# Springer Tracts in Civil Engineering

**Series Editors**

Giovanni Solari, Wind Engineering and Structural Dynamics Research Group, University of Genoa, Genova, Italy

Sheng-Hong Chen, School of Water Resources and Hydropower Engineering, Wuhan University, Wuhan, China

Marco di Prisco, Politecnico di Milano, Milano, Italy

Ioannis Vayas, Institute of Steel Structures, National Technical University of Athens, Athens, Greece

**Springer Tracts in Civil Engineering** (STCE) publishes the latest developments in Civil Engineering - quickly, informally and in top quality. The series scope includes monographs, professional books, graduate textbooks and edited volumes, as well as outstanding PhD theses. Its goal is to cover all the main branches of civil engineering, both theoretical and applied, including:

- Construction and Structural Mechanics
- Building Materials
- Concrete, Steel and Timber Structures
- Geotechnical Engineering
- Earthquake Engineering
- Coastal Engineering; Ocean and Offshore Engineering
- Hydraulics, Hydrology and Water Resources Engineering
- Environmental Engineering and Sustainability
- Structural Health and Monitoring
- Surveying and Geographical Information Systems
- Heating, Ventilation and Air Conditioning (HVAC)
- Transportation and Traffic
- Risk Analysis
- Safety and Security

**Indexed by Scopus**

To submit a proposal or request further information, please contact:
Pierpaolo Riva at Pierpaolo.Riva@springer.com (Europe and Americas) Mengchu Huang at mengchu.huang@springer.com (China)

More information about this series at http://www.springer.com/series/15088

Sandro Longo · Maria Giovanna Tanda ·
Luca Chiapponi

# Problems in Hydraulics
# and Fluid Mechanics

 Springer

Sandro Longo
Department of Engineering and Architecture
(DIA)
University of Parma
Parma, Italy

Maria Giovanna Tanda
Department of Engineering and Architecture
(DIA)
University of Parma
Parma, Italy

Luca Chiapponi
Department of Engineering and Architecture
(DIA)
University of Parma
Parma, Italy

ISSN 2366-259X          ISSN 2366-2603   (electronic)
Springer Tracts in Civil Engineering
ISBN 978-3-030-51389-4          ISBN 978-3-030-51387-0   (eBook)
https://doi.org/10.1007/978-3-030-51387-0

This Springer imprint is published by the registered company Springer Nature Switzerland AG
The registered company address is: Gewerbestrasse 11, 6330 Cham, Switzerland

*To my Family and to Vincenzo and Gilda (S.L.)*
*To my Family (M.G.T.)*
*To Irene and Tommaso (L.C.)*

# Preface

The exercises of this textbook, fully worked, can be a useful complement to the Lessons and are suitable to accustom the students to perform the calculations applied to situations of technical interest. In some exercises, we propose two different methods of solving, to highlight that the level of complexity of the calculations is often related to the choice of method. In general, we have preferred the simplest method. Chapters 1 and 2 deal with exercises on forces on flat and curved surfaces and humps. Chapter 3 is entirely dedicated to floating bodies. Chapter 4 deals with some classic exercises that require the application of balance of linear and angular momentum, in inertial and non-inertial references. Chapter 5 analyzes pipeline systems, with particular applications to industrial plants in Chap. 6. Chapter 7 deals with hydraulic systems with machines (pumps and turbines). This is followed by Chap. 8, dedicated to transient phenomena in closed pipelines. Chapter 9 deals with flows in open channels. The Appendices contain some data and formulas of practical interest.

The book is addressed to undergraduates and graduates in Engineering Sciences. In many exercises, some parameters are given in terms of $C_u$ and $C_{pu}$, that are, for example, the last and second-last digit of the registration number. This diversifies the calculations of the students during the written tests.

Parma, Italy  
March 2020

Sandro Longo  
Maria Giovanna Tanda  
Luca Chiapponi

# Introduction

Fluid Mechanics deals with the behaviour of fluid under the action of forces, at rest or in motion. Fluids are classified into liquids, gases and vapours. Liquids are difficult to compress and generally the range of pressure is limited so as to consider them as incompressible, although in some practical applications their compressibility is invoked to explain the behaviour of elastic waves in water hammer. Gases can be easily compressed, although the change of pressure during flow is often so modest that no significant change of volume takes place, and the motion is defined isochoric: in isochoric flow, gases behave like liquids. Vapours are gases which can be condensed by increasing the pressure without lowering temperature. Hence, in isothermal or almost isothermal flows, vapours can afford to phase transition with the coexistence of a liquid and a gas phase. Their analysis belongs to multi-phase systems.

Hydraulics deals with water, although concepts derived for water are broadly adopted for the Fluid Dynamics of liquids similar to water, like oil and kerosene.

In a fluid at rest the internal stresses reduce to pressure, a normal force per unit surface acting normally to any surface, and the shear stress is null. In a fluid in motion the internal stresses include also shear stress, but no torque per unit surface is considered. A special category of fluids, called polar fluids, is described by including also torque per unit surface. A polar fluid is capable of transmitting stress couples and being subjected to body torques.

Fluids are continua and differ from solid since in solids a stress determines a strain proportional (linearly or not) to the stress, whereas in fluids a stress determines a rate of strain proportional (linearly or not) to the stress. Linear proportionality between strain rate and stress is a characteristic of Newtonian fluids. Non-Newtonian fluids show a more complex relation between strain rate and stress, in some cases involving also stress varying with strain and in time.

The fundamental equations adopted for solving problems are mass conservation, linear momentum and angular momentum balances, energy conservation. These equations are the consequence of some principles (e.g. linear momentum balance equation is derived from the hypothesis of homogeneity of space), and require boundary and initial conditions. In several real cases, some material properties are necessary to model, e.g. the reaction of a fluid to a stress state or the heat flux due to an imposed thermal field. These are described by constitutive equations, mechanical and thermal, which contain some material parameters, like viscosity, which is evaluated through experiments. The constitutive equations must have a non-dimensional formulation, satisfy coordinate indifference (have a tensorial formulation), frame-reference indifference (material indifference) and satisfy the second principle of thermodynamics.

## SI Units and Dimensions

The *Système International d'Unités* (SI) has been introduced with the aim of a common language for units, although in several countries former systems are still used. The SI is based on seven fundamental quantities: mass, length, time, electric current, thermodynamic temperature, amount of substance and luminous intensity. Fundamental quantities are independent and are sufficient to describe all other quantities, defined as derived quantities. The fundamental units are listed in Table 1.

**Table 1** Fundamental quantities and units of measurements in the SI

| Fundamental quantities | Symbol | Denomination of the unit | Symbol of the unit |
|---|---|---|---|
| Length | L | metre | m |
| Mass | M | kilogram | kg |
| Time | T | second | s |
| Thermodynamic temperature | $\Theta$ | kelvin | K |
| Electric current | I | ampere | A |
| Luminous intensity | C | candela | cd |
| Amount of substance | mol | mole | mol |

However, recently (in 2019), the fundamental units in SI have been defined in terms of seven dimensional constants. The conversions have been introduced once the required accuracy in measuring these constants has been achieved. Although the seven dimensional constants could be adopted as fundamental, the past seven fundamental units are still in use, but they are now defined in terms of (1) the caesium hyperfine frequency, $\Delta v_{Cs}$; (2) the speed of light in vacuum, $c$; (3) the Planck constant, $h$; (4) elementary charge, $e$; (5) Boltzmann constant, $k$; (6) Avogadro constant, $N_A$, and (7) the luminous efficacy of a defined visible radiation, $K_{cd}$.

For instance, the caesium-133 hyperfine frequency is equal to

$$\Delta v_{Cs} = \Delta v \left(^{133}Cs\right)_{hfs} = 9\ 192\ 631\ 770\ \text{Hz},$$

and the time unit is expressed as

**Fig. 1** Logo of the SI constants (internal ring), in force since 20 May 2019 to define the fundamental units (external ring)

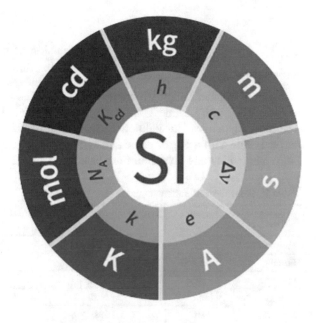

$$1\,\text{s} = \frac{9\ 192\ 631\ 770}{\Delta v_{\text{Cs}}}.$$

We could say that the time of rotation of the Earth about its axis is $0.79 \times 10^{15} \Delta v_{\text{Cs}}^{-1}$, but it is a lot more in everyday life to say that it is 86 400 s.

Figure 1 shows the logo used to disseminate the new way by the *Bureau International des Poids et Mesures* (BIPM). The BIPM is an international organization established by the Metre Convention, through which Member States act together on matters related to measurement science and measurement standards.

Units are necessary to express the size of a measured quantity with respect to a known quantity: if a man is 1.75 m tall, we are simply saying that the ratio between his height and the fundamental unit (metre) is 1.75. The use of a different system of units can modify the number, but not the intrinsic height of the man, in the sense that the same man is defined as 5 3/4 inches tall in the Imperial System, not being different from the 1.75 m tall man in SI units.

In order to avoid misunderstanding, it is convenient to think in terms of mass, length, time, force, etc., instead of thinking in terms of the units. The use of dimensions instead of the units renders the analysis independent on the systems used, and is helpful in checking the correctness of the equations. Thus

$$\text{velocity} = \frac{\text{distance}}{\text{time}},$$

hence

$$\text{dimensions of velocity} = \frac{\text{dimension of distance}}{\text{dimension of time}} = \frac{L}{T}.$$

In Fluid Mechanics and Hydraulics, all quantities are a monomial combination of mass M, length L, time T and temperature $\Theta$. This last dimension is typical of Thermofluidynamics. In addition to the fundamental units, there are also some derived units with a proper name and symbols. The most common derived units in Fluid Mechanics and Hydraulics are listed in Table 2.

**Table 2** Some derived units with name

| Quantity | Name | Symbol | Expression in derived SI units | Expression in fundamental SI units |
|----------|------|--------|-------------------------------|-----------------------------------|
| Frequency | hertz | Hz | | $s^{-1}$ |
| Force | newton | N | | $m \ kg \ s^{-2}$ |
| Pressure | pascal | Pa | $N/m^2$ | $m^{-1} \ kg \ s^{-2}$ |
| Energy, work | joule | J | N m | $m^2 \ kg \ s^{-2}$ |
| Power, energy flux | watt | W | $J/s$ | $m^2 \ kg \ s^{-3}$ |

A necessary but not sufficient condition for the correctness of the equations is the dimensional homogeneity. For instance, the head is the sum of three terms, representative of potential, pressure and kinetic energy, hence in its expression $H = z + p/\gamma + V^2/2g$ the dimension of all terms is a length (energy per unit weight). Table 3 lists the most common and frequently used dimensions for the quantities encountered in Fluid Dynamics and Hydraulics.

The dimensions are helpful in evaluating the coefficient of conversions between different systems (which are still in use). In Hydraulics, some classical formulas, where some coefficients are dimensional, are still common. The Chézy formula for discharge in a channel, $Q = \chi \Omega \sqrt{R_h i_b}$, adopts different formulations for the dimensional coefficient $\chi$, which in SI units is defined, e.g. according to Bazin as $\chi = 87/(1 + \gamma/\sqrt{R_h})$, where $\gamma$ is the Bazin coefficient of roughness which must be expressed in $m^{1/2}$.

## The Scientific Notation with Physical Measurements

The expression of measurements (a ratio between the measured variable and the unit of measurement for that variable) is a way to communicate also the uncertainty of the value. The number of significant figures in a measurement is related to the overall uncertainty of the entire procedure: the more accurate is the procedure to obtain a measure, the greater the number of significant figures it can report. Zeros do not contribute to the number of significant figures, unless they are between non-zero numbers or unless they are underlined. 0.000 123 and 452, and 121.0 have three significant figures, but 0.001 204 and 1201, and 303.1, and 862.$\underline{000}$ 0 have four significant figures. To avoid ambiguities, it is preferable to use the scientific notation: the number 121.0, with three significant figures, can be written as $1.21 \times 10^2$, 862.$\underline{000}$ 0 can be written as $8.620 \times 10^2$ and 0.001 204 can be written as $1.204 \times 10^{-3}$. The number before the power of ten should be preferably between 1 and 10 and should contain all the significant figures, without the need to underline zeros which are significant.

In combining values of measurements or data with different uncertainties, (i) for multiplication and division the number of significant digits in the result can be no greater than the number of significant digits in the least-precise measured value; (ii) for addition and subtractions the result should have the same number places (tens place, ones place, tenths place, etc.) as the least-precise starting value: if you have 1.012 kg of salt (uncertainty of 1 thousandths of kilogram), and then you buy 5.4 kg of salt (uncertainty 0.1 kg), you have 1.012 + 5.4 = 6.4 kg of salt.

**Table 3** Dimensions of quantities commonly used in Fluid Mechanics and Hydraulics

| Quantity | | Dimensions |
|---|---|---|
| Length | all linear measurements | L |
| Area | length × length | $L^2$ |
| Volume | area × length | $L^3$ |
| First moment of area | area × length | $L^3$ |
| Second moment of area | area × length$^2$ | $L^4$ |
| Angle | arc/radius | 1 |
| Strain | a ratio | 1 |
| Head | energy/weight | L |
| Energy gradient | head/length | 1 |
| Time | | T |
| Velocity | distance/time | $LT^{-1}$ |
| Angular velocity | angle/time | $T^{-1}$ |
| Acceleration | velocity/time | $LT^{-2}$ |
| Angular acceleration | angular velocity/time | $T^{-2}$ |
| Volume discharged | volume/time | $L^3T^{-1}$ |
| Kinematic viscosity | dynamic viscosity/mass density | $L^2T^{-1}$ |
| Mass | | M |
| Force | mass × acceleration | $MLT^{-2}$ |
| Weight | force | $MLT^{-2}$ |
| Mass density | mass/volume | $ML^{-3}$ |
| Specific weight | weight/volume | $ML^{-2}T^{-2}$ |
| Mass discharged | mass/time | $MT^{-3}$ |
| Pressure (intensity) | force/area | $ML^{-1}T^{-2}$ |

(continued)

**Table 3** (continued)

| Quantity | | Dimensions |
|---|---|---|
| Shear stress | force/area | $ML^{-1}T^{-2}$ |
| Elastic modulus | stress/strain | $ML^{-1}T^{-2}$ |
| Bulk modulus | stress/strain | $ML^{-1}T^{-2}$ |
| Impulse | force × time | $MLT^{-1}$ |
| Momentum | mass × velocity | $MLT^{-1}$ |
| Work, energy | force × distance | $ML^{2}T^{-2}$ |
| Power | work/time | $ML^{2}T^{-3}$ |
| Moment of force | force × distance | $ML^{2}T^{-2}$ |
| Dynamic viscosity | shear stress/velocity gradient | $ML^{-1}T^{-1}$ |
| Surface tension | energy/area | $MT^{-2}$ |

Table 4 lists a series of number with different numbers of figures and notations. Units derived by means of the basic units must have a numerical unit coefficient and the multiples and submultiples of the units of measurement must be expressed as integer exponent powers of ten (Table 5).

# Writing Rules

In technical writing, in order to make it easier to understand and to avoid misinterpretation it is advisable to follow some basic rules.

**Table 4** Notations for significant figures

| Number | Significant Figures | |
|---|---|---|
| 3.651 | 4 | There are no zeros and all numbers are significant |
| 1010.56 | 6 | The two zeros are significant here because they occur between other significant figures |
| 0.219 8 | 4 | The first zero is only a placeholder for the decimal point and is not significant |
| 0.000 044 2 | 3 | The first five zeros are placeholders needed to report the data to the hundred-thousandths place |
| 33.100 | 3 | With no underlines or scientific notation, the last two zeros are placeholders and are not significant |
| 11 891 <u>00</u>0 | 7 | The two underlined zeros are significant, while the last zero is not, as it is not underlined |
| $5.457 \times 10^{13}$ | 4 | In scientific notation, all numbers reported before the power of ten are significant |

(continued)

**Table 4** (continued)

| Number | Significant Figures | |
|---|---|---|
| $6.520 \times 10^{-23}$ | 4 | In scientific notation, all numbers reported before the power of ten, including zeros, are significant |
| $0.320 \times 10^{-2}$ | 3 | In scientific notation, all numbers reported before the power of ten, including zeros (but not before decimal point), are significant |

Units of measurement expressed in symbolic form always begin with a lowercase letter, unless they are derived from a name of person. For example, 1 s and not 1 S, 12 A (from the name of André-Marie Ampère) and not 12 a. In addition, a space between the number and the symbol (23 m and not 23m) is always required and symbols should never be indicated in italics or bold: 1 s and not 1 *s* or 1 **s** (the use of units in bold in some results of the exercises of the present book is an exception adopted to highlight the relevant values).

**Table 5** Multiples and submultiples in SI

| Coefficient | Name | Symbol | Coefficient | Name | Symbol |
|---|---|---|---|---|---|
| $10^1$ | deca | da | $10^{-1}$ | deci | d |
| $10^2$ | etto | h | $10^{-2}$ | centi | c |
| $10^3$ | kilo | k | $10^{-3}$ | milli | m |
| $10^6$ | mega | M | $10^{-6}$ | micro | $\mu$ |
| $10^9$ | giga | G | $10^{-9}$ | nano | n |
| $10^{12}$ | tera | T | $10^{-12}$ | pico | p |
| $10^{15}$ | peta | P | $10^{-15}$ | femto | f |
| $10^{18}$ | exa | E | $10^{-18}$ | atto | a |
| $10^{21}$ | zetta | Z | $10^{-21}$ | zepto | z |
| $10^{24}$ | yotta | Y | $10^{-24}$ | yocto | y |

If you need to write the unit of measure in full in the text, you will always use lowercase characters, even if the unit is derived from a person name: ampère and not Ampère, newton and not Newton.

A unit of measure symbol consisting of the product of two or more units can be written either by interposing a point or by leaving a space: $13.2 \, N \cdot m$ or $13.2 \, N \, m$.

In the case of the quotient between two units of measurement you can write, for example, $3.8 \, m/s$ or $3.8 \, ms^{-1}$ or $3.8 \, m \cdot s^{-1}$ or $3.8 \, \frac{m}{s}$. The second form is the advisable one.

For multiples or submultiples prefixes, only those greater than $10^6$ are shown with a capital letter; therefore, 1.5 MJ and not 1.5 mJ, 22 kg and not 22 Kg. Notice, in this regard, that the prefix 'm' (milli-) indicates $10^{-3}$, while the prefix 'M' (Mega-) indicates $10^6$. Again, the multiple or submultiple symbol is placed next to the unit of measure symbol, without space: 13.2 mW and not 13.2 m W.

In the scientific notation it is necessary that the units are the basic ones: write, then, $3.2 \times 10^5$ m and not $3.2 \times 10^2$ km. In the notation with the prefixes it is also advisable to choose the prefix so that the number is between 0.1 and 1000, then 7.8 MJ and not 7800 kJ. Double prefixes are not allowed, so 1.2 µF (microfarad) and not 1.2 mmF (millimillifarad).

When writing numbers containing more than four digits in sequence, a spacing is appropriate, grouping the digits in groups of three to the left and to the right of the decimal point; therefore, 12 000 and not 12000, then 13.224 32 and not 13.22432.

Notice, finally, that the International Organization for Standardization (ISO) suggests the comma as the decimal separator, while, in English-speaking countries, the comma is the separator of the thousands and the point is the decimal separator. Therefore, to avoid confusion between the SI notation and the Anglo-Saxon notation, it is not advisable to use a point or comma to separate the thousands. Since 2003, the use of the decimal point is also allowed in English texts.

# Contents

# Chapter 1
# Hydrostatic Forces on Submerged Plane Surfaces

A fluid at rest in contact with a surface exerts only a normal force per unit area, called pressure. The pressure is isotropic (at a given point it is the same regardless of the orientation of the infinitesimal surface containing the point and on which it is evaluated) and once the pressure field has been determined, it is possible to calculate the total force acting on a surface of finite dimensions. For planar surfaces the analysis is simplified because the direction of the hydrostatic force is normal to the surface, and the force always enters the surface. The magnitude of the force can be calculated by integration and, *only for homogeneous fluids*, also as a product of the pressure in the centroid of the surface and the surface area. The force is applied to the pressure centre, which generally differs from the centroid of the surface, and which can be calculated by imposing the equivalence of the moments of the force and the vectorial sum of the elementary moments of the elementary forces. Pressure centre is always below the centroid, with respect to the water line, except for horizontal surfaces.

In this chapter there are some exercises for calculating the pressure distributions, mainly referring to U-tube manometers. Then forces on surfaces in homogeneous or stratified fluid are calculated, possibly with the calculation of the equilibrium condition for isostatic system. The relevant geometric properties for some common shapes of plane surfaces are listed in the Appendix.

**Exercise 1.1**  Consider a mercury (Hg) U-tube manometer used to measure the pressure of a liquid in a sphere, see Fig. 1.1.

– Calculate the pressure in A, if $h_1 = (20 + C_u)$ cm and $h_2 = (10 + C_{pu})$ cm.

  Assume $\gamma_w = 9806\,\mathrm{N\,m^{-3}}$, $\gamma_{Hg}/\gamma_w = 13.6$.

  **Solution**  The pressure in B, in the left limb, is

---

$C_u$ and $C_{pu}$, that are two integer numbers between 0 and 9, for example, the last and second-last digit of the registration number.

© Springer Nature Switzerland AG 2021
S. Longo et al., *Problems in Hydraulics and Fluid Mechanics*, Springer Tracts
in Civil Engineering, https://doi.org/10.1007/978-3-030-51387-0_1

**Fig. 1.1**  Mercury U-tube
manometer

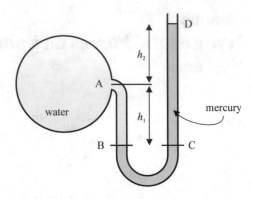

$$p_B \equiv p_C = p_D + \gamma_{Hg}(h_1 + h_2),$$

where C, in the right limb, is at the same level of B. The pressure in A is

$$p_A = p_B - \gamma_w h_1.$$

Hence,

$$p_A = p_D + \gamma_{Hg}(h_1 + h_2) - \gamma_w h_1.$$

For $C_u = C_{pu} = 0$ it results $h_1 = 20$ cm, $h_2 = 10$ cm. Assuming that in D the pressure equals the atmospheric pressure (the zero is the value of the average atmospheric relative pressure at the sea level), results

$$\therefore \quad p_A = p_D + \gamma_{Hg}(h_1 + h_2) - \gamma_w h_1 =$$
$$0 + 13.6 \times 9806 \times (0.2 + 0.1) - 9806 \times 0.2 = \mathbf{38\,050\ Pa} \quad (\text{gage}).$$

---

**Exercise 1.2**  Consider a differential inverted U-tube manometer used to measure the difference of pressure between two taps A and B, see Fig. 1.2.

– Calculate the difference of pressure $p_A - p_B$ if the valve in C is open and the tube is at contact with the atmosphere.
– Calculate the new values of $h_1$ and $h_2$ if the valve in C is open and the tube is at contact with an ambient where $p_C = -10^2$ Pa.

Assume $\gamma_w = 9806\,\mathrm{N\,m^{-3}}$, $h_1 = (15 + C_u)$ cm and $h_2 = (10 + C_{pu})$ cm.

**Solution**  The pressure in A, in the left limb, is

$$p_A = p_D + \gamma_w h_1,$$

**Fig. 1.2** Differential
inverted U-tube manometer

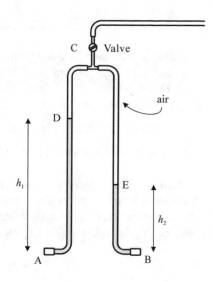

and the pressure in B is

$$p_B = p_E + \gamma_w h_2.$$

Since $p_D = p_E = p_C$, results

$$p_A - p_B = \gamma_w(h_1 - h_2).$$

If the pressure in C is reduced, the difference of levels between D and E does
not change, and both levels change of $p_C/\gamma_w$, hence $h_1' = h_1 - p_C/\gamma_w$ and $h_2' = h_2 - p_C/\gamma_w$. Decreasing/increasing $p_C$, both levels move up/down. The pressure in
C can be used to fit the two menisci D and E within the length of the U-tube.

For $C_u = C_{pu} = 0$ it results $h_1 = 15\,\text{cm}$, $h_2 = 10\,\text{cm}$.
If $p_C = 0$ (gage), then

$$\therefore \qquad p_A - p_B = \gamma_w(h_1 - h_2) = 9806 \times (0.15 - 0.10) = \textbf{490 Pa}.$$

If $p_C = -10^2\,\text{Pa}$,

$$\therefore \qquad h_1' = h_1 - p_C/\gamma_w = 0.15 + \frac{10^2}{9806} = \textbf{16 cm},$$

$$\therefore \qquad h_2' = h_2 + p_C/\gamma_w = 0.10 + \frac{10^2}{9806} = \textbf{11 cm},$$

and

$$p_A - p_B = \gamma_w(h_1' - h_2') = 9806 \times (0.16 - 0.11) = \textbf{490 Pa}.$$

**Exercise 1.3** Consider a differential U-tube manometer with enlarged ends and water and oil in the left and right limb, respectively, see Fig. 1.3. If $p_1 = p_2$ the levels in the two ends are different. In a second step, impose $p_1' = (30 + C_{pu})$ Pa and $p_2 = 0$ (gage), see Fig. 1.4.

– Calculate the new level of the interface between water and oil.

Assume $\gamma_w = 9806\,\mathrm{N\,m^{-3}}$, $s_o \equiv \gamma_o/\gamma_w = 0.95$, $A_1 = 2A_2 = 50a$.

**Solution** If $p_1 = p_2$ results also $\gamma_w h_1 = \gamma_o h_2$. Imposing a new value of pressure $p_1'$ on the left end, requires a new equilibrium condition. If $p_1' > p_1$ the air-water interface, at the left end, shifts downwards by a $\Delta_1$ distance; the water-oil and oil-air interfaces, at the right end, shift upwards. The pressure in A on the left limb is

$$p_A = p_1' + \gamma_w(h_1 - \Delta_1),$$

and, on the right limb,

$$p_A = p_2 + \gamma_o(h_2 - \delta + \Delta_2) + \gamma_w \delta.$$

**Fig. 1.3** Differential U-tube manometer with enlarged ends and $p_1 = p_2 = 0$

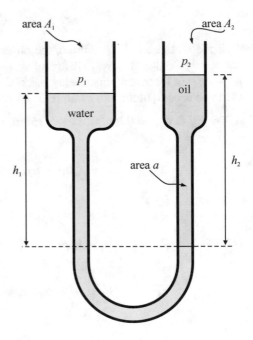

**Fig. 1.4** Differential U-tube manometer with enlarged ends and $p_1' > p_2$

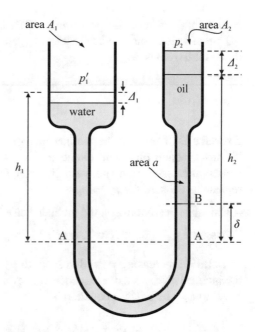

Hence

$$p_1' - p_2 = \gamma_o(h_2 - \delta + \Delta_2) + \gamma_w\delta - \gamma_w(h_1 - \Delta_1),$$

or

$$p_1' - p_2 = \gamma_o(-\delta + \Delta_2) + \gamma_w\delta + \gamma_w\Delta_1,$$

since $\gamma_w h_1 = \gamma_o h_2$ from the initial condition. Mass conservation also requires that $\Delta_1 A_1 = \Delta_2 A_2 = \delta a$. In terms of $\delta$ it results

$$\delta = \frac{p_1' - p_2}{\gamma_w\left[\left(1 + \dfrac{a}{A_1}\right) - s_0\left(1 - \dfrac{a}{A_2}\right)\right]},$$

where $s_0 = \gamma_o/\gamma_w$.

For $C_u = C_{pu} = 0$ it results $p_1' = 30\,\text{Pa}$,

$$\therefore \quad \delta = \frac{p_1' - p_2}{\gamma_w\left[\left(1 + \dfrac{a}{A_1}\right) - s_0\left(1 - \dfrac{a}{A_2}\right)\right]} =$$

$$\frac{30}{9806\left[\left(1 + \dfrac{1}{50}\right) - 0.95\left(1 - \dfrac{1}{25}\right)\right]} = \mathbf{2.83\,cm}.$$

Notice that a differential pressure of 30 Pa is equivalent to a head of $30/9806 \approx$ 0.3 cm of water. The gain due to the enlarged ends and to the use of a second lighter fluid is $2.83/0.3 \approx 9$.

---

**Exercise 1.4** Consider the inclined manometer in Fig. 1.5, realized with a circular cross-section pipe with diameter $d = 0.5$ cm. The manometer is connected to a circular cross-section tank with diameter $D = 20$ cm. The dashed horizontal line represents the zero for $p_a = p_b$.

– Calculate the reading on the inclined scale if $p_a - p_b = 100$ Pa.

Assume $\gamma_m = 8200\,\mathrm{N\,m^{-3}}$, $\alpha = 10°$.

**Solution** Increasing $p_a$ with respect to $p_b$, forces the manometric fluid to rise in the inclined pipe. As a consequence, the level in the left tank drops with a reduction equal to $h_a$. The balance equation is

$$p_a = p_b + \gamma_m(h_b + h_a) \equiv p_b + \gamma_m(h'_b \sin\alpha + h_a),$$

where $h'_b$ is the reading on the inclined scale.

Mass conservation requires that

$$\frac{\pi D^2}{4}h_a = \frac{\pi d^2}{4}h'_b \rightarrow h_a = h'_b\frac{d^2}{D^2},$$

hence

$$p_a = p_b + \gamma_m h'_b\left(\sin\alpha + \frac{d^2}{D^2}\right) \rightarrow h'_b = \frac{p_a - p_b}{\gamma_m\left(\sin\alpha + d^2/D^2\right)}.$$

Inserting the numerical values, yields

$$\therefore \qquad h'_b = \frac{p_a - p_b}{\gamma_m\left(\sin\alpha + d^2/D^2\right)} = \frac{100}{8200 \times \left(\sin 10° + 0.5^2/20^2\right)} = \mathbf{7\,cm}.$$

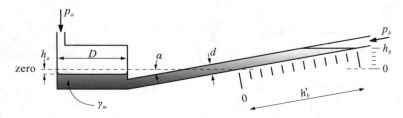

**Fig. 1.5** Inclined manometer

Notice that the vertical reading is $h_b = h'_b \sin \alpha = 7 \sin 10° = 1.2\,\text{cm}$. The gain due to the inclination is $7/1.2 \approx 6$. If $D \gg d$, the reading on the inclined scale is $h'_b \approx (p_a - p_b)/(\gamma_m \sin \alpha)$ and the gain could be increased with a reduction of $\alpha$. However, meniscus errors become dominant for very small angles. In practical applications the angle is seldom less than 10°.

---

**Exercise 1.5**  Consider the differential manometer with multiple U-tubes in Fig. 1.6. The manometric fluid is mercury, all other limbs are filled with air.

– Calculate the sum of the readings $\Delta h_1$ and $\Delta h_2$ if $p_a - p_b = 15\,000\,\text{Pa}$.

   Assume $\gamma_m = 133\,400\,\text{N}\,\text{m}^{-3}$.

   **Solution** The balance equation is

$$p_A = p_B + \gamma_m \Delta h_1,$$

and

$$p_C = p_D + \gamma_m \Delta h_2.$$

Neglecting the specific weight of the air, yields

$$p_A = p_b, \quad p_B = p_C, \quad p_D = p_b,$$

**Fig. 1.6**  Multiple U-tubes manometer

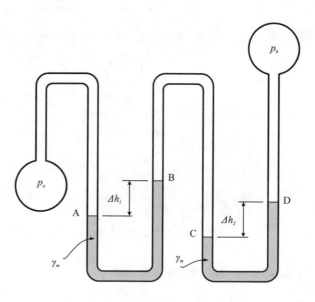

hence
$$p_a - p_b = \gamma_m(\Delta h_1 + \Delta h_2) \to (\Delta h_1 + \Delta h_2) = \frac{p_a - p_b}{\gamma_m}.$$

Inserting the numerical values, yields

$$\therefore \qquad (\Delta h_1 + \Delta h_2) = \frac{p_a - p_b}{\gamma_m} = \frac{15\,000}{133\,400} = \mathbf{11.2\,cm}.$$

With this pressure gauge it is possible to measure large differences in pressure, appropriately increasing the number of limbs: the air simply transfers the pressure between the menisci. However, the accuracy of the measurement is reduced as it is necessary to read separately the difference in level for each branch and to add them up, with uncertainties that add up.

---

**Exercise 1.6**  An isosceles triangle-shaped plate is pivoted on the horizontal top side, see Fig. 1.7. The tank contains fluid concrete, with specific gravity $s = 2.4$.

– Calculate the force on the plate and the point of application of the force.
– Calculate the minimum horizontal force, orthogonal to the plate, applied in D, required to prevent the plate from rotating and opening.

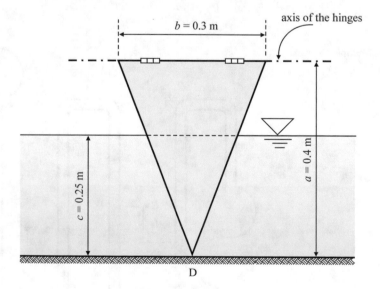

**Fig. 1.7**  Schematic of the triangular plate

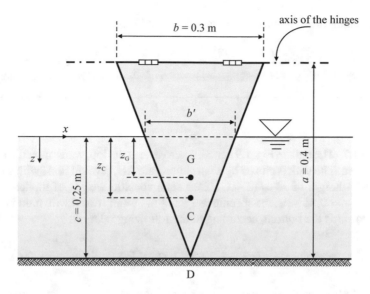

**Fig. 1.8** Coordinate system adopted for the calculation

**Solution** The magnitude of the force is equal to the product of the pressure in the centroid of the submerged surface by the area of the submerged surface. In the coordinate system shown in Fig. 1.8, the position of the centroid is $z_G = c/3$. The force is

$$\therefore \qquad F = \gamma_c \frac{c}{3} \frac{b'c}{2} = 2.4 \times 9806 \times \frac{0.25}{3} \times \frac{0.1875 \times 0.25}{2} = \mathbf{46\,N},$$

where $b'$ is the base of the triangle at $z = 0$, calculated with similarity of triangles:

$$b' = \frac{bc}{a} = \frac{0.3 \times 0.25}{0.4} = \mathbf{0.1875\,m}.$$

The point of application of the force (center of pressure) has a coordinate equal to:

$$\therefore \qquad z_C = \frac{I_{xx}}{S_x} = \frac{\dfrac{b'c^3}{12}}{\dfrac{b'c^2}{6}} = \frac{c}{2} = \mathbf{0.125\,m},$$

where $S_x$ and $I_{xx}$ are the first moment of area and the second moment of area (area moment of inertia) of the submerged surface with respect to the $x$−axis, respectively. The values of the two moments of area are given in Appendix.

By equating the moments about the axis of the hinges, yields:

$$\therefore \quad R = \frac{F\,(a - c/2)}{a} = \frac{46 \times (0.4 - 0.125)}{0.4} = \mathbf{31.625\,N.}$$

---

**Exercise 1.7** The tank in Fig. 1.9 contains kerosene at the top, water mixed with mud at the bottom. The tank is closed by a flat sluice gate $\overline{AB}$, pivoted in A, with a unitary depth and a height of $H = (2 + C_u/2)$ m. The specific weight of the kerosene is equal to $\gamma_{ker} = 0.81 \times \gamma_w$, the specific weight of the water mixed with mud increases linearly towards the bottom, according to the following relation:

$$\gamma(z') = \gamma_w + (1 + C_u)\,\frac{\gamma_w}{49}\frac{z'}{h_2}.$$

The fluids depths are equal to $h_1 = (2 + C_{pu}/2)$ m, $h_2 = (8 + C_{pu}/2)$ m.

– Calculate the force exerted on the sluice gate.
– Calculate the position of the centre of pressure.

Assume $\gamma_w = 9800\,\mathrm{N\,m^{-3}}$.

**Fig. 1.9** Schematic of the tank containing stratified fluids

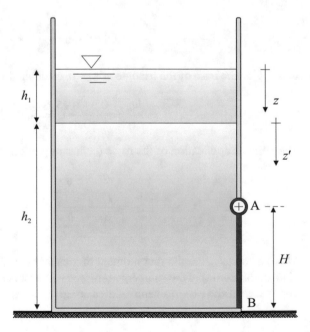

**Solution** The pressure in the kerosene, for $0 < z < h_1$, varies linearly and is equal to $p(z) = \gamma_{ker} z \equiv 0.81 \times \gamma_w z$. At the interface between kerosene and water mixed with mud, the pressure is equal to:

$$p(h_1) = \gamma_{ker} h_1.$$

The pressure in the water mixed with mud, for $h_1 < z < h_1 + h_2$, i.e. between $z' = 0$ and $z' = h_2$, is calculated by using the indefinite form of hydrostatic equation written in the $z'$ coordinate system:

$$\frac{dp}{dz'} = \gamma(z') \equiv \gamma_w + (1 + C_u)\frac{\gamma_w}{49}\frac{z'}{h_2} \rightarrow \int_{p(0)}^{p(z')} dp = \int_0^{z'}\left[\gamma_w + (1 + C_u)\frac{\gamma_w}{49}\frac{z'}{h_2}\right]dz'.$$

Hence

$$p(z') = p(0) + \gamma_w z' + (1 + C_u)\frac{\gamma_w}{98}\frac{z'^2}{h_2} = \gamma_{ker}h_1 + \gamma_w z' + (1 + C_u)\frac{\gamma_w}{98}\frac{z'^2}{h_2}.$$

The force exerted on the sluice gate is calculated by integrating the elementary forces associated with the pressure in the fluid on the surface of the gate, of unitary depth:

$$F = \int_{h_2-H}^{h_2}\left[\gamma_{ker}h_1 + \gamma_w z' + (1 + C_u)\frac{\gamma_w}{98}\frac{z'^2}{h_2}\right]dz'$$

$$= \gamma_{ker}h_1 z'\Big|_{h_2-H}^{h_2} + \gamma_w\frac{z'^2}{2}\Big|_{h_2-H}^{h_2} + (1 + C_u)\frac{\gamma_w}{294}\frac{z'^3}{h_2}\Big|_{h_2-H}^{h_2}$$

$$= \gamma_{ker}h_1 H + \gamma_w\frac{h_2^2 - (h_2 - H)^2}{2} + (1 + C_u)\frac{\gamma_w}{294}\left[\frac{h_2^3}{h_2} - \frac{(h_2 - H)^3}{h_2}\right].$$

The centre of pressure is calculated by imposing that the moment about an axis (for example, the trace of the interface between the two liquids) due to the distribution of elementary forces, equates the moment of the resultant of these forces calculated about the same axis:

$$F \times z_C' = \int_{h_2-H}^{h_2}\left[\gamma_{ker}h_1 + \gamma_w z' + (1 + C_u)\frac{\gamma_w}{98}\frac{z'^2}{h_2}\right]z'\,dz' \rightarrow$$

$$z'_C = \frac{\left[\gamma_{ker}h_1\dfrac{z'^2}{2} + \gamma_w\dfrac{z'^3}{3} + (1+C_u)\dfrac{\gamma_w}{392}\dfrac{z'^4}{h_2}\right]\Big|_{h_2-H}^{h_2}}{F}$$

$$= \frac{\gamma_{ker}h_1\dfrac{h_2^2}{2} + \gamma_w\dfrac{h_2^3}{3} + (1+C_u)\dfrac{\gamma_w}{392}\dfrac{h_2^4}{h_2}}{F}$$

$$- \frac{\gamma_{ker}h_1\dfrac{(h_2-H)^2}{2} + \gamma_w\dfrac{(h_2-H)^3}{3} + (1+C_u)\dfrac{\gamma_w}{392}\dfrac{(h_2-H)^4}{h_2}}{F}.$$

For $C_u = C_{pu} = 0$ it results $\gamma = 9800 + 200(z'/h)\,\mathrm{N\,m^{-3}}$, $h_1 = 2\,\mathrm{m}$, $h_2 = 8\,\mathrm{m}$, $H = 2\,\mathrm{m}$.

$$p(z') = 15\,876 + 9800z' + 100\frac{z'^2}{h_2}\,\mathrm{Pa}\ (z',\ h_2\ \text{in metres}).$$

$$F = \gamma_{ker}h_1 H + \gamma_w\frac{h_2^2 - (h_2-H)^2}{2} + \frac{\gamma_w}{294}\left[\frac{h_2^3}{h_2} - \frac{(h_2-H)^3}{h_2}\right] \rightarrow$$

$$\therefore \qquad F = 0.81 \times 9800 \times 2 \times 2 + 9800 \times \frac{8^2 - (8-2)^2}{2}$$

$$+ \frac{9800}{294} \times \left[\frac{8^3}{8} - \frac{(8-2)^3}{8}\right] = \mathbf{170.2\ kN}.$$

$$z'_C = \frac{\gamma_{ker}h_1\dfrac{h_2^2}{2} + \gamma_w\dfrac{h_2^3}{3} + \dfrac{\gamma_w}{392}\dfrac{h_2^4}{h_2}}{F}$$

$$- \frac{\gamma_{ker}h_1\dfrac{(h_2-H)^2}{2} + \gamma_w\dfrac{(h_2-H)^3}{3} + \dfrac{\gamma_w}{392}\dfrac{(h_2-H)^4}{h_2}}{F} \rightarrow$$

$$\therefore \qquad z'_C = \frac{0.81 \times 9800 \times 2 \times \dfrac{8^2}{2} + 9800 \times \dfrac{8^3}{3} + \dfrac{9800}{392} \times \dfrac{8^4}{8}}{170\,200}$$

$$- \frac{0.81 \times 9800 \times 2 \times \dfrac{(8-2)^2}{2} + 9800 \times \dfrac{(8-2)^3}{3} + \dfrac{9800}{392} \times \dfrac{(8-2)^4}{8}}{170\,200} = \mathbf{7.05\ m}.$$

**Fig. 1.10** Pressure diagram

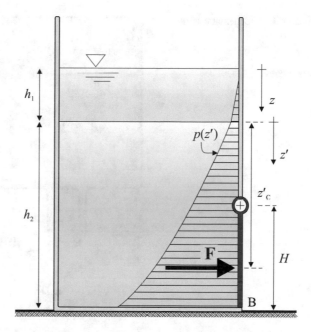

Figure 1.10 shows the pressure diagram along the vertical. In this case, if the fluid had been of uniform density, the centre of pressure would have been closer to the free surface.

---

**Exercise 1.8** The tank in Fig. 1.11 is separated in two tanks by an inclined septum with a circular opening of diameter $D = (0.50 + C_{pu}/20)$ m closed by a sluice gate. The differential manometer shows a difference in level of $\Delta h = (0.10 + C_u/10)$ m. The manometric fluid is mercury (Hg).

- Calculate the direction and magnitude of the force exerted on the circular gate.
- Calculate the centre of pressure.

Assume $\gamma_w = 9800\,\mathrm{N\,m^{-3}}$, $\gamma_{Hg}/\gamma_w = 13.6$.

**Solution** The pressure head for the left tank is higher than the pressure head for the right tank (Fig. 1.12), with a difference equal to

$$\delta = \frac{(\gamma_{Hg} - \gamma_w)\,\Delta h}{\gamma_w}.$$

The force exerted on the gate by the fluid in the left tank is orthogonal to the plane of the gate and is equal to

**Fig. 1.11** Schematic of the tank

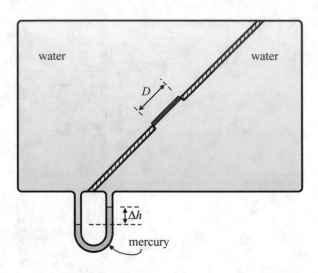

$$F_l = \gamma_w z_{Glx} \frac{\pi D^2}{4},$$

and the force exerted by the fluid in the right tank, also orthogonal to the plane of the gate, is equal to

$$F_r = \gamma_w z'_{Grx} \frac{\pi D^2}{4},$$

where $z_{Glx}$ and $z'_{Grx}$ are the coordinates of the centroid of the gate with respect to the piezometric level of the fluid in the left and in the right tank, respectively. The total force is

$$F = F_l - F_r = \gamma_w \left(z_{Glx} - z'_{Grx}\right) \frac{\pi D^2}{4} = \gamma_w \delta \frac{\pi D^2}{4} = \left(\gamma_{Hg} - \gamma_w\right) \Delta h \frac{\pi D^2}{4},$$

is orthogonal to the circular gate and is applied in the centroid, **F** being due to the action of a uniform pressure originating from the difference between two trapezoidal pressure diagrams with equal inclination (see Fig. 1.12). Therefore, the centre of pressure and the centroid coincide.

For $C_u = C_{pu} = 0$ it results $D = 0.50\,\text{m}$, $h = 0.10\,\text{m}$ and

$$\therefore \qquad F = \left(\gamma_{Hg} - \gamma_w\right) \Delta h \frac{\pi D^2}{4}$$

$$= (13.6 - 1) \times 9800 \times 0.10 \times \frac{\pi \times 0.5^2}{4} = \mathbf{2424\,N.}$$

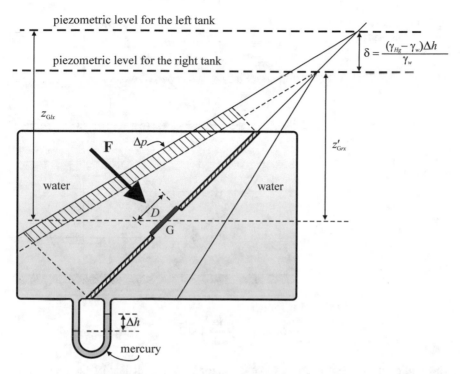

**Fig. 1.12** Piezometric levels (gage pressure) for the left and the right tanks, respectively, and pressure diagram

---

**Exercise 1.9**  In the system in Fig. 1.13 the sluice gate, of depth $L = (2 + C_u/2)$ m, is pivoted in A, at a height of $a = (4 + C_{pu})$ m from the horizontal flat bottom. The two liquids have specific weight $\gamma_1 = 10\,000\,\mathrm{N\,m^{-3}}$ and $\gamma_2 = 12\,000\,\mathrm{N\,m^{-3}}$, respectively. The level of the liquid on the left is $h_1 = (3 + C_{pu}/2)$ m.

- Calculate the force exerted by the liquid on the left.
- Calculate the centre of pressure.
- Calculate the $h_2$ level corresponding to an incipient opening of the gate.

Neglect the weight of the gate.

**Solution**  As shown in the diagram in Fig. 1.14, the force exerted on the flat sluice gate by the liquid on the left is orthogonal to the surface and has a magnitude equal to:

$$S_1 = \gamma_1 \frac{h_1}{2} L h_1 \sqrt{2},$$

and is applied at a distance (parallel to the gate) equal to $h_1\sqrt{2}/3$ from the bottom. The arm of this force with respect to the pivot in A is equal to

**Fig. 1.13** Schematic of the sluice gate pivoted in A

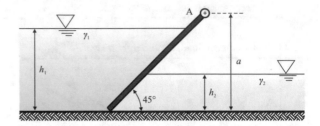

**Fig. 1.14** System of forces acting on the gate

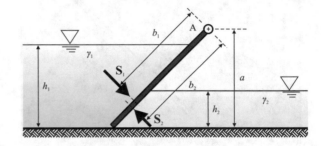

$$b_1 = \left(a - \frac{h_1}{3}\right)\sqrt{2}.$$

The liquid on the right exerts a force orthogonal to the surface of the gate with a magnitude equal to:

$$S_2 = \gamma_2 \frac{h_2}{2} L h_2 \sqrt{2} \ \ \text{if} \ \ h_2 \le a,$$

applied at a distance (parallel to the gate) equal to $h_2\sqrt{2}/3$ from the bottom. The arm of this force with respect to the pivot in A is

$$b_2 = \left(a - \frac{h_2}{3}\right)\sqrt{2}.$$

By imposing the balance of moments about the axis trough A, in conditions of incipient rotation yields

$$S_1 b_1 = S_2 b_2 \ \rightarrow \ \gamma_1 \frac{h_1}{2} L h_1 \sqrt{2} \left(a - \frac{h_1}{3}\right)\sqrt{2} = \gamma_2 \frac{h_2}{2} L h_2 \sqrt{2} \left(a - \frac{h_2}{3}\right)\sqrt{2},$$

equivalent to the following 3rd order equation in the unknown $h_2$,

$$h_2^3 - 3a h_2^2 + 3\frac{\gamma_1}{\gamma_2} h_1^2 \left(a - \frac{h_1}{3}\right) = 0.$$

This equation admits at least one real solution. It can be demonstrated that if $h_1 < a$ and $\gamma_1 < \gamma_2$, it also results $h_2 < h_1 < a$.

For $C_u = C_{pu} = 0$ it results $L = 2\,\mathrm{m}$, $a = 4\,\mathrm{m}$, $h_1 = 3\,\mathrm{m}$,

$$\therefore \qquad S_1 = \gamma_1 \frac{h_1}{2} L h_1 \sqrt{2} = 10\,000 \times \frac{3}{2} \times 2 \times 3 \times \sqrt{2} = \mathbf{127.3\ kN},$$

$$\therefore \qquad b_1 = \left(a - \frac{h_1}{3}\right)\sqrt{2} = \left(4 - \frac{3}{3}\right) \times \sqrt{2} = \mathbf{4.24\ m}.$$

The computation of the value of $h_2$ requires the solution of the following equation:

$$h_2^3 - 3ah_2^2 + 3\frac{\gamma_1}{\gamma_2}h_1^2\left(a - \frac{h_1}{3}\right) = 0 \rightarrow$$

$$h_2^3 - 3 \times 4h_2^2 + 3 \times \frac{10\,000}{12\,000} \times 3^2 \times \left(4 - \frac{3}{3}\right) = 0 \rightarrow$$

$$h_2^3 - 12h_2^2 + 67.5 = 0.$$

The solutions can be obtained with a numerical procedure or by applying the complex analytical formula due to Cardano (1501–1576), resulting in:

$$\therefore \qquad \begin{aligned} h_2 &= \mathbf{2.69\,m}, \\ h_2 &= 11.49\,\mathrm{m}, \\ h_2 &= -2.19\,\mathrm{m}, \end{aligned}$$

of which only the first one is acceptable.

---

**Exercise 1.10** In the system in Fig. 1.15 a rectangular sluice gate of length $L$ and depth $b = 3\,\mathrm{m}$ (the latter dimension is orthogonal to the drawing plane) is pivoted at the top side and separates two tanks containing water. The water level in the left tank is $h = (2 + C_u/20)\,\mathrm{m}$ and the weight of the sluice gate is $P = (6 + C_{pu}/20) \times 10^4\,\mathrm{N}$.

- Calculate the force exerted on the sluice gate by the water in the tank on the left, and its centre of pressure.
- Calculate the maximum level of the water in the right tank to prevent the sluice gate from opening.

Assume $\gamma_w = 9800\ \mathrm{N\,m^{-3}}$.

**Fig. 1.15** Schematic of the tank and of the upper pivoted sluice gate

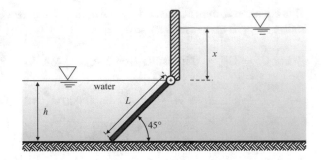

**Fig. 1.16** System of forces acting on the sluice gate and pressure diagram

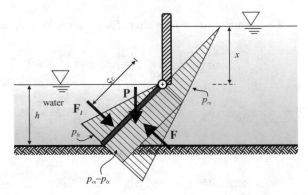

**Solution** The force exerted on the sluice gate by the water in the left tank is orthogonal to the sluice gate and has a magnitude equal to:

$$F_l = \gamma_w \frac{h}{2} bL,$$

where $L$ is the length of the sluice gate, equal to

$$L = h\sqrt{2}.$$

This force is applied at a distance $\varsigma_C = 2L/3$ from the free surface level of the water, measured parallel to the sluice gate.

The system of acting forces is shown in Fig. 1.16.

Considering the hydrostatic forces due to the water in the left and right tanks, we can see that two different pressure distributions act on the sluice gate:

- from left, a pressure with triangular diagram;
- from right, a pressure with trapezoidal diagram.

Since the liquid is the same for the two tanks, the slope of the two diagrams is the same and since the elementary forces due to the pressure on the two sides have opposing direction, only the rectangular part of the pressure diagram acts on the sluice gate, from right to left. The sluice gate is ultimately subjected to the action of a uniform pressure equal to $\gamma_w x$, exerting a clockwise torque about the pivot. Imposing the balance equation of moments of forces yields:

$$\gamma_w x b L \frac{L}{2} - P \frac{h}{2},$$

where $h/2$ is the arm of the weight. The condition that prevents the sluice gate from opening is

$$x \leq \frac{Ph}{\gamma_w b L^2}.$$

For $C_u = C_{pu} = 0$ it results $b = 3\,\mathrm{m}$, $h = 2\,\mathrm{m}$, $P = 6 \times 10^4\,\mathrm{N}$, and

$$L = h\sqrt{2} = 2 \times \sqrt{2} = 2.83\,\mathbf{m},$$

$$\therefore \qquad F_l = \gamma_w \frac{h}{2} b L = 9800 \times \frac{2}{2} \times 3 \times 2.83 = \mathbf{83.2\,kN},$$

$$\therefore \qquad \varsigma_C = \frac{2}{3} L = \frac{2}{3} \times 2.83 = \mathbf{1.89\,m},$$

$$\therefore \qquad x \leq \frac{Ph}{\gamma_w b L^2} \equiv \frac{6 \times 10^4 \times 2}{9800 \times 3 \times 2.82^2} - \mathbf{0.51\,m}.$$

---

**Exercise 1.11** In the system in Fig. 1.17 a sluice gate pivoted at the top side, $b = 2.0\,\mathrm{m}$ width orthogonal to the drawing, separates two tanks, containing water in the left and oil in the right. The oil level in the right tank is $h_2 = (2 + C_u/20)\,\mathrm{m}$, the weight of the sluice gate is $P = (6 + C_{pu}/20) \times 10^4\,\mathrm{N}$, and $a = (1 + C_u/10)\,\mathrm{m}$.

- Calculate the force exerted on the sluice gate by the oil in the right tank.
- Calculate the centre of pressure of the force.
- Calculate the minimum value for the water depth in the left tank to prevent opening of the gate.

Assume $\gamma_w = 9806\,\mathrm{N\,m^{-3}}$, $\gamma_o = 0.8 \times 9806\,\mathrm{N\,m^{-3}}$.

**Solution** Considering the schematic in Fig. 1.18, the force exerted by the oil in the right tank is normal to the plane of the gate and has a magnitude equal to

$$F_o = \gamma_o (h_2 - a/2) bL \quad \text{(we assume that } h_2 > a \text{ is always satisfied),} \qquad (1.1)$$

**Fig. 1.17** Schematic of the sluice gate and of the tanks

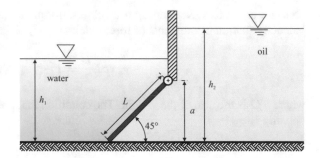

**Fig. 1.18** System of forces acting on the gate

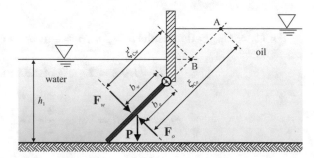

where $L = a\sqrt{2}$ is the length of the gate.

The centre of pressure of $F_o$, measured along the sluice gate from the free surface level of the oil, is calculated as follows:

$$\xi_{Co} = \xi_{Go} + \frac{I_{Gxx}}{S_x} \equiv \left(h_2\sqrt{2} - L/2\right) + \frac{\frac{1}{12}bL^3}{bL\left(h_2\sqrt{2} - L/2\right)},$$

where $I_{Gxx}$ is second moment of inertia of the sluice gate with respect to its centroid and $S_x$ is the moment of inertia of the sluice gate with respect to A. If the torque due to the force exerted by the oil and to the weight of the gate, calculated about the pivot, is counter-clockwise, the gate will not rotate even if it is $h_1 = 0$.

This condition of stability of the gate (independent from the value of $h_1$) is

$$F_o b_o - P\frac{a}{2} < 0, \tag{1.2}$$

where $b_o$ is the arm of the force due to the oil with respect to the pivot, equal to:

$$b_o = \xi_{Co} - (h_2\sqrt{2} - L) \equiv \left[\frac{\frac{1}{12}bL^3}{bL\left(h_2\sqrt{2} - L/2\right)} + \frac{L}{2}\right], \tag{1.3}$$

and $a/2$ is the arm of the weight of the gate. Substituting Eqs. (1.1–1.3) into Eq. (1.2), yields:

$$\gamma_o \left(h_2 - a/2\right) bl \left[\frac{\frac{1}{12}bL^3}{bL\left(h_2\sqrt{2} - L/2\right)} + \frac{L}{2}\right] - P\frac{a}{2} < 0.$$

If this latter condition is not met, the minimum water level required to prevent the gate from opening shall be calculated by requiring that the moment of all forces about the pivot be zero. Choosing a coordinate system with a positive moment in the clockwise direction, the condition for rotational equilibrium is:

$$F_o b_o - F_w b_w - P\frac{a}{2} = 0.$$

The force exerted by the water in the left tank is orthogonal to the sluice gate and has a magnitude equal to:

$$\begin{cases} F_w = \gamma_w \left(h_1 - a/2\right) bL \equiv \gamma_w \frac{\sqrt{2}}{2}\left(h_1\sqrt{2} - L/2\right) bL, & \text{if } h_1 \geq a, \\ F_w = \gamma_w \frac{h_1^2\sqrt{2}}{2}b, & \text{if } h_1 < a. \end{cases}$$

This force is applied at a distance from the free surface level of the water, measured parallel to the sluice gate, equal to:

$$\begin{cases} \xi'_{Cw} = \xi'_{Gw} + \frac{I'_{Gxx}}{S'_x} \equiv \left(h_1\sqrt{2} - L/2\right) + \frac{\frac{1}{12}bL^3}{bL\left(h_1\sqrt{2} - L/2\right)}, & \text{if } h_1 \geq a, \\ \xi'_{Cw} = \frac{2}{3}h_1\sqrt{2}, & \text{if } h_1 < a, \end{cases}$$

where $I'_{Gxx}$ is the second moment of inertia of the sluice gate with respect to its centroid G, and $S'_x$ is the moment of inertia of the sluice gate with respect to the water line. The arm of the force exerted by the water, with respect to the pivot, is equal to:

$$\begin{cases} b_w = \xi'_{Cw} - (h_1\sqrt{2} - L), & \text{if } h_1 \geq a, \\ b_w = \frac{2}{3}h_1\sqrt{2} + (L - h_1\sqrt{2}), & \text{if } h_1 < a. \end{cases}$$

The torque equilibrium condition is

$$\gamma_o \left(h_2 - a/2\right) bL \left[ \frac{\frac{1}{12}bL^3}{bL\left(h_2\sqrt{2} - L/2\right)} + \frac{L}{2} \right] -$$

$$\gamma_w \frac{\sqrt{2}}{2}\left(h_1\sqrt{2} - L/2\right) bL \left[ \frac{\frac{1}{12}bL^3}{bL\left(h_1\sqrt{2} - L/2\right)} + \frac{L}{2} \right] - P\frac{a}{2} \leq 0, \text{ if } h_1 \geq a,$$

$$\gamma_o \left(h_2 - a/2\right) bL \left[ \frac{\frac{1}{12}bL^3}{bL\left(h_2\sqrt{2} - L/2\right)} + \frac{L}{2} \right] -$$

$$\gamma_w \frac{h_1^2\sqrt{2}}{2} b \left[ \frac{2}{3}h_1\sqrt{2} + (L - h_1\sqrt{2}) \right] - P\frac{a}{2} \leq 0, \text{ if } h_1 < a.$$

In the first case ($h_1 \geq a$), substituting $\left(h_1\sqrt{2} - L/2\right) = x$ yields

$$\underbrace{\gamma_o \left(h_2 - a/2\right) bL \left[ \frac{\frac{1}{12}bL^3}{bL\left(h_2\sqrt{2} - L/2\right)} + \frac{L}{2} \right]}_{M_o} -$$

$$\gamma_w \frac{\sqrt{2}}{24}bL^3 - \gamma_w \frac{\sqrt{2}}{4}xbL^2 - P\frac{a}{2} \leq 0, \text{ if } h_1 \geq a,$$

hence

$$x \geq \frac{M_o - P\frac{a}{2} - \gamma_w \frac{\sqrt{2}}{24}bL^3}{\gamma_w \frac{\sqrt{2}}{4}bL^2},$$

or

$$h_1 \geq \frac{L}{2\sqrt{2}} + \frac{M_o - P\frac{a}{2} - \gamma_w \frac{\sqrt{2}}{24}bL^3}{\gamma_w \frac{1}{2}bL^2}.$$

In the second case ($h_1 < a$), a third degree equation is obtained which always allows a real positive solution, provided that

$$F_o b_o - P\frac{a}{2} \geq 0.$$

We notice that this last condition is the same condition stated at the beginning of the exercise and related to the dependence of the torque balance on the water level $h_1$.

For $C_u = C_{pu} = 0$ it results $b = 2.0$ m, $h_2 = 2$ m, $P = 6 \times 10^4$ N, $a = 1$ m, $\gamma_w = 9806$ N m$^{-3}$, $\gamma_o = 7845$ N m$^{-3}$. Hence

$$\therefore \qquad L = a\sqrt{2} = 1 \times \sqrt{2} = \mathbf{1.41\ m},$$

$$\therefore \qquad F_o = \gamma_o \left(h_2 - a/2\right) bL = 7845 \times (2 - 1/2) \times 2 \times 1 \times \sqrt{2} = \mathbf{33.28\ kN},$$

$$\therefore \qquad b_o = \frac{\frac{1}{12}bL^3}{bL\left(h_2\sqrt{2} - L/2\right)} + \frac{L}{2}$$

$$= \frac{\frac{1}{12} \times 2 \times \left(1 \times \sqrt{2}\right)^3}{2 \times 1 \times \sqrt{2} \times \left(2 \times \sqrt{2} - 1 \times \sqrt{2}/2\right)} + \frac{1 \times \sqrt{2}}{2} = \mathbf{0.78\ m},$$

$$\therefore \qquad F_o b_o - P\frac{a}{2} \rightarrow 33\,280 \times 0.78 - 6 \times 10^4 \times \frac{1}{2} < \mathbf{0}.$$

Hence, the gate is always closed, even for $h_1 = 0$.

---

**Exercise 1.12** In the tank in Fig. 1.19, a rectangular gate $\overline{AB}$ separates two fluids of specific weight $\gamma_1 = 8000$ N m$^{-3}$ and $\gamma_2 = 9800$ N m$^{-3}$, respectively. The fluid depths are $h_1 = 3.5$ m and $h_2 = 4.5$ m and the height of the rectangular gate is $H = 1.5$ m.

– Calculate the force per unit depth acting on the gate.
– Calculate the direction and the pressure centre of the force.

**Solution** By considering the schematic shown in Fig. 1.20, the fluid 1 exerts a horizontal force (per unit depth), pointing to the right equal to

$$F_1 = \gamma_1 \left(h_1 - \frac{H}{2}\right) H = 8000 \times \left(3.5 - \frac{1.5}{2}\right) \times 1.5 = \mathbf{33.0\ kN},$$

applied at a distance from the free surface of fluid 1 equal to

**Fig. 1.19** Schematic of the tank with the flat sluice gate pivoted in A

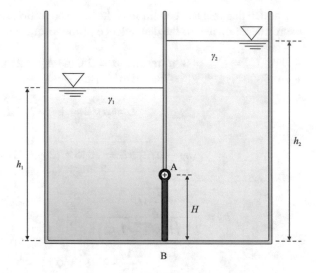

**Fig. 1.20** Schematic for the calculation of forces and centres of pressure

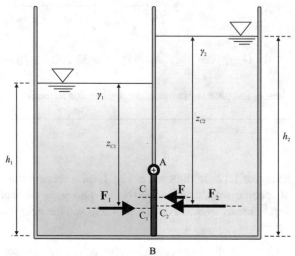

$$z_{C1} = z_{G1} + \frac{I_{G1xx}}{S_x} = \left(h_1 - \frac{H}{2}\right) + \frac{\frac{1}{12}H^3}{H\left(h_1 - H/2\right)} =$$

$$\left(3.5 - \frac{1.5}{2}\right) + \frac{\frac{1}{12} \times 1.5^3}{1.5 \times (3.5 - 1.5/2)} = \mathbf{2.82\ m}.$$

Fluid 2 exerts a horizontal force (per unit depth) pointing to the left equal to

$$F_2 = -\gamma_2 \left( h_2 - \frac{H}{2} \right) H = -9800 \times \left( 4.5 - \frac{1.5}{2} \right) \times 1.5 = -55.1\,\text{kN},$$

applied at a distance from the free surface of fluid 2 equal to

$$z_{C2} = z_{G2} + \frac{I_{G2x'x'}}{S_{x'}} = \left( h_2 - \frac{H}{2} \right) + \frac{\frac{1}{12}H^3}{H\,(h_2 - H/2)} =$$

$$\left( 4.5 - \frac{1.5}{2} \right) + \frac{\frac{1}{12} \times 1.5^3}{1.5 \times (4.5 - 1.5/2)} = \textbf{3.80 m}.$$

The resulting force is pointing to the left and is equal to

$$\therefore \qquad F = F_1 + F_2 = 33\,000 - 55\,125 = -\,\textbf{22.1 kN}.$$

The balance of moment of forces about the axis A yields

$$F_1 \times \overline{AC_1} + F_2 \times \overline{AC_2} = F \times \overline{AC} \rightarrow \overline{AC} = \frac{F_1 \times \overline{AC_1} + F_2 \times \overline{AC_2}}{F},$$

where $\overline{AC_1} = z_{C1} - (h_1 - H)$, $\overline{AC_2} = z_{C2} - (h_2 - H)$, and $\overline{AC}$ are the arms of the forces exerted by the two fluids and of the resulting force with respect to the pivot in A, see Fig. 1.20. Hence,

$$\therefore \qquad \overline{AC} = \frac{33.0 \times (2.82 - 3.5 + 1.5) - 55.1 \times (3.80 - 4.5 + 1.5)}{-22.1} = \textbf{0.77 m}.$$

---

**Exercise 1.13** In the system in Fig. 1.21 the flat sluice gate, of unitary depth, is pivoted in A and is inclined at an angle to the horizontal $\alpha = (45 + C_{pu})°$. The upper fluid is oil, the lower fluid, consisting of water and mud, has specific weight increasing downwards according to the following equation:

$$\gamma = \gamma_w + \frac{\gamma_w}{98} \frac{z'}{b}.$$

- Calculate the total force exerted by water and mud and oil on the gate.
- Calculate the centre of pressure of the force.

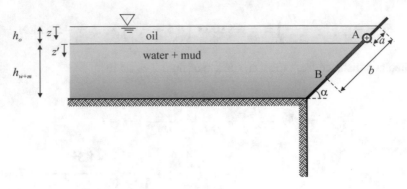

**Fig. 1.21** Schematic of the gate with a stratified fluid

Assume $a = (2 + C_u/10)$ m, $b = (4 + C_{pu}/10)$ m, $h_o = (2.5 + C_u/10)$ m, $\gamma_w = 9800\,\mathrm{N\,m^{-3}}$, $\gamma_o = 0.8\gamma_w$. Notice that the pivot in A is at a higher level than the separation plane between oil and water plus mud.

**Solution** It is convenient to introduce the two new $Ox$ and $Ox'$ coordinate systems shown in Fig. 1.22, in addition to the $z$ and $z'$ coordinate systems. The pressure in the oil, for $0 < z < h_o$, varies linearly and, in the coordinate system $z$, it is equal to $p = \gamma_o z \equiv 0.8\gamma_w z$. In the $Ox$ coordinate system, in the oil $(0 < x < h_o/\sin\alpha)$, it results:

$$p(x) = (\gamma_o \sin\alpha)\, x.$$

The pressure in the fluid water plus mud, for $h_o < z < h_o + z'$, is calculated by integrating the hydrostatic equation. In the $z'$ coordinate system, it results:

$$\frac{dp}{dz'} = \gamma(z') \equiv \gamma_w + \frac{\gamma_w}{98}\frac{z'}{b} \rightarrow \int\limits_{p(0)}^{p(z')} dp = \int\limits_0^{z'} \left(\gamma_w + \frac{\gamma_w}{98}\frac{z'}{b}\right) dz' \rightarrow$$

$$p(z') = p(0) + \gamma_w z' + \frac{\gamma_w}{98}\frac{z'^2}{2b} = \gamma_o h_o + \gamma_w z' + \frac{\gamma_w}{98}\frac{z'^2}{2b}.$$

In the $O'x'$ coordinate system and in the domain occupied by the mixture of water and mud, it results:

$$p(x') = \gamma_o h_o + (\gamma_w \sin\alpha)\, x' + \left(\frac{\gamma_w}{98}\sin^2\alpha\right)\frac{x'^2}{2b}.$$

The qualitative diagram of the pressure acting on the flat gate is shown in Fig. 1.23. The force is calculated by integrating the elementary forces associated with the pressure on the surface of the gate, (the domain of integration is the shaded grey area in Fig. 1.23), separating the integral in the two contributions due to the oil and to the mixture of water and mud, respectively:

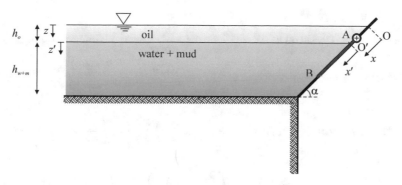

**Fig. 1.22** Coordinate systems used for the calculation of forces

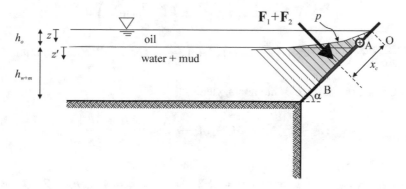

**Fig. 1.23** Diagram of the pressure acting on the flat gate

$$F_1 = \int\limits_{a}^{h_o/\sin\alpha} (\gamma_o \sin\alpha)\, x\, dx = \frac{\gamma_o}{2} \sin\alpha \left( \frac{h_o^2}{\sin^2\alpha} - a^2 \right) \quad \text{(integration in } Ox),$$

$$F_2 = \int\limits_{0}^{c} \left[ \gamma_o h_o + (\gamma_w \sin\alpha)\, x' + \left( \frac{\gamma_w}{98} \sin^2\alpha \right) \frac{x'^2}{2b} \right] dx'$$

$$= \gamma_o h_o c + \left( \frac{\gamma_w}{2} \sin\alpha \right) c^2 + \left( \frac{\gamma_w}{98b} \sin^2\alpha \right) \frac{c^3}{6} \quad \text{(integration in } O'x'),$$

where $c = b - (h_o/\sin\alpha)$ represents the length of the gate at contact with the mixture of water plus mud.

The centre of pressure is calculated by imposing that the moment about an axis (for example, the axis orthogonal to the sheet and passing through O) of the distribution of forces, coincides with the moment (calculated with respect to the same axis) of the resultant of these forces. The two forces exert, with respect to the axis previously defined, two counter-clockwise moments equal to:

$$M_1 = \int\limits_{a}^{h_o/\sin\alpha} (\gamma_o \sin\alpha)\, x^2\, dx = \frac{\gamma_o}{3}\sin\alpha \left(\frac{h_o^3}{\sin^3\alpha} - a^3\right) \quad \text{(integration in } Ox\text{)},$$

and

$$M_2 = \int\limits_{0}^{c} \left[\gamma_o h_o + (\gamma_w \sin\alpha)\, x' + \left(\frac{\gamma_w}{98}\sin^2\alpha\right)\frac{x'^2}{2b}\right]\left(\frac{h_o}{\sin\alpha} + x'\right) dx' \rightarrow$$

$$M_2 = \left[(\gamma_o h_o)\, c + \left(\frac{\gamma_w}{2}\sin\alpha\right) c^2 + \left(\frac{\gamma_w}{98}\frac{1}{6b}\sin^2\alpha\right) c^3\right]\frac{h_o}{\sin\alpha} +$$

$$\left(\frac{\gamma_o}{2}h_o\right) c^2 + \left(\frac{\gamma_w}{3}\sin\alpha\right) c^3 + \left(\frac{\gamma_w}{98}\frac{1}{8b}\sin^2\alpha\right) c^4$$

$$= F_2\frac{h_o}{\sin\alpha} + \left(\frac{\gamma_o}{2}h_o\right) c^2 + \left(\frac{\gamma_w}{3}\sin\alpha\right) c^3 + \left(\frac{\gamma_w}{98}\frac{1}{8b}\sin^2\alpha\right) c^4 \quad \text{(integration in } O'x'\text{)}.$$

By imposing the balance of moments, yields

$$(F_1 + F_2)\, x_C = M_1 + M_2 \rightarrow x_C = \frac{M_1 + M_2}{F_1 + F_2}.$$

For $C_u = C_{pu} = 0$ it results $\alpha = 45°$, $a = 2\,\text{m}$, $b = 4\,\text{m}$, $h_o = 2.5\,\text{m}$, $\gamma_o = 7840\,\text{N}\,\text{m}^{-3}$,

$$\therefore \qquad c = b - \frac{h_o}{\sin\alpha} = 4 - \frac{2.5}{\sin 45°} = 0.46\,\text{m},$$

$$\gamma = 9800 + 25z'\,\text{N}\,\text{m}^{-3} \quad (z' \text{ in metres}),$$

$$p(x) = (\gamma_o \sin\alpha)\, x = 5543.7x\,\text{Pa} \quad (x \text{ in metres}),$$

$$p(x') = 7840h_o + (9800\sin\alpha)x' + (100\sin^2\alpha)\frac{x'^2}{2b} =$$
$$19\,600 + 6930x' + 6.25x'^2\,\text{Pa} \quad (x' \text{ in metres}),$$

$$\therefore \qquad F_1 = \frac{\gamma_o}{2}\sin\alpha \left(\frac{h_o^2}{\sin^2\alpha} - a^2\right)$$
$$= \frac{7840}{2} \times \sin 45° \times \left(\frac{2.5^2}{\sin^2 45°} - 2^2\right) = 23\,560\,\text{N},$$

$$\therefore \quad F_2 = (\gamma_o h_o)\,c + \left(\frac{\gamma_w}{2}\sin\alpha\right)c^2 + \left(\frac{\gamma_w}{98b}\sin^2\alpha\right)\frac{c^3}{6}$$

$$= 7840 \times 2.5 \times 0.46 + \left(\frac{9800}{2}\times\sin 45°\right)\times 0.46^2$$

$$+ \left(\frac{9800}{98\times 4}\times\sin^2 45°\right)\times\frac{0.46^3}{6} = 9749\ \text{N},$$

$$\therefore \quad M_1 = \frac{\gamma_o}{3}\sin\alpha\left(\frac{h_o^3}{\sin^3\alpha} - a^3\right)$$

$$= \frac{7840}{3}\times\sin 45° \times\left(\frac{2.5^3}{\sin^3 45°} - 2^3\right) = \mathbf{66\,883\ N\,m},$$

$$\therefore \quad M_2 = F_2\frac{h_o}{\sin\alpha} + \left(\frac{\gamma_o}{2}h_o\right)c^2 + \left(\frac{\gamma_w}{3}\sin\alpha\right)c^3 + \left(\frac{\gamma_w}{98}\frac{1}{8b}\sin^2\alpha\right)c^4$$

$$= 9749 \times \frac{2.5}{\sin 45°} + \left(\frac{7840}{2}\times 2.5\right)\times 0.46^2 + \left(\frac{9800}{3}\times\sin 45°\right)\times 0.46^3$$

$$+ \left(\frac{9800}{98}\times\frac{1}{8\times 4}\times\sin^2 45°\right)\times 0.46^4 = \mathbf{36\,768\ N\,m},$$

$$\therefore \quad x_C = \frac{M_1 + M_2}{F_1 + F_2} = \frac{66\,883 + 36\,768}{23\,560 + 9749} = \mathbf{3.11\ m}.$$

**Exercise 1.14** A flat sluice gate of mass $M = 2000\,\text{kg}$ is pivoted along the lower side, pivot in A in Fig. 1.24. The length of the gate in the direction orthogonal to the drawing is $l = 8\,\text{m}$.

– Calculate the size $b$ of the gate if the system is in equilibrium in the configuration shown in Fig. 1.24.

The liquid is water of specific weight $\gamma_w = 9800\,\text{N\,m}^{-3}$.

**Solution** As shown in Fig. 1.25, the forces acting on the gate are the weight **P** and the force **S** of the fluid. If the system is in equilibrium, the resulting torque with respect to the pivot A is null, i.e.

$$\mathbf{M_S + M_P = 0}.$$

The torque $\mathbf{M_S}$ is clockwise and has a magnitude equal to:

**Fig. 1.24** Schematic of the flat sluice gate pivoted in A

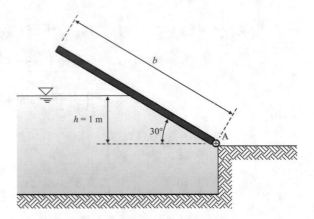

**Fig. 1.25** Schematic of the forces and the torques acting on the gate

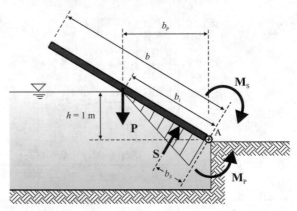

$$|\mathbf{M_S}| = Sb_s,$$

where $S$ is the force exerted by water, $b_s$ is the action arm.

The torque $\mathbf{M_P}$ is counter-clockwise and is due to the weight $P = Mg$ acting with an arm equal to $b_P = (b/2)\cos 30°$. In equilibrium condition, it results:

$$Sb_S = Mgb_P \rightarrow Sb_S = Mg\frac{b}{2}\cos 30° \rightarrow b = \frac{2Sb_S}{\cos 30° Mg}.$$

The portion of the gate at contact with water has length:

$$b_i = h/\sin 30° = \mathbf{2\,m}.$$

The magnitude of the force exerted by the water is equal to:

$$S = \gamma_w z_G b_i l = 9806 \times 0.5 \times 2 \times 8 = \mathbf{78\,448\,N},$$

where $z_G$ is the depth of the centroid of the immersed portion of the gate. The action arm with respect to the pivot is equal to

$$b_S = b_i/3 \approx 0.66\,\text{m}.$$

Hence

$$\therefore \qquad b = \frac{2Sb_S}{\cos 30° Mg} = \frac{2 \times 78\,448 \times 0.66}{\cos 30° \times 2000 \times 9.806} = 6.15\,\text{m}.$$

---

**Exercise 1.15** The wooden shuttering for concrete constructional work of a staircase in Fig. 1.26 is filled with fluid concrete with $\gamma_c/\gamma_w = 2.4$, where $\gamma_w = 9806\,\text{N}\,\text{m}^{-3}$ is the specific weight of water. The weight of the shuttering is $P_{sh} = 370\,\text{N}$ and the step of the staircase is $l = 0.90\,\text{m}$ wide.

– Calculate the ballast required to hold the shuttering in place, assuming that the rotational equilibrium is always guaranteed.

**Solution** The vertical force on the shuttering is calculated as the product of the pressure acting on the shuttering that delimits the treads of the steps by the area of the surface of the treads themselves. With reference to the diagram in Fig. 1.27, results:

$$p_1 = \gamma_c 2h, \quad p_2 = \gamma_c h.$$

Hence,

$$F_{1z} = -p_1 bl = -\gamma_c 2hbl,$$

**Fig. 1.26** schematic of the shuttering containing fluid concrete

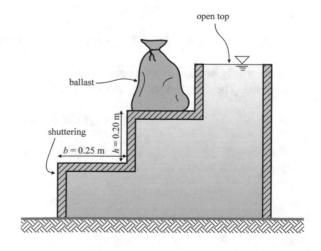

**Fig. 1.27** Diagram of the
pressures and vertical forces
acting on the shuttering

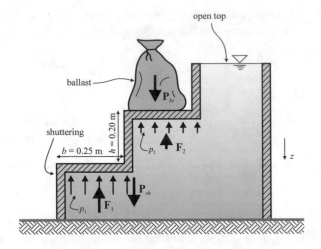

and

$$F_{2z} = -p_2bl = -\gamma_c hbl.$$

The equilibrium condition in the vertical, yields

$$P_{bt} + P_{sh} + F_{1z} + F_{2z} = 0 \rightarrow P_{bt} = -F_{1z} - F_{2z} - P_{sh}.$$

Inserting the values of the variables, yields

$$P_{sh} = \textbf{370 N},$$

$$F_{1z} = -\gamma_c 2hbl = -2.4 \times 9806 \times 2 \times 0.2 \times 0.25 \times 0.9 = \textbf{-2118 N},$$

$$F_{2z} = -\gamma_c hbl = -2.4 \times 9806 \times 0.2 \times 0.25 \times 0.9 = \textbf{-1059 N},$$

$$\therefore \qquad P_{bt} = -F_{1z} - F_{2z} - P_{sh} = 2118 + 1059 - 370 = \textbf{2807 N}.$$

Alternatively, we can apply a global method, valid for surfaces that are not necessarily flat. The equilibrium equation of the dashed control volume in Fig. 1.28,
is

$$\textbf{G} + \boldsymbol{\Pi}_0 + \boldsymbol{\Pi}_1 + \boldsymbol{\Pi}_2 = \textbf{0}.$$

The physical meaning of the various forces is clarified in Fig. 1.29. In terms of
relative pressure, results $\boldsymbol{\Pi}_1 = \textbf{0}$. Also:

$$\Pi_{2z} = -3p_fbl,$$

**Fig. 1.28** Control volume
for the application of a
global method

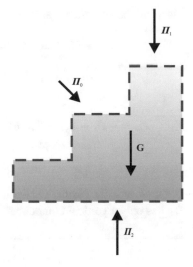

**Fig. 1.29** Schematic of the
forces described in the global
method

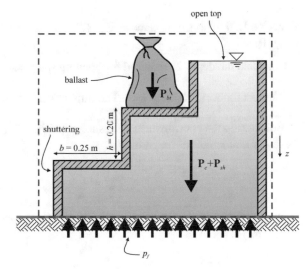

where $p_f$ is the pressure at the bottom, and $l$ is the step width. The other forces are

$$G = P_c = \gamma_c 6bhl$$

and

$$\Pi_{0z} = P_{bt} + P_{sh}.$$

Hence:

$$P_c + P_{bt} + P_{sh} - 3p_f bl = 0 \rightarrow P_{bt} = -P_c - P_{sh} + 3p_f bl.$$

Inserting the numerical values yields

$$P_c = \gamma_c 6bhl = 2.4 \times 9806 \times 6 \times 0.25 \times 0.20 \times 0.9 = \textbf{6354 N},$$

$$p_f = 3\gamma_c h = 3 \times 2.4 \times 9806 \times 0.20 = \textbf{14 120 Pa},$$

$$\therefore \qquad P_{bt} = -P_c - P_{sh} + 3p_f bl$$
$$= -6354 - 370 + 3 \times 14\ 120 \times 0.25 \times 0.9 = \textbf{2807 N}.$$

Obviously, the results obtained with the two methods are the same. Notice that $\Pi_0$ has only a vertical component, since there is no horizontal force to balance.

---

**Exercise 1.16** The tilting gate in Fig. 1.30 is pivoted at the edge, with $b = (2 + C_{pu}/10)$ m.

– Calculate the depth $D$ required for opening the gate.

Neglect the weight of the gate. The fluid is water with $\gamma_w = 9806\,\mathrm{N\,m^{-3}}$.

**Solution** In the coordinate system shown in Fig. 1.31, the fluid exerts a horizontal force per unit length of magnitude

$$F_x = \frac{1}{2}\gamma_w D^2,$$

applied at a distance of $D/3$ from the pivot. The moment of this force (positive if clockwise) is equal to

**Fig. 1.30** Tilting gate

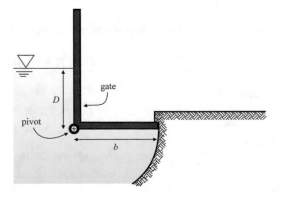

**Fig. 1.31** Schematic for the
calculation of the forces

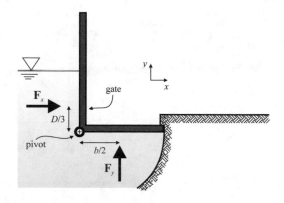

$$M_{F_x} = \frac{1}{6}\gamma_w D^3.$$

The vertical force per unit length is equal to

$$F_y = \gamma_w D b,$$

and is applied at a distance $b/2$ from the pivot.

The moment of this force (negative, since it is counter clockwise) is equal to

$$M_{F_y} = -\gamma_w D \frac{b^2}{2}.$$

Neglecting the moment of the weight of the gate, the gate opens if the total moment is positive (clockwise):

$$M_{F_x} + M_{F_y} > 0 \rightarrow \frac{1}{6}\gamma_w D^3 - \gamma_w D \frac{b^2}{2} \equiv \frac{\gamma_w D}{2}\left(\frac{1}{3}D^2 - b^2\right) > 0,$$

i.e. if $D > b\sqrt{3}$.

For $C_u = C_{pu} = 0$ it results $b = 2$ m,

$$\therefore \qquad D > b\sqrt{3} = 2\sqrt{3} = \mathbf{3.46\,m}.$$

# Chapter 2
# Hydrostatic Forces on Submerged Curved Surfaces

If the surface at contact with the fluid is curved, the action of the fluid can be computed by integrating the elementary forces due to the local pressure multiplied by the infinitesimal area of the surface. In practice, it is convenient to adopt integral balance equation for a control volume partly (or completely) delimited by the surface, and then to balance the forces acting on the external surface and body forces. It is also convenient to resolve the total force acting on the surface into horizontal and vertical components. A horizontal component of the force passes through the centroid of the vertical projection of the surface on a plane normal to the horizontal direction considered. By considering two directions in the horizontal plane, the corresponding two horizontal components can be determined and vector composed to calculate the total horizontal force acting on the surface, resulting, in general, in a single horizontal vector plus a moment which is required to render them coplanar. The vertical force passes through the centroid of the vertical (axis) cylinder of fluid with director represented by the contour of the surface. The horizontal and vertical force components can be vector composed, giving in general a single vector plus a moment.

Some complex computations are required in specific cases where part of the volume delimited by the surface is subject to the force of Archimedes.

Specific attention should be paid for multi-layered fluid of different densities and for stratified fluids.

**Exercise 2.1** The cylindrical gate $\overset{\frown}{\text{AC}}$ in Fig. 2.1, that is a semicircle of diameter $D$ and length $L$ orthogonal to the sheet, is pivoted in C and is in equilibrium due to the horizontal force **S** applied in A, which contrasts the pressure action of water and oil on the left. The specific weight of the liquids, the size of the gate and the depths of the water and oil (equal to $D/2$) are known.

– Calculate the magnitude of the horizontal force **S** to be applied in A to maintain the gate in equilibrium.

---

$C_u$ and $C_{pu}$, that are two integer numbers between 0 and 9, for example, the last and second-last digit of the registration number.

© Springer Nature Switzerland AG 2021
S. Longo et al., *Problems in Hydraulics and Fluid Mechanics*, Springer Tracts in Civil Engineering, https://doi.org/10.1007/978-3-030-51387-0_2

**Fig. 2.1** Cylindrical gate
pivoted in C and subject to
oil and water pressure on the
left side

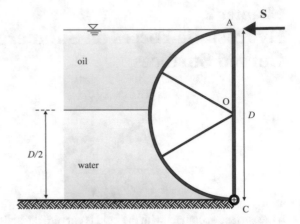

**Fig. 2.2** Pressure diagram
and virtual volume for the
application of the global
method of equilibrium

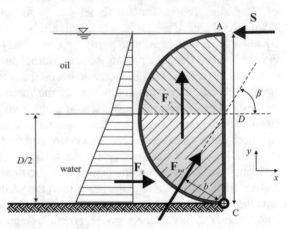

Numerical data: $D = (3 + 0.2 \times C_u)$ m, $L = (4 + 0.1 \times C_{pu})$ m, $\gamma_w = 9806\,\mathrm{N\,m^{-3}}$,
$\gamma_o = 0.8\gamma_w$.

**Solution** The cylindrical pivot is capable of developing any constraining reaction
through the trace axis C (orthogonal to the sheet), but is not capable of developing
resistant torques parallel to its axis. Therefore, the system of acting forces (hydrostatic
and external) must have zero moment with respect to the pivot.

The cylindrical gate is a curved surface with a contour line contained in a plane.
Applying the global equation of static equilibrium to the fluid volume enclosed
between the curved surface and the vertical plane AC (Fig. 2.2), a horizontal force
is calculated equal to the sum of the horizontal force of the oil, pointing to the right
and with magnitude

$$F_{xo} = \gamma_o \frac{D^2}{8} L,$$

and the horizontal force of the water, pointing to the right and with magnitude:

$$F_{xw} = (2\gamma_o + \gamma_w) \frac{D^2}{8} L.$$

The resulting horizontal force has magnitude:

$$F_x = \gamma_o \frac{D^2}{8} L + (2\gamma_o + \gamma_w) \frac{D^2}{8} L.$$

The vertical force is a vector with magnitude equal to the weight of fluid contained in the volume delimited by the curved surface and the vertical surface of the AC trace, but with the opposite orientation:

$$F_y = \gamma_o \frac{\pi D^2}{16} L + \gamma_w \frac{\pi D^2}{16} L.$$

The total force, which has a magnitude equal to

$$F_{tot} = \sqrt{F_x^2 + F_y^2},$$

crosses the axis O of the gate (since it is the sum of elementary forces all acting normally to the surface, hence pointing to O), and forms an angle to the horizontal equal to:

$$\beta = \tan^{-1}\left(\frac{F_y}{F_x}\right).$$

The arm with respect to C is equal to $b = (D/2) \cos \beta$. By imposing the equilibrium of the moment of forces about the axis C, $S$ is calculated as:

$$S = \frac{F_{tot}b}{D} \equiv \frac{F_x}{2}.$$

For $C_u = C_{pu} = 0$ it results $D = 3\,\mathrm{m}$, $L = 4\,\mathrm{m}$, $\gamma_w = 9806\,\mathrm{N\,m^{-3}}$, $\gamma_o = 7840\,\mathrm{N\,m^{-3}}$.

$$F_{xo} = \gamma_o \frac{D^2}{8} L = 7840 \times \frac{3^2}{8} \times 4 = \mathbf{35\,280\,N},$$

$$F_{xw} = (2\gamma_o + \gamma_w) \frac{D^2}{8} L = (2 \times 7840 + 9806) \times \frac{3^2}{8} \times 4 = \mathbf{114\,690\,N},$$

$$F_x = F_{xo} + F_{xw} = 35\,280 + 114\,690 = \mathbf{149\,970\,N},$$

$$F_y = \gamma_o \frac{\pi D^2}{16} L + \gamma_w \frac{\pi D^2}{16} L =$$

$$7840 \times \frac{\pi \times 3^2}{16} \times 4 + 9806 \times \frac{\pi \times 3^2}{16} \times 4 = \mathbf{124\,730\,N},$$

$$F_{tot} = \sqrt{F_x^2 + F_y^2} = \sqrt{149\,970^2 + 124\,730^2} = \mathbf{195\,060\,N},$$

$$\beta = \tan^{-1}\left(\frac{F_y}{F_x}\right) = \tan^{-1}\left(\frac{124\,730}{149\,970}\right) = \mathbf{39°\,45'},$$

$$b = \frac{D}{2}\cos\beta = \frac{3}{2} \times \cos 39°\,45' = \mathbf{1.153\,m},$$

$$\therefore \qquad S = \frac{F_{tot}b}{D} \equiv \frac{F_x}{2} = \frac{149\,970}{2} = \mathbf{74\,985\,N}.$$

**Exercise 2.2** The hemispherical cup in Fig. 2.3 is held in position by an external force **F**.

– Calculate the magnitude, the direction and the centre of pressure of the force **F**.

Data: $\gamma_w = 9806\,\mathrm{N\,m^{-3}}$, $R = (0.10 + C_u/100)\,\mathrm{m}$, $H_1 = R/3$, $H_2 = R/4$.

**Solution** The external force **F** must balance the forces due to the fluids inside and outside the cup. For the symmetry of the configuration, the force will not have horizontal component but only vertical component. Moreover, on the curved surface ADCB (see Fig. 2.4) the force is null due to the presence of water inside and outside the surface (the cup has null thickness by hypothesis) since, point by point, the elementary forces are equal and opposite.

It remains the force on the inner side of the surface DEVFC due to the air trapped in the cup, and the force on the curved outer surface DE-FC due to the water. The external surface EVF is in contact with the air at gage pressure and, therefore, the acting force is null. Applying the global equilibrium equation to the DEVFC volume filled with air, we observe that the force $S_{air}$ (neglecting the weight **G** of the air contained) is equal to the force on the flat surface $\overline{DC}$ with a pressure equal to:

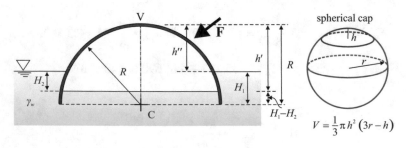

**Fig. 2.3** Hemispherical cup in water

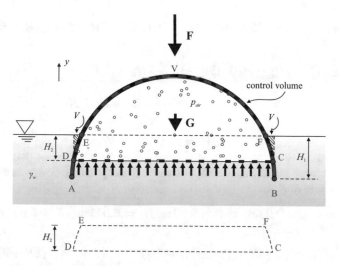

**Fig. 2.4** Control volume and schematic of the acting forces

$$p_{air} = \gamma_w H_2.$$

Hence,

$$S_{air} = p_{air}\pi \frac{\overline{DC}^2}{4} = \gamma_w H_2 \pi \frac{\overline{DC}^2}{4}, \tag{2.1}$$

pointing upwards, positive in the coordinate system fixed in Fig. 2.4.

The force of the water outside the cup can be evaluated very simply: it is equal to the weight of the dashed volume $V$ in Fig. 2.4 and is pointing downwards:

$$S_w = -\gamma_w V.$$

The resulting force is

$$S_{tot} = S_{air} + S_w = \gamma_w H_2 \pi \frac{\overline{DC}^2}{4} - \gamma_w V. \tag{2.2}$$

The Equation (2.2) has an effective geometric interpretation since the first term (coincident with 2.1) corresponds to the weight of liquid contained within a cylindrical volume of circular base of diameter $\overline{DC}$ and height equal to $H_2$. Therefore, Eq. (2.2) is equivalent to the calculation of the weight of the volume of liquid DE-FC obtained by subtracting the volume $V$ from the former cylindrical volume.

The volume of the solid of revolution DEFC, called spherical segment, can be calculated by subtracting the volume of the small height cap EVF, with height $h = R - H_1$, from the volume of the spherical cap DVC, with height $h = R - (H_1 - H_2)$:

$$V_{EVF} = \frac{1}{3}\pi(R - H_1)^2 [3R - (R - H_1)] \equiv \frac{1}{3}\pi(R - H_1)^2 (2R + H_1),$$

$$V_{DVC} = \frac{1}{3}\pi[R - (H_1 - H_2)]^2 \{3R - [R - (H_1 - H_2)]\} \equiv$$

$$\frac{1}{3}\pi(R - H_1 + H_2)^2 (2R + H_1 - H_2),$$

$$S_{tot} = \gamma_w V' = \gamma_w (V_{DVC} - V_{EVF}).$$

The required force **F** that balances $S_{tot}$ is vertical and pointing downwards, with the same magnitude of $S_{tot}$.

For $C_u = C_{pu} = 0$ it results $R = 0.10$ m, $H_1 = 0.033$ m, $H_2 = 0.025$ m.

$$V' = \left[ \frac{1}{3}\pi(R - H_1 + H_2)^2 (2R + H_1 - H_2) - \frac{1}{3}\pi(R - H_1)^2 (2R + H_1) \right] =$$

$$\frac{1}{3} \times \pi \times \left[ \begin{array}{l} (0.10 - 0.033 + 0.025)^2 \times (2 \times 0.10 + 0.033 - 0.025) - \\ (0.10 - 0.033)^2 \times (2 \times 0.10 + 0.033) \end{array} \right] =$$

$$\mathbf{7.483 \times 10^{-4} \ m^3}.$$

$$\therefore \qquad F = -S_{tot} = -\gamma_w V' = -9806 \times 7.483 \times 10^{-4} = -\mathbf{7.34\,N}.$$

---

**Exercise 2.3** Figure 2.5 shows a cylindrical sluice gate of unit length and radius $R = (0.50 + C_{pu}/20)$ m, pivoted in A, which opens rotating about A when the level in the tank is equal to $h_w = (3.50 + C_u/10)$ m.

- Determine the magnitude, the direction and the line of application of the force exerted by the water, for incipient opening condition.
- Determine the weight of the sluice gate.

Assume $\gamma_w = 9.8$ kN m$^{-3}$.

**Solution** Fixing the coordinate system shown in Fig. 2.6, the water exerts a horizontal force pointing to the right equal to

$$F_x = \gamma_w \left( h_w - \frac{R}{2} \right) R,$$

and a vertical upward-pointing force of magnitude equal to the weight of the dashed volume in Fig. 2.6:

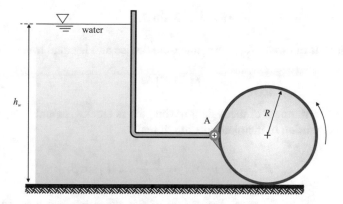

**Fig. 2.5** Cylindrical sluice gate pivoted in A

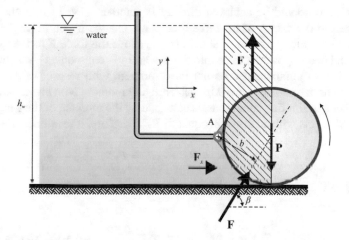

**Fig. 2.6** Coordinate system and schematic of the acting forces

$$F_y = \gamma_w \left[ (h_w - R) R + \frac{\pi R^2}{4} \right].$$

The resulting force has magnitude

$$|\mathbf{F}| = \sqrt{F_x^2 + F_y^2},$$

and crosses the axis of the cylindrical circular gate at an angle to the horizontal equal to

$$\beta = \tan^{-1} \left( \frac{F_y}{F_x} \right).$$

The arm with respect to the rotation axis A is equal to

$$b = R \sin \beta.$$

The moment of the water force is counter-clockwise and is equal to

$$M = |\mathbf{F}| b.$$

The stabilizing moment of the weight of the gate is clockwise and is equal to $PR$. In limiting condition of incipient rotation, it results

$$PR = |\mathbf{F}| b \rightarrow P = \frac{|\mathbf{F}| b}{R} \equiv |\mathbf{F}| \sin \beta.$$

$|\mathbf{F}| \sin \beta$ is equal to $F_y$, hence the weight of the sluice gate must be equal to the vertical component of the water force.

This solution could be achieved directly in another way. In fact, the moment with respect to the axis A of the force $\mathbf{F}$ does not change if you move the point of application of $\mathbf{F}$ along its line of action. If you move the force $\mathbf{F}$ to the axis of the cylinder and decompose the force into the $x-$ and $y-$ components, it is immediate to check that $x-$component does not have moment with respect to A, while the $y-$component has counter-clockwise destabilising moment with an arm equal to $R$. This moment, in limiting conditions, is balanced by the moment of the weight of the gate $\mathbf{P}$, which acts with an arm still equal to $R$. Therefore, it results

$$F_y R = PR \rightarrow P = F_y.$$

For $\mathbf{C}_u = \mathbf{C}_{pu} = 0$ it results $R = 0.50\,\mathrm{m}$, $h_w = 3.50\,\mathrm{m}$.

$$F_x = \gamma_w \left( h_w - \frac{R}{2} \right) R = 9800 \times \left( 3.50 - \frac{0.50}{2} \right) \times 0.50 = \mathbf{15\,925\,N},$$

$$F_y = \gamma_w \left[ (h_w - R) R + \frac{\pi R^2}{4} \right] =$$

$$9800 \times \left[ (3.50 - 0.50) \times 0.50 + \frac{\pi \times 0.50^2}{4} \right] = \mathbf{16\,625\,N},$$

$$\therefore \qquad |\mathbf{F}| = \sqrt{F_x^2 + F_y^2} = \sqrt{15\,925^2 + 16\,625^2} = \mathbf{23\,020\,N},$$

$$\therefore \qquad \beta = \tan^{-1} \left( \frac{F_y}{F_x} \right) = \tan^{-1} \left( \frac{16\,625}{15\,925} \right) = \mathbf{46° \; 14'},$$

$$b = R \sin \beta = 0.5 \times \sin 46° \; 14' = \mathbf{0.36\,m},$$

$$\therefore \qquad P = |\mathbf{F}| \sin \beta \equiv F_y = 23\,020 \times \sin 46° \; 14' = \mathbf{16\,625\,N}.$$

**Exercise 2.4** The hemispherical cup in Fig. 2.7 contains partly air, partly oil and partly water, and is held in position by an external force $\mathbf{F}$ acting on it. Given the geometric dimensions listed below and the characteristics of the fluids, calculate:

– the magnitude, the direction and the line of application of the force $\mathbf{F}$;
– the pressure of the air inside the cup.

Data: $\gamma_w = 9800\,\mathrm{N\,m^{-3}}$, $\gamma_o = 0.8\gamma_w$, $R = (0.10 + C_u/100)\,\mathrm{m}$, $h_1 = R/3$, $h_2 = R/4$, $h_3 = R/2$.

**Solution** The external force $\mathbf{F}$ must balance the hydrostatic force due to the fluids inside and outside the cup. By symmetry, the hydrostatic force will not have horizontal component but only a vertical component. On the curved surface DA-CB (Fig. 2.8) the force is null because the internal and external forces exerted by the water are balanced (the thickness of the cup is negligible). It is necessary to calculate the force due to the water pressing on the external surface DE-FC and the force due to the air and oil on the internal side DEGVHFC. The external surface DE-FC receives a vertical force, downwards (as is evident, given the direction of the elementary force) and equal to the weight of the dashed volume in Fig. 2.8 filled with water. This volume can be calculated as the difference between the cylindrical volume having a circular base with radius $r_2$ (Fig. 2.7) and the spherical segment with bases of radius $r_2$ and $r_3$:

$$S_{ext} = -\gamma_w V = -\gamma_w \left[ \pi r_2^2 \,(h_3 - h_2) - \frac{\pi\,(h_3 - h_2)}{2}\,(r_2^2 + r_3^2) - \frac{\pi\,(h_3 - h_2)^3}{6} \right].$$

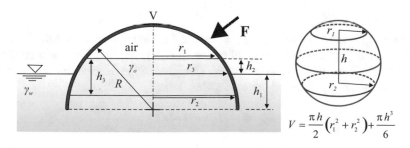

$$V = \frac{\pi h}{2}\left(r_1^2 + r_2^2\right) + \frac{\pi h^3}{6}$$

**Fig. 2.7** Hemispherical cup subject to pressure of water, oil and air

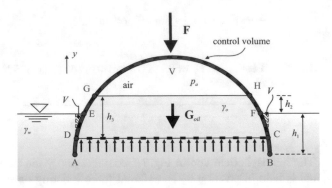

**Fig. 2.8**  Control volume and schematic of forces

**Fig. 2.9**  Schematic for the
calculation of the forces
acting on the internal surface
of the cup

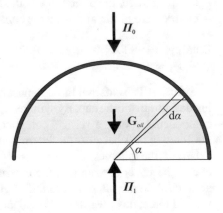

The force acting on the internal side DEGVHFC, due to air and oil, can be calculated
with different methodologies, but the most convenient is based on the method of the
global static equilibrium.

Isolating the volume of interest (see Fig. 2.9), which is partially filled with oil
and air, considering all surface and mass forces acting on the control volume, and
neglecting the air weight, it results:

$$\boldsymbol{\Pi}_0 + \boldsymbol{\Pi}_1 + \mathbf{G}_{oil} = \mathbf{0},$$

in which $\boldsymbol{\Pi}_0$ is the force on the curved surface due to the surrounding fluids, $\boldsymbol{\Pi}_1$ is
the force on the circular flat surface of radius $r_2$, $\mathbf{G}_{oil}$ is the weight of the oil that fills
the spherical segment with radii $r_2$ and $r_1$. The force of the internal fluids $\mathbf{S}_{int}$ on the
curved surface is equal to and opposed to $\boldsymbol{\Pi}_0$. Hence,

$$\mathbf{S}_{int} \equiv -\boldsymbol{\Pi}_0 = \boldsymbol{\Pi}_1 + \mathbf{G}_{oil}.$$

Since $\boldsymbol{\Pi}_1$ and $\mathbf{G}_{oil}$ are vertical, also $\mathbf{S}_{int}$ is vertical. The calculus indicates that:

$$\Pi_1 = pA = \gamma_w \, (h_3 - h_2) \, \pi r_2^2,$$

$$G_{oil} = -\gamma_o V_{oil} = -\gamma_o \left[ \frac{\pi h_3}{2} \left( r_1^2 + r_2^2 \right) + \frac{\pi h_3^3}{6} \right].$$

Hence,

$$S_{int} = \Pi_1 + G_{oil} = \gamma_w \, (h_3 - h_2) \, \pi r_2^2 - \gamma_o \left[ \frac{\pi h_3}{2} \left( r_1^2 + r_2^2 \right) + \frac{\pi h_3^3}{6} \right].$$

Finally, the total force is equal to the algebraic sum of the internal and external forces:

$$S_{tot} = S_{int} + S_{ext}, \quad F = -S_{tot}.$$

The air pressure is calculated by applying the hydrostatic equation:

$$p_a = \gamma_w \, (h_3 - h_2) - \gamma_o h_3.$$

Alternatively, it is possible to perform the calculation by integrating the elementary forces due to the pressure acting inside and outside the cup. Figure 2.10 shows the pressure diagram and the radial coordinate system $R - \alpha$ chosen for an easy calculation. The relative air pressure in the cap is uniform and acts on the spherical cap GVH; acting on the elementary surface of area

$$dA = 2\pi R^2 \cos \alpha \, d\alpha,$$

it generates an elementary vertical (only) force equal to

$$dF_{ya} = p_a 2\pi R^2 \cos \alpha \sin \alpha \, d\alpha.$$

The horizontal component is self-balanced and has the only effect of modifying the tensional state of the cup material. The vertical force of the air in the shell is obtained by integration:

$$F_{ya} = \int_{\alpha_0}^{\pi/2} p_a 2\pi R^2 \sin \alpha \cos \alpha \, d\alpha = p_a \pi R^2 \cos^2 \alpha \Big|_{\pi/2}^{\alpha_0} = p_a \pi R^2 \cos^2 \alpha_0.$$

The initial angle of integration is equal to:

$$\alpha_0 = \sin^{-1} \left( \frac{h_1 + h_2}{R} \right).$$

Notice that the $R \cos \alpha_0$ coincides with the radius $r_1$. Therefore, ignoring the weight of the air, the vertical force of the air on the shell is equal to the force of the

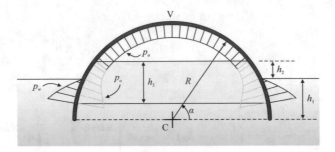

**Fig. 2.10**  Radial distribution of internal and external pressure

air on the circular base surface of the shell. The oil pressure is increasing downwards, according to the following relationship:

$$p_o = p_a + \gamma_o \left( h_1 + h_2 - R \sin \alpha \right),$$

and generates only an elementary vertical force equal to:

$$\mathrm{d} F_{yo} = p_o 2\pi R^2 \cos \alpha \sin \alpha \mathrm{d}\alpha.$$

The vertical force of the oil is obtained by integration:

$$F_{yo} = \int_{\alpha_1}^{\alpha_0} \left[ p_a + \gamma_o \left( h_1 + h_2 - R \sin \alpha \right) \right] 2\pi R^2 \sin \alpha \cos \alpha \, \mathrm{d}\alpha =$$

$$\left[ -p_a \pi R^2 \cos^2\alpha - \gamma_o \left( h_1 + h_2 \right) \pi R^2 \cos^2\alpha - \frac{2}{3}\pi R^3 \gamma_o \sin^3\alpha \right]\Bigg|_{\alpha_1}^{\alpha_0} =$$

$$- \pi p_a h_3 (h_3 - 2h_1 - 2h_2) - \pi \gamma_o \frac{h_3^2}{3}(2h_3 - 3h_1 - 3h_2).$$

The limits of integration are equal to:

$$\alpha_1 = \sin^{-1} \left( \frac{h_1 + h_2 - h_3}{R} \right), \quad \alpha_0.$$

Notice that, with the values fixed for $h_1$, $h_2$ and $h_3$ and for the predetermined ratio between the specific weight of the oil and water, by mere coincidence the vertical force of the oil is zero. The water pressure outside, acting on the surface ED and FC, generates a self-balanced horizontal force and a vertical force downwards. The water pressure increases downwards according to the following relationship:

$$p_w = \gamma_w \left( h_1 - R \sin \alpha \right),$$

and generates a vertical (only) force equal to:

$$F_{yw} = \int_0^{\alpha_1} p_w 2\pi R^2 \sin\alpha \cos\alpha \, d\alpha = \left[ -\gamma_w h_1 \pi R^2 \cos^2\alpha - \frac{2}{3}\pi R^3 \gamma_w \sin^3\alpha \right]\Big|_0^{|\alpha_1|} =$$

$$- \gamma_w \pi \frac{(h_2 - h_3)^2}{3}(3h_1 - 2h_3 + 2h_2).$$

In summary, the calculations performed by integrating elementary forces are much more complex than those required by using the global equation of statics.

For $C_u = C_{pu} = 0$ it results $R = 0.10\,\text{m}$, $h_1 = 0.033\,\text{m}$, $h_2 = 0.025\,\text{m}$, $h_3 = 0.05\,\text{m}$, $\gamma_w = 9800\,\text{N}\,\text{m}^{-3}$, $\gamma_o = 7840\,\text{N}\,\text{m}^{-3}$.

$$r_1 = \sqrt{R^2 - (h_1 + h_2)^2} = \sqrt{0.10^2 - (0.033 + 0.025)^2} = \mathbf{0.0815\,m},$$

$$r_2 = \sqrt{R^2 - (h_1 + h_2 - h_3)^2} =$$
$$\sqrt{0.10^2 - (0.033 + 0.025 - 0.05)^2} = \mathbf{0.0997\,m},$$

$$r_3 = \sqrt{R^2 - h_1^2} = \sqrt{0.10^2 - 0.033^2} = \mathbf{0.0944\,m},$$

$$V = \pi r_2^2 (h_3 - h_2) - \frac{\pi (h_3 - h_2)}{2}(r_2^2 + r_3^2) - \frac{\pi (h_3 - h_2)^3}{6} =$$
$$\pi \times 0.0997^2 \times (0.05 - 0.025) - \frac{\pi \times (0.05 - 0.025)}{2} \times (0.0997^2 + 0.0944^2)$$
$$- \frac{\pi \times (0.05 - 0.025)^3}{6} = \mathbf{3.2217 \times 10^{-5}\,m^3},$$

$$V_{oil} = \frac{\pi h_3}{2}(r_1^2 + r_2^2) + \frac{\pi h_3^3}{6} =$$
$$\frac{\pi \times 0.05}{2} \times (0.0815^2 + 0.0997^2) + \frac{\pi \times 0.05^3}{6} = \mathbf{1.3678 \times 10^{-3}\,m^3},$$

$$\therefore \quad F = -\left[ \gamma_w (h_3 - h_2) \pi r_2^2 - \gamma_o V_{oil} - \gamma_w V \right]$$
$$= - \begin{bmatrix} 9800 \times (0.05 - 0.025) \times \pi \times 0.0997^2 \\ -7840 \times 1.3678 \times 10^{-3} \\ -9800 \times 4.0393 \times 10^{-5} \end{bmatrix} = \mathbf{3.47\,N}.$$

This force is directed upwards. The air pressure is

$$\therefore p_a = \gamma_w \, (h_3 - h_2) - \gamma_o h_3 =$$
$$9800 \times (0.05 - 0.025) - 7840 \times 0.05 = -\mathbf{147\,Pa}.$$

---

**Exercise 2.5**  The structure in Fig. 2.11 is supported by an internal pressure equal to the atmospheric pressure. It has a hemispherical dome ABD whose diametral plane is inclined at an angle $\alpha$ to the horizontal.

– Determine the magnitude and the horizontal inclination of the total hydrostatic force (due to internal air and external liquids) on the hemispherical shell.

The geometrical data and other useful values are $H_1 = (10 + C_{pu})$ m, $H_2 = (5 + C_{pu})$ m, $h = (3 + C_u/20)$ m, $\alpha = (60 + C_u)°$, $\gamma_1 = 10\,100\,\mathrm{N\,m^{-3}}$, $\gamma_2 = 6500\,\mathrm{N\,m^{-3}}$.

**Solution**  The net force due to the air at atmospheric pressure inside the dome is zero because the atmospheric pressure also acts on the surface of the oil and therefore, on the whole, the effect on the surface of interest is null. Applying the global equation of static equilibrium to the ideal-minded hemispherical volume filled with liquid (global equation method, Fig. 2.12), yields

$$\boldsymbol{\Pi}_0 + \boldsymbol{\Pi}_1 + \mathbf{G}_s = \mathbf{0},$$

**Fig. 2.11**  Schematic of the structure

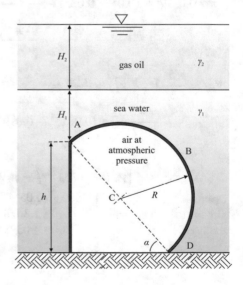

**Fig. 2.12** Schematic for the application of the method of the global equilibrium equation

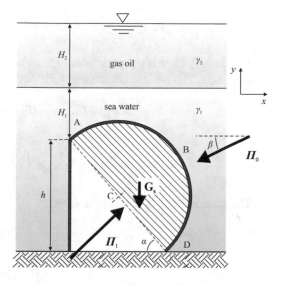

where $\boldsymbol{\Pi}_1$ is the force on the flat circular surface and centre C, $\mathbf{G_s}$ is the weight of seawater that ideally fills the volume of the hemisphere, $\boldsymbol{\Pi}_0$ is the force on the lateral surface of the hemisphere. The hemisphere has a radius $R$, obtained from the trigonometric relation:

$$R = \frac{h}{2 \sin \alpha}.$$

Since the hydrostatic force coincides with $\boldsymbol{\Pi}_0$, it results:

$$\mathbf{F} \equiv \boldsymbol{\Pi}_0 = -\boldsymbol{\Pi}_1 - \mathbf{G_s}.$$

The force $\boldsymbol{\Pi}_1$ is calculated as the product of the pressure in C and the area of the maximum circle of the hemisphere:

$$|\boldsymbol{\Pi}_1| = \left[ \gamma_2 H_2 + \gamma_1 \left( H_1 + \frac{h}{2} \right) \right] \pi R^2,$$

while the weight of the liquid in the hemisphere is equal to:

$$|\mathbf{G_s}| = \gamma_1 \frac{2}{3} \pi R^3.$$

Since the vectors $\mathbf{G_s}$ and $\boldsymbol{\Pi}_1$ are vertical and inclined at an angle $\alpha$, respectively, it is easy to deduce that the hydrostatic force on the hemisphere has a component $F_x$ equal to the $x$-component of the vector $-\boldsymbol{\Pi}_1$, hence

$$F_x = -\pi R^2 \sin \alpha \left[ \gamma_2 H_2 + \gamma_1 \left( H_1 + \frac{h}{2} \right) \right].$$

The horizontal force is pointing from right to left, opposite to the positive direction of the $x$-axis. The vertical component $F_y$ is equal to:

$$F_y \equiv \Pi_{0y} = -\Pi_{1y} - G_{sy},$$

that is

$$F_y = -\pi R^2 \cos \alpha \left[ \gamma_2 H_2 + \gamma_1 \left( H_1 + \frac{h}{2} \right) \right] + \gamma_1 \frac{2}{3} \pi R^3.$$

The vertical force is upward-pointing if

$$F_y > 0 \rightarrow \alpha > \cos^{-1} \left[ \frac{2R}{3 \left( \dfrac{\gamma_2 H_2}{\gamma_1} + H_1 + \dfrac{h}{2} \right)} \right].$$

The resulting force has a magnitude equal to:

$$|\mathbf{F}| = \sqrt{F_x^2 + F_y^2},$$

it has direction through the centre of the sphere (since it is the composition of elementary forces due to pressure, all orthogonal to the surface and, therefore, concurrent in the centre of the sphere), it is contained in the vertical plane $x - y$ by symmetry, and it has an inclination to the horizontal equal to:

$$\beta = \tan^{-1} \left( \frac{F_y}{F_x} \right).$$

To better understand these results, see the schematic in Fig. 2.13, where on the left there is the section of the volumes obtained with a vertical plane passing through the centre of the hemisphere, on the right there is an assonometric view. The vertical force acting on the surface is equal to the weight of the volume of fluid (two-layer) $V_1 + V_2$ above the cap, downwards directed, and to the buoyancy force relative to the volume $V_3$, upwards directed.

The exploded view in Fig. 2.14 shows the volumes required for the calculation of the two forces. The two forces are applied in the centre of gravity of the fluid masses contained in $V_1$ and $V_2$, and in the centre of the hull of volume $V_3$, respectively. However, calculations with this scheme introduce some analytical complexities, due to the presence of an intersection volume between an elliptical base cylinder with vertical axis, and a spherical surface (volume $V_2$). It is convenient to add and subtract the volume $V'$ which simplifies the calculus to the volume of a cylinder intersected by a plane, and of a hemisphere, according to the scheme shown in Fig. 2.15.

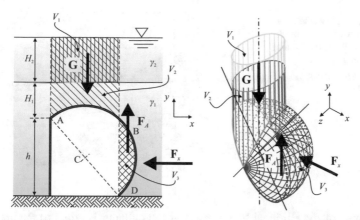

**Fig. 2.13** Section and assonometric view of the volumes defined for the calculation of the forces

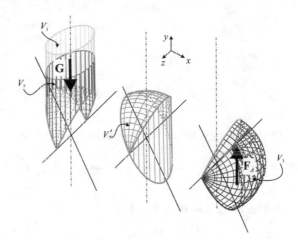

**Fig. 2.14** Exploded in assonometry of the volumes defined for the calculation of the vertical forces

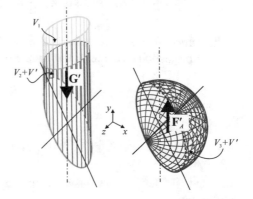

**Fig. 2.15** Exploded in assonometry of the volumes defined for the simplified calculation of the vertical forces

**Fig. 2.16** Exploded in assonometry of the volumes defined for the calculation of the horizontal forces

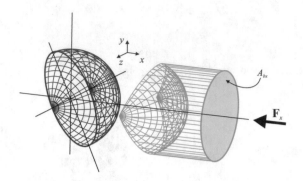

The area of the elliptical base surface of the cylinder is equal to the area of the projection along the $y$-axis of the maximum circle of the hemisphere:

$$A_{by} = \pi R^2 \cos \alpha,$$

with $R = h/(2 \sin \alpha)$. Hence

$$G' = -\gamma_2 H_2 \pi R^2 \cos \alpha - \gamma_1 \left( H_1 + \frac{h}{2} \right) \pi R^2 \cos \alpha.$$

The buoyancy force on the hemisphere is equal to:

$$F_A' = \gamma_1 \frac{2}{3} \pi R^3.$$

The vertical force is equal to:

$$F_y = -\gamma_2 H_2 \pi R^2 \cos \alpha - \gamma_1 \left( H_1 + \frac{h}{2} \right) \pi R^2 \cos \alpha + \gamma_1 \frac{2}{3} \pi R^3.$$

With reference to Fig. 2.16, the horizontal force is equal to the hydrostatic force acting on the $x$-projection of the maximum circle of the hemisphere, having a surface area equal to

$$A_{bx} = \pi R^2 \sin \alpha.$$

Hence,

$$F_x = -\pi R^2 \sin \alpha \left[ \gamma_2 H_2 + \gamma_1 \left( H_1 + \frac{h}{2} \right) \right],$$

pointing to the left, as previously obtained by applying the global equation of static equilibrium.

For $C_u = C_{pu} = 0$ it results $H_1 = 10$ m, $H_2 = 5$ m, $h = 3$ m, $\alpha = 60°$, $\gamma_1 = 10\,100$ N m$^{-3}$, $\gamma_2 = 6500$ N m$^{-3}$,

$$R = \frac{h}{2 \sin \alpha} = \frac{3}{2 \times \sin 60°} = 1.73\,\text{m},$$

$$F_y = -\gamma_2 H_2 \pi R^2 \cos \alpha - \gamma_1 \left( H_1 + \frac{h}{2} \right) \pi R^2 \cos \alpha + \gamma_1 \frac{2}{3} \pi R^3 =$$

$$- 6500 \times 5 \times \pi \times 1.73^2 \times \cos 60° - 10\,100 \times \left( 10 + \frac{3}{2} \right) \times \pi \times 1.73^2 \times \cos 60°$$

$$+ 10\,100 \times \frac{2}{3} \times \pi \times 1.73^3 = - \mathbf{589.3\,kN},$$

downwards directed.

$$F_x = \left[ -\gamma_2 H_2 - \gamma_1 \left( H_1 + \frac{h}{2} \right) \right] \pi R^2 \sin \alpha =$$

$$\left[ -6500 \times 5 - 10\,100 \times \left( 10 + \frac{3}{2} \right) \right] \times \pi \times 1.73^2 \times \sin 60° = - \mathbf{1210.4\,kN},$$

$$\therefore \qquad F = \sqrt{F_x^2 + F_y^2} = \sqrt{1210.4^2 + 589.3^2} = \mathbf{1346.2\,kN},$$

$$\therefore \qquad \beta = \tan^{-1} \left( \frac{F_y}{F_x} \right) = \tan^{-1} \left( \frac{-589.3}{-1210.4} \right) = \mathbf{26°}.$$

**Exercise 2.6** The tank in Fig. 2.17 has a hemisphere inserted in the wall and is only partially filled with water. The relative air pressure is indicated by the pressure gauge and is negative. Determine the magnitude and horizontal inclination of the hydrostatic force on the hemisphere in the two cases:

– water level at a distance $h$ above point C (situation illustrated in Fig. 2.17);
– water level passing through point C.

The numerical values are $p_m = -(0.49 + C_u/20) \times 10^5\,\text{Pa}$, $R = (0.25 + C_u/100)\,\text{m}$, $h = (1 + C_{pu}/10)\,\text{m}$, $\gamma_w = 9806\,\text{N\,m}^{-3}$.

**Solution** Case *(a): Water level at a distance h from C.*
Applying the global equation of static equilibrium to the ideal-minded hemispherical volume filled with liquid (global equation method, Fig. 2.18), yields

$$\boldsymbol{\Pi}_0 + \boldsymbol{\Pi}_1 + \mathbf{G}_w = 0,$$

**Fig. 2.17** Schematic of the
tank with hemispherical
surface

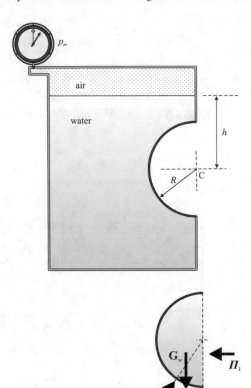

**Fig. 2.18** Control volume
and schematic of the forces
for the application of the
global equation method in
the case of a hemisphere
completely immersed in
water

where $\boldsymbol{\Pi}_1$ is the force on the flat circular surface, $\mathbf{G}_w$ is the weight of water that
ideally fills the volume of the hemisphere, $\boldsymbol{\Pi}_0$ is the force on the lateral surface of
the hemisphere. The unknown hydrostatic force $\mathbf{F}$ coincides with $\boldsymbol{\Pi}_0$, hence

$$\mathbf{F} \equiv \boldsymbol{\Pi}_0 = -\boldsymbol{\Pi}_1 - \mathbf{G}_w.$$

Since the vectors $\mathbf{G}_w$ and $\boldsymbol{\Pi}_1$ are vertical and horizontal, respectively, it follows that
the hydrostatic force on the hemisphere has component $F_x$ equal to the magnitude
of $\boldsymbol{\Pi}_1$, i.e.

$$F_x = (\gamma_w h + p_m) \frac{\pi D^2}{4},$$

where $p_m$ is the relative air pressure indicated by the pressure gauge. The direction
of the horizontal force depends on the sign of the relative pressure in the centroid of
the flat surface: if the pressure is negative, the force is pointing to the left (towards
the negative of the $x$-axis). The vertical component $F_y$ is pointing upwards and has
a magnitude equal to the magnitude of $\mathbf{G}_w$:

**Fig. 2.19** Control volume and schematic of the forces for the application of the global equation method in the case of a hemisphere partially immersed in water

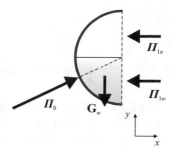

$$F_y = \gamma_w \frac{\pi D^3}{6} \frac{1}{2}.$$

The magnitude of the force is equal to

$$|\mathbf{F}| = \sqrt{F_x^2 + F_y^2},$$

and the inclination to the horizontal is calculated as

$$\alpha = \tan^{-1}\left(\frac{F_y}{F_x}\right).$$

Case *(b): Water level through C.*

The analysis is identical to that carried out for case *a*), but the force is due both to the water, which acts on the lower quarter, and to the air, which acts on the upper quarter of the hemisphere. It is not required to separate the two forces, which can therefore be evaluated by applying the global equation to the hemisphere filled with liquid only in the lower quarter (Fig. 2.19).

By neglecting the weight of the air, the global equation can be written as:

$$\boldsymbol{\Pi}_0 + \mathbf{G}_w + \boldsymbol{\Pi}_{1a} + \boldsymbol{\Pi}_{1w} = \mathbf{0},$$

where $\mathbf{G}_w$ is the weight of the quarter of a sphere filled with liquid, $\boldsymbol{\Pi}_{1w}$ is the force due to the liquid that presses on the flat surface of semicircular shape, $\boldsymbol{\Pi}_{1a}$ is the force due to the air on the flat portion of the upper semicircular shape and $\boldsymbol{\Pi}_0$ is the force on the lateral surface of the hemisphere. The unknown force $\mathbf{F}$ coincides with $\boldsymbol{\Pi}_0$, hence

$$\mathbf{F} = -\mathbf{G}_w - \boldsymbol{\Pi}_{1a} - \boldsymbol{\Pi}_{1w},$$

where

$$G_w = \gamma_w \frac{\pi D^3}{6} \frac{1}{4},$$

$$\Pi_{1a} = p_m \frac{\pi D^2}{8},$$

$$\Pi_{1w} = p_{Gw} \frac{\pi D^2}{8} = \left( \gamma_w \frac{4}{3\pi} \frac{D}{2} + p_m \right) \frac{\pi D^2}{8},$$

since the centroid of the semicircle is located at a distance $4D/(6\pi)$ from the centre of the circle, see in Appendix. The horizontal component is still given by the force on the flat surface. It turns out, then:

$$F_x = -\Pi_{1a} - \Pi_{1w}. \tag{2.3}$$

The sign still depends on the value of the algebraic sum in Eq. (2.3).

The vertical component $F_y$ is directed upwards and is equal to the magnitude of $\mathbf{G}_w$:

$$F_y = \gamma_w \frac{\pi D^3}{6} \frac{1}{4}.$$

For $C_u = C_{pu} = 0$ it results $p_m = -0.49 \times 10^5$ Pa, $R = 0.25$ m, $h = 1.0$ m.

Case (a): Water level at a distance h from C.

$$F_x = (\gamma_w h + p_m) \frac{\pi D^2}{4} = (9806 \times 1.0 - 0.49 \times 10^5) \times \frac{\pi \times 0.5^2}{4} = -7700\,\text{N},$$

$$F_y = \gamma_w \frac{\pi D^3}{6} \frac{1}{2} = 9806 \times \frac{\pi \times 0.5^3}{6} \times \frac{1}{2} = 320\,\text{N},$$

$$\therefore \qquad |\mathbf{F}| = \sqrt{F_x^2 + F_y^2} = \sqrt{7700^2 + 320^2} = 7706\,\text{N},$$

$$\therefore \qquad \alpha = \tan^{-1}\left( \frac{F_y}{F_x} \right) = \tan^{-1}\left( \frac{320}{-7700} \right) = -2°\,24'.$$

Case (b): Water level through C.

$$F_x = \left( \gamma_w \frac{4}{3\pi} \frac{D}{2} + 2p_m \right) \frac{\pi D^2}{8} =$$

$$\left( 9806 \times \frac{4}{3 \times \pi} \times \frac{0.5}{2} - 2 \times 0.49 \times 10^5 \right) \times \frac{\pi \times 0.5^2}{8} = -9525\,\text{N},$$

$$F_y = \gamma_w \frac{\pi D^3}{6} \frac{1}{4} = 9806 \times \frac{\pi \times 0.5^3}{6} \times \frac{1}{4} = 160\,\text{N},$$

$$\therefore \qquad |\mathbf{F}| = \sqrt{F_x^2 + F_y^2} = \sqrt{9525^2 + 160^2} = \mathbf{9526\,N},$$

$$\therefore \qquad \alpha = \tan^{-1}\left(\frac{F_y}{F_x}\right) = \tan^{-1}\left(\frac{160}{-9525}\right) = \mathbf{-1°}.$$

**Exercise 2.7** In the schematic in Fig. 2.20, the cylindrical circular gate, with a length $L = 2$ m and a radius $R = (4 + C_{pu}/2)$ m, is pivoted in O. The gate is symmetrical to the horizontal and $\alpha = (120 + 5 \times C_u)°$.

- Calculate the force on the gate and its point of application, if the fluid is homogeneous with specific weight $\gamma = 9800\,\mathrm{N\,m^{-3}}$.
- Perform the same calculations, if the fluid is stratified with variable specific weight according to the following relationship:

$$\gamma_s = \left(\gamma_0 + \gamma'\frac{z}{R}\right)\mathrm{N\,m^{-3}},$$

where $\gamma_0 = 9800\,\mathrm{N\,m^{-3}}$ and $\gamma' = 1000\,\mathrm{N\,m^{-3}}$.

**Solution** The fluid level is equal to

$$h = 2R\sin\frac{\alpha}{2}.$$

In the coordinate system shown in Fig. 2.21, if the fluid is homogeneous, the horizontal force is equal to

$$F_{h,x} = \frac{1}{2}\gamma h^2 L,$$

**Fig. 2.20** Cylindrical circular gate pivoted in O

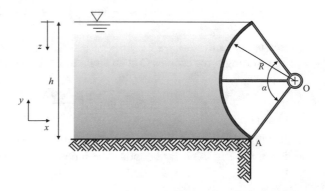

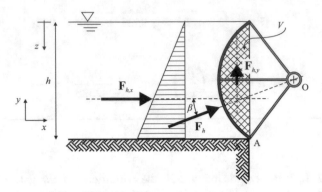

**Fig. 2.21** Pressure diagram and virtual volume $V$ (dashed) for the application of the global equation method in the case of homogeneous fluid

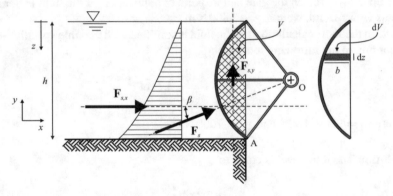

**Fig. 2.22** Pressure diagram and virtual volume $V$ (dashed) for the application of the global equation method in the case of variable density stratified fluid

and the vertical force is the buoyancy that acts on the dashed volume $V$ in Fig. 2.21, that is equal to

$$F_{h,y} = \gamma \left( \frac{\alpha}{2} R^2 - R^2 \sin \frac{\alpha}{2} \cos \frac{\alpha}{2} \right) L \equiv \frac{1}{2} \gamma R^2 L \, (\alpha - \sin \alpha) \,,$$

pointing upwards.

The resulting force crosses O at an angle to the horizontal equal to:

$$\beta_h = \tan^{-1} \left( \frac{F_{h,y}}{F_{h,x}} \right) .$$

If the fluid is stratified, the pressure distribution increases more than linearly with the depth, see Fig. 2.22. By adopting the indefinite equation of static equilibrium, yields

**Fig. 2.23** Schematic for the calculation of $b$

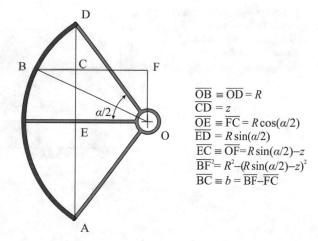

$$\overline{OB} \equiv \overline{OD} = R$$
$$\overline{CD} = z$$
$$\overline{OE} \equiv \overline{FC} = R\cos(\alpha/2)$$
$$\overline{ED} = R\sin(\alpha/2)$$
$$\overline{EC} \equiv \overline{OF} = R\sin(\alpha/2) - z$$
$$\overline{BF}^2 = R^2 - (R\sin(\alpha/2) - z)^2$$
$$\overline{BC} \equiv b = \overline{BF} - \overline{FC}$$

$$p(z) = \int\limits_0^z \gamma_s(z)\mathrm{d}z = \int\limits_0^z \left(\gamma_0 + \gamma'\frac{z}{R}\right)\mathrm{d}z = \gamma_0 z + \frac{\gamma'}{2}\frac{z^2}{R},$$

and the horizontal force is equal to

$$F_{s,x} = \int\limits_0^h p(z)\, L\,\mathrm{d}z = \int\limits_0^h \left(\gamma_0 z + \frac{\gamma'}{2}\frac{z^2}{R}\right) L\,\mathrm{d}z - \frac{\gamma_0}{2}h^2 L + \frac{\gamma'}{6}\frac{h^3}{R}L.$$

The vertical force is upward-pointing and has magnitude equal to the weight of the fluid (stratified) contained in volume $V$ in Fig. 2.22. This weight is calculated by integration as follows.

The elementary volume $\mathrm{d}V$ per unit of depth is a parallelepiped with base equal to

$$b = \sqrt{R^2 - \left(R\sin\frac{\alpha}{2} - z\right)^2} - R\cos\frac{\alpha}{2}$$

(see Fig. 2.23) and height $\mathrm{d}z$. Thus, the vertical force is equal to

$$F_{s,y} = \int\limits_0^h \left(\gamma_0 + \gamma'\frac{z}{R}\right) L \left[\sqrt{R^2 - \left(R\sin\frac{\alpha}{2} - z\right)^2} - R\cos\frac{\alpha}{2}\right]\mathrm{d}z.$$

As an alternative, it is possible to calculate the forces by integrating the elementary force acting on the infinitesimal surface $R\,\mathrm{d}\theta$ of unit depth, see Fig. 2.24. As an example, for the force in the vertical direction, results

**Fig. 2.24** Pressure diagram
and coordinate system for
the integration of elementary
forces acting on the gate

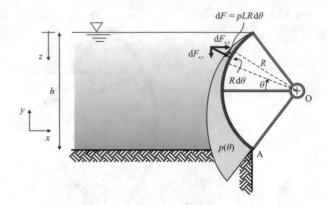

$$F_{s,y} = \int\limits_{-\alpha/2}^{\alpha/2} \left( \gamma_0 z + \frac{\gamma'}{2} \frac{z^2}{R} \right) L R \sin\theta \, d\theta. \tag{2.4}$$

The integral in Eq. (2.4) can be calculated analytically by expressing $z$ as a function of $R$ and $\theta$. The definite integral, calculated between $-\alpha/2$ and $+\alpha/2$, has the following value:

$$F_{s,y}\Big|_{-\alpha/2}^{\alpha/2} = \underbrace{\frac{1}{2}\gamma_0 R^2 L \left( \alpha - \sin\alpha \right)}_{\text{basic term}} + \underbrace{\frac{1}{2}\gamma' \sin\frac{\alpha}{2} R^2 L \left( \alpha - \sin\alpha \right)}_{\text{density variation contribution}} \equiv$$

$$\frac{1}{2} R^2 L \left( \gamma_0 + \gamma' \sin\frac{\alpha}{2} \right) (\alpha - \sin\alpha).$$

The force is pointing upwards.

The resulting force passes through O and forms an angle with respect to the horizontal, equal to:

$$\beta_s = \tan^{-1} \left( \frac{F_{s,y}}{F_{s,x}} \right).$$

For $C_u = C_{pu} = 0$ it results $L = 2$ m, $R = 4$ m, $\alpha = 120°$.

*Homogeneous Fluid:*

$$F_{h,x} = \frac{1}{2}\gamma h^2 L = \frac{1}{2} \times 9800 \times 6.93^2 \times 2 = \mathbf{470.40 \text{ kN}},$$

$$F_{h,y} = \gamma \left( \frac{\alpha}{2} R^2 - R^2 \sin\frac{\alpha}{2} \cos\frac{\alpha}{2} \right) L =$$

$$9800 \times \left( \frac{120° \times \pi}{2 \times 180°} \times 4^2 - 4^2 \times \sin\frac{120°}{2} \times \cos\frac{120°}{2} \right) \times 2 = \mathbf{192.61 \text{ kN}},$$

$$\therefore \qquad |\mathbf{F}_h| = \sqrt{F_{h,x}^2 + F_{h,y}^2} = \sqrt{470.40^2 + 192.61^2} = \mathbf{508.31\,kN},$$

$$\therefore \qquad \beta_h = \tan^{-1}\left(\frac{F_{h,y}}{F_{h,x}}\right) = \tan^{-1}\left(\frac{192.61}{470.40}\right) = \mathbf{22°\ 16'}.$$

*Stratified Fluid:*

$$F_{s,x} = \frac{9800}{2}h^2 L + \frac{1000}{3}\frac{h^3}{R}L =$$

$$\frac{9800}{2} \times 6.93^2 \times 2 + \frac{1000}{6} \times \frac{6.93^3}{2} \times 2 = \mathbf{478.65\,kN},$$

$$F_{s,y}\big|_{-\alpha/2}^{\alpha/2} = \frac{1}{2}R^2 L\left(\gamma_0 + \gamma' \sin\frac{\alpha}{2}\right)(\alpha - \sin\alpha) =$$

$$\frac{1}{2} \times 4^2 \times 2 \times \left(9800 + 1000 \times \sin\frac{120°}{2}\right) \times \left(\frac{120°}{180°}\pi - \sin 120°\right) = \mathbf{209.63\,kN},$$

$$\therefore \qquad |\mathbf{F}_s| = \sqrt{F_{s,x}^2 + F_{s,y}^2} = \sqrt{478.65^2 + 209.63^2} = \mathbf{522.54\,kN},$$

$$\therefore \qquad \beta_s = \tan^{-1}\left(\frac{F_{s,y}}{F_{s,x}}\right) = \tan^{-1}\left(\frac{209.63}{478.65}\right) = \mathbf{23°\ 39'}.$$

---

**Exercise 2.8** The cylinder in Fig. 2.25 is immersed in water and has a length, orthogonally to the drawing, equal to $L = 1.50$ m.

- Calculate the horizontal and the vertical components of the force on the cylinder.
- Calculate the direction of force.

Assume $\gamma_w = 9800\,\mathrm{N\,m^{-3}}$, $R = (0.50 + C_u/10)$ m, $\alpha = (45 + C_{pu})°$.

**Solution** The horizontal component of the force is equal to the force on the curved surface $\overgroup{AC}$, which delimits the portion of the dashed cylinder in Fig. 2.26:

$$F_x = \gamma_w \frac{R^2}{2}(1 + \cos\alpha)^2 L,$$

since the forces acting on the symmetric curved surface $\overgroup{AEB}$ are self-balanced.

**Fig. 2.25** Schematic of the cylinder in water

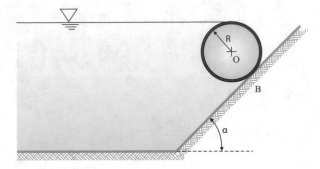

**Fig. 2.26** Coordinate system and schematics for calculating the horizontal force component

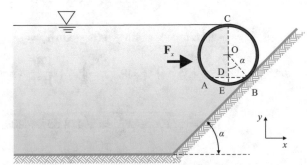

**Fig. 2.27** Coordinate system and schematic for calculating the vertical force component

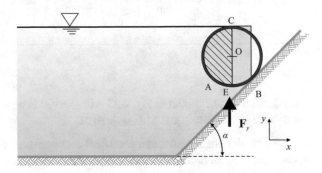

The vertical component of the force is pointing upwards and is equal to the weight of the yellow dashed volume in Fig. 2.27. This volume may be decomposed into the volume of the semi-cylinder $\overset{\frown}{\text{CAE}}$ and the vertical solid delimited below by the arc $\overset{\frown}{\text{EB}}$. It results

$$F_y = \gamma_w \frac{\pi R^2}{2} L + \gamma_w \left[ R^2 \left( \frac{2 + \cos \alpha}{2} \right) \sin \alpha + \frac{\alpha}{2} R^2 \right] L.$$

The resultant of the forces crosses the axis of the cylinder, trace O, and is inclined to the horizontal at an angle $\beta = \tan^{-1}\left(F_y/F_x\right)$.

For $C_u = C_{pu} = 0$ it results $R = 0.50$ m, $\alpha = 45°$, $L = 1.50$ m.

$$\therefore \qquad F_x = \gamma_w \frac{R^2}{2}(1 + \cos\alpha)^2 L = 9800 \times \frac{0.5^2}{2} \times (1 + \cos 45°)^2 \times 1.50 = \mathbf{5.35\,kN},$$

$$\therefore \qquad F_y = \gamma_w \frac{\pi R^2}{2}L + \gamma_w\left[R^2\left(\frac{2 + \cos\alpha}{2}\right)\sin\alpha + \frac{\alpha}{2}R^2\right]L =$$

$$9800 \times \frac{\pi \times 0.5^2}{2} \times 1.50$$

$$+\, 9800 \times \left[0.5^2 \times \left(\frac{2 + \cos 45°}{2}\right) \times \sin 45° + \frac{\pi}{8} \times 0.5^2\right] \times 1.50$$

$$=\mathbf{10.73\,kN},$$

$$\therefore \qquad \beta = \tan^{-1}\left(\frac{F_y}{F_x}\right) = \tan^{-1}\left(\frac{10.73}{5.35}\right) = \mathbf{63°30'}.$$

---

**Exercise 2.9** In the system shown in Fig. 2.28, the sealed tank has a unitary depth and contains pressurized air.

- Determine the magnitude, direction and line of application of the force exerted by air, oil and water on the quarter-cylinder surface.

Assume $\gamma_o = 8000\,\mathrm{N\,m^{-3}}$, $\gamma_w = 9800\,\mathrm{N\,m^{-3}}$, $\gamma_a = 0$, $h_1 = (3.00 + C_u/10)$ m, $h_2 = (2.00 + C_u/10)$ m, $R = 1.50$ m, $p_m = (0.20 + C_{pu}/20) \times 10^5$ Pa.

**Solution** The force exerted by the oil in the container must account for the presence of the air under pressure above it; as a consequence, the isobaric plane at atmospheric pressure for the oil is above the interface air-oil by the quantity:

**Fig. 2.28** Schematic of the pressurized tank

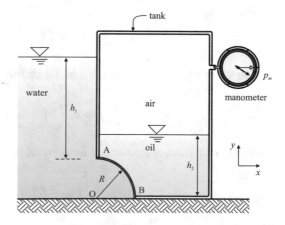

**Fig. 2.29** Schematic for the
calculation of the internal
forces

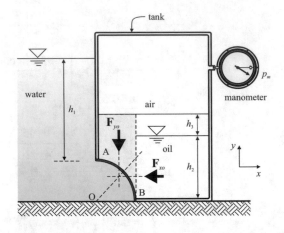

$$h_3 = \frac{p_m}{\gamma_o},$$

as shown in Fig. 2.29. The oil force on the curved surface (per unit length) can be
calculated, for example, by using the components method. The horizontal component
is equal to:

$$F_{xo} = -\gamma_o \left[ h_3 + \left( h_2 - \frac{R}{2} \right) \right] R,$$

and is negative in the adopted coordinate system.

The vertical component is equal to the weight of the volume (dashed in Fig. 2.29)
between the curved surface and the plane of the hydrostatic oil loads. Therefore:

$$F_{yo} = -\gamma_o \left[ R (h_2 + h_3) - \frac{\pi R^2}{4} \right].$$

The resultant of the forces exerted by the oil intersects the axis O at an angle to the
horizontal equal to:

$$\alpha_o = \tan^{-1} \left( \frac{F_{yo}}{F_{xo}} \right).$$

The water on the left side exerts a horizontal force equal to

$$F_{xw} = \gamma_w \left( h_1 + \frac{R}{2} \right) R,$$

pointing to the right, and a vertical force equal to

$$F_{yw} = \gamma_w \left[ (h_1 + R) R - \frac{\pi R^2}{4} \right],$$

**Fig. 2.30** Schematic for the calculation of the external forces

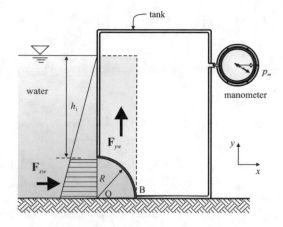

pointing upwards. This vertical force coincides, in magnitude, with the weight of the dashed virtual volume in Fig. 2.30. The resultant of the forces exerted by the water intersects the axis O at an angle to the horizontal equal to

$$\alpha_w = \tan^{-1}\left(\frac{F_{yw}}{F_{xw}}\right).$$

The total force has a horizontal component equal to

$$F_x = F_{xo} + F_{xw},$$

and a vertical component equal to

$$F_y = F_{yo} + F_{yw}.$$

The total force crosses the centre O at an angle to the horizontal equal to

$$\alpha = \tan^{-1}\left(\frac{F_y}{F_x}\right).$$

For $C_u = C_{pu} = 0$ it results $\gamma_o = 8000\,\mathrm{N\,m^{-3}}$, $\gamma_w = 9800\,\mathrm{N\,m^{-3}}$, $\gamma_a = 0$, $h_1 = 3.00\,\mathrm{m}$, $h_2 = 2.00\,\mathrm{m}$, $R = 1.50\,\mathrm{m}$, $p_m = 0.20 \times 10^5\,\mathrm{Pa}$.

$$h_3 = \frac{p_m}{\gamma_o} = \frac{0.20 \times 10^5}{8000} = 2.5\,\mathrm{m},$$

$$F_{xo} = -\gamma_o \left[ h_3 + \left( h_2 - \frac{R}{2} \right) \right] R =$$

$$- 8000 \times \left[ 2.5 + \left( 2 - \frac{1.5}{2} \right) \right] \times 1.5 = -45.0 \, \text{kN},$$

$$F_{yo} = -\gamma_o \left[ R \, (h_2 + h_3) - \frac{\pi R^2}{4} \right] =$$

$$- 8000 \times \left[ 1.5 \times (2 + 2.5) - \frac{\pi \times 1.5^2}{4} \right] = -39.9 \, \text{kN},$$

$$F_{xw} = \gamma_w \left( h_1 + \frac{R}{2} \right) R = 9800 \times \left( 3 + \frac{1.5}{2} \right) \times 1.5 = 55.1 \, \text{kN},$$

$$F_{yw} = \gamma_w \left[ (h_1 + R) \, R - \frac{\pi R^2}{4} \right] =$$

$$9800 \times \left[ (3 + 1.5) \times 1.5 - \frac{\pi \times 1.5^2}{4} \right]$$

$$= 48.8 \, \text{kN},$$

$$\therefore \qquad F_x = F_{xo} + F_{xw} = -45.0 + 55.1 = 100.1 \, \text{kN},$$

$$\therefore \qquad F_y = F_{yo} + F_{yw} = -39.9 + 48.8 = 88.7 \, \text{kN},$$

$$\therefore \qquad \alpha = \tan^{-1} \left( \frac{F_y}{F_x} \right) = \tan^{-1} \left( \frac{88.7}{100.1} \right) = 41° \, 32'.$$

---

**Exercise 2.10** In the system in Fig. 2.31 the cylindrical gate, of depth $L = (2 + C_u/2)$ m and of radius $R = (4 + C_{pu}/2)$ m, with axis of cylindrical symmetry in C, is pivoted in O. The angle $O\widehat{C}A = 45°$ and the distance $\overline{OC} = R/2$. The pivot does not coincide with the axis of cylindrical symmetry, and the gate has a sealing in A.

- Calculate the water force on the gate.
- Calculate the horizontal force **F** that must be applied in B to open the gate.

Neglect the weight of the gate. Assume $\gamma_w = 9806 \, \text{N m}^{-3}$.

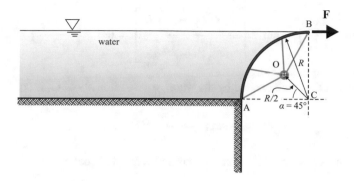

**Fig. 2.31** Cylindrical gate subject to water pressure

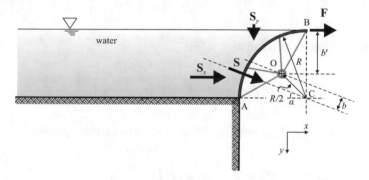

**Fig. 2.32** Schematic for calculating the forces acting on the cylindrical gate

**Solution** In the coordinate system in Fig. 2.32, the water exerts a horizontal force to the right equal to

$$S_x = \frac{1}{2}\gamma_w R^2 L,$$

and a vertical downward force equal to the weight of the volume of fluid above the gate,

$$S_y = \gamma_w R^2 \left(1 - \frac{\pi}{4}\right) L.$$

The magnitude of the total force is equal to

$$|\mathbf{S}| = \sqrt{S_x^2 + S_y^2} = \gamma_w R^2 L \sqrt{\frac{1}{4} + \left(\frac{4-\pi}{4}\right)^2}.$$

The force exerted by the water crosses C at an angle to the horizontal equal to

$$\alpha = \tan^{-1}\left(\frac{S_y}{S_x}\right) = \tan^{-1}\left[\frac{\gamma_w R^2\left(1 - \frac{\pi}{4}\right)L}{\frac{1}{2}\gamma_w R^2 L}\right] = \tan^{-1}\left(\frac{4 - \pi}{2}\right) = 23° \; 13'.$$

The arm of the force exerted by the water, for rotation about O, is equal to

$$b = \frac{R}{2}\sin\left(\frac{\pi}{4} - \alpha\right) \approx 0.185\,R,$$

The moment is counter-clockwise and has a magnitude equal to

$$M_S = |S|\,b = \left(\sqrt{S_x^2 + S_y^2}\right)b \approx \gamma_w R^2 L\left[\sqrt{\frac{1}{4} + \left(\frac{4 - \pi}{4}\right)^2}\right] \times 0.185R.$$

The arm of force **F** is equal to

$$b' = R - \frac{R}{2}\cos\frac{\pi}{4} = R\left(\frac{4 - \sqrt{2}}{4}\right),$$

and the moment about O is clockwise if **F** is pointing to the right, and has a magnitude equal to

$$M_F = |F|\,b' = |F|\,R\left(\frac{4 - \sqrt{2}}{4}\right).$$

Imposing the balance of moments about O yields

$$M_S = M_F \rightarrow |F| \approx \frac{\gamma_w R^2 L\sqrt{\frac{1}{4} + \left(\frac{4 - \pi}{4}\right)^2} \times 0.185R}{R\left(\frac{4 - \sqrt{2}}{4}\right)}.$$

For $C_u = C_{pu} = 0$ it results $L = 2\,$m, $R = 4\,$m.

$$S_x = \frac{1}{2}\gamma_w R^2 L = \frac{1}{2} \times 9806 \times 4^2 \times 2 = 156.90\,\text{kN},$$

$$S_y = \gamma_w R^2\left(1 - \frac{\pi}{4}\right)L = 9806 \times 4^2 \times \left(1 - \frac{\pi}{4}\right) \times 2 = 67.34\,\text{kN},$$

$$|S| = \sqrt{S_x^2 + S_y^2} = \sqrt{156.90^2 + 67.34^2} = 170.74\,\text{kN},$$

$$\therefore \qquad |\mathbf{F}| \approx \frac{170.74 \times 0.185 \times 4}{4 \times \left(\dfrac{4 - \sqrt{2}}{4}\right)} = 49\,\text{kN}.$$

**Exercise 2.11** In the system in Fig. 2.33 the cylindrical gate, of depth $L = (2 + C_u/2)$ m and of radius $R = (4 + C_{pu}/2)$ m, with axis of cylindrical symmetry of trace C, is pivoted in A. The two liquids on the left have specific weight $\gamma_1 = 11\,000\,\text{N}\,\text{m}^{-3}$ and $\gamma_2 = 10\,000\,\text{N}\,\text{m}^{-3}$, and their depths are $h_1 = (1 + C_{pu}/4)$ m and $h_2 = (3 + C_{pu}/4)$ m, respectively.

- Calculate the total force exerted by the liquids on the gate.
- Calculate the inclination to the horizontal of the force.
- Calculate the horizontal force **F** that must be applied in B to keep the gate in place.

 Neglect the weight of the gate.

 **Solution** In the coordinate system shown in Fig. 2.34, the upper layer liquid exerts a horizontal force equal to

$$S_{2x} = \frac{1}{2}\gamma_2 h_2^2 L,$$

and a vertical force equal to the weight of volume $V_2$:

$$S_{2y} = \gamma_2 V_2 = \gamma_2 \left[ R h_2 \sin \alpha_2 - \frac{R^2}{2}\alpha_2 + \frac{R(R - h_2)}{2} \sin \alpha_2 \right] L,$$

where

$$\alpha_2 = \frac{\pi}{2} - \alpha = \cos^{-1}\left(\frac{R - h_2}{R}\right).$$

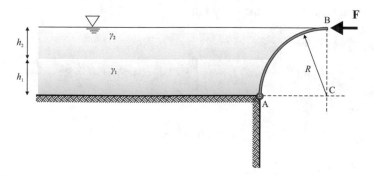

**Fig. 2.33** Cylindrical gate subject to stratified fluids pressure

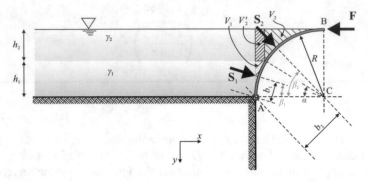

**Fig. 2.34** Schematic for calculating the forces

The resulting force has magnitude

$$|\mathbf{S}_2| = \sqrt{S_{2x}^2 + S_{2y}^2},$$

and crosses the centre of curvature C at an angle to the horizontal equal to

$$\beta_2 = \tan^{-1}\left(\frac{S_{2y}}{S_{2x}}\right).$$

The arm with respect to the pivot in A is equal to

$$b_2 = R \sin \beta_2,$$

and the clockwise moment has magnitude $|\mathbf{M}_2| = b_2 \, |\mathbf{S}_2|$.
    The lower layer liquid exerts a horizontal force equal to

$$S_{1x} = \frac{1}{2} \left(\gamma_1 h_1 + 2\gamma_2 h_2\right) h_1 L,$$

and a vertical force equal to the weight of the volume $V_2' + V_1$:

$$S_{1y} = \gamma_2 V_2' + \gamma_1 V_1 =$$
$$\gamma_2 h_2 R \left(1 - \cos\alpha\right) L + \gamma_1 \left[R \left(1 - \cos\alpha\right) h_1 - \frac{\alpha}{2} R^2 + \frac{R}{2} h_1 \cos\alpha\right] L,$$

where

$$\alpha = \sin^{-1}\left(\frac{h_1}{R}\right).$$

The resulting force has magnitude $|\mathbf{S}_1| = \sqrt{S_{1x}^2 + S_{1y}^2}$ and crosses the centre of curvature C, at an angle to the horizontal equal to

$$\beta_1 = \tan^{-1}\left(\frac{S_{1y}}{S_{1x}}\right).$$

The arm with respect to the pivot in A is equal to $b_1 = R \sin \beta_1$ and the clockwise moment has magnitude $|\mathbf{M}_1| = b_1 |\mathbf{S}_1|$.

The total force has magnitude

$$|\mathbf{S}| = \sqrt{(S_{1x} + S_{2x})^2 + (S_{1y} + S_{2y})^2}$$

and still crosses the centre of curvature C, at an angle to the horizontal equal to

$$\beta = \tan^{-1}\left(\frac{S_{1y} + S_{2y}}{S_{1x} + S_{2x}}\right).$$

The magnitude of force $\mathbf{F}$ is obtained by imposing the balance of moments about the pivot A:

$$RF + b_1 S_1 + b_2 S_2 = 0 \rightarrow F = -\frac{b_1 S_1 + b_2 S_2}{R}.$$

For $C_u = C_{pu} = 0$ it results $L = 2\,\text{m}$, $R = 4\,\text{m}$, $h_1 = 1\,\text{m}$, $h_2 = 3\,\text{m}$,

$$S_{2x} = \frac{1}{2}\gamma_2 h_2^2 L = \frac{1}{2} \times 10\,000 \times 3^2 \times 2 = \mathbf{90.00\,kN},$$

$$\alpha_2 = \cos^{-1}\left(\frac{R - h_2}{R}\right) = \cos^{-1}\left(\frac{4-3}{4}\right) = \mathbf{75° \, 31'},$$

$$S_{2y} = \gamma_2 V_2 = \gamma_2 \left[ Rh_2 \sin\alpha_2 - \frac{R^2}{2}\alpha_2 + \frac{R(R - h_2)}{2}\sin\alpha_2 \right] L =$$

$$10\,000 \times \left[ \begin{array}{c} 4.0 \times 3.0 \times \sin 75°31' - \dfrac{4.0^2}{2} \times \dfrac{75°31'}{180°} \times \pi + \\[2mm] \dfrac{4.0\,(4.0 - 3.0)}{2} \times \sin 75°31' \end{array} \right] \times 2.0$$

$$= \mathbf{60.21\,kN},$$

$$|\mathbf{S}_2| = \sqrt{S_{2x}^2 + S_{2y}^2} = \sqrt{90.00^2 + 60.21^2} = \mathbf{108.28\,kN},$$

$$\beta_2 = \tan^{-1}\left(\frac{S_{2y}}{S_{2x}}\right) = \tan^{-1}\left(\frac{60.21}{90.00}\right) = \mathbf{33° \, 47'},$$

$$b_2 = R \sin \beta_2 = 4.0 \times \sin 33° \; 47' = \mathbf{2.22\,m},$$

$$S_{1x} = \frac{1}{2} \left( \gamma_1 h_1 + 2\gamma_2 h_2 \right) h_1 L =$$

$$\frac{1}{2} \left( 11\,000 \times 1 + 2 \times 10\,000 \times 3 \right) \times 1.0 \times 2.0$$

$$= \mathbf{71.00\,kN},$$

$$\alpha = \sin^{-1} \left( \frac{h_1}{R} \right) = \sin^{-1} \left( \frac{1.0}{4.0} \right) = \mathbf{14° \; 28'},$$

$$S_{1y} = \gamma_2 h_2 R \left( 1 - \cos \alpha \right) L + \gamma_1 \left[ R \left( 1 - \cos \alpha \right) h_1 - \frac{\alpha_1}{2} R^2 + \frac{R}{2} h_1 \cos \alpha \right] L =$$

$$10\,000 \times 3.0 \times 4.0 \times \left( 1 - \cos 14°28' \right) \times 2.0 +$$

$$11\,000 \times \left[ \begin{array}{l} 4.0 \times \left( 1 - \cos 14°28' \right) \times 1.0 - \\ \dfrac{14°28'}{180°} \times \pi \times \dfrac{4.0^2}{2} + \dfrac{4.0}{2} \times 1.0 \times \cos 14°28' \end{array} \right] \times 2.0 = \mathbf{8.55\,kN},$$

$$|S_1| = \sqrt{S_{1x}^2 + S_{1y}^2} = \sqrt{71.00^2 + 8.55^2} = \mathbf{71.50\,kN},$$

$$\beta_1 = \tan^{-1} \left( \frac{S_{1y}}{S_{1x}} \right) = \tan^{-1} \left( \frac{8.55}{71.00} \right) = \mathbf{6° \; 52'},$$

$$b_1 = R \sin \beta_1 = 4.0 \times \sin 6°52' = \mathbf{0.48\,m}.$$

The total force has components

$$S_x = S_{1x} + S_{2x} = 71.00 + 90.00 = \mathbf{161.00\,kN},$$

$$S_y = S_{1y} + S_{2y} = 8.55 + 60.21 = \mathbf{68.76\,kN},$$

it has magnitude

$$\therefore \qquad |S| = \sqrt{S_x^2 + S_y^2} = \sqrt{161.00^2 + 68.76^2} = \mathbf{175.10\,kN},$$

and it is inclined at an angle to the horizontal equal to

$$\therefore \qquad \beta = \tan^{-1} \left( \frac{S_y}{S_x} \right) = \tan^{-1} \left( \frac{68.76}{161.00} \right) = \mathbf{23° \; 8'}.$$

The horizontal force required for equilibrium has magnitude equal to

$$\therefore \quad F = -\frac{b_1 S_1 + b_2 S_2}{R} \cdot = -\frac{0.48 \times 71.5 + 2.22 \times 108.28}{4.0} = -68.80\,\text{kN}.$$

This force must be pointing to the left in the schematic in Fig. 2.34.

---

**Exercise 2.12** In the system in Fig. 2.35 the cylindrical gate, with depth $L = (2 + C_u/2)$ m and radius $R = (4 + C_{pu}/2)$ m, is pivoted in A.

- Calculate the horizontal force **P** required to keep the gate in the position.
- Calculate the force if the gate was pivoted in B.

Neglect the weight of the gate, assume $\gamma_w = 9806\,\text{N m}^{-3}$.

**Solution** In the coordinate system shown in Fig. 2.36, the horizontal component of the water force is equal to

$$F_x = \frac{1}{2}\gamma_w R^2 L,$$

and the vertical component of the force is equal to

$$F_y = \gamma_w \frac{\pi R^2}{4} L.$$

The force magnitude is equal to

$$|\mathbf{F}| = \sqrt{F_x^2 + F_y^2}.$$

Since the elementary forces are normal to the surface and cross B, the integral force also crosses B at an angle to the horizontal equal to

$$\alpha = \tan^{-1}\left(\frac{F_y}{F_x}\right).$$

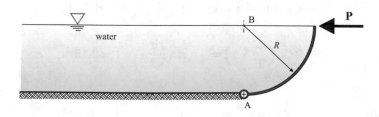

**Fig. 2.35** Cylindrical gate pivoted in A

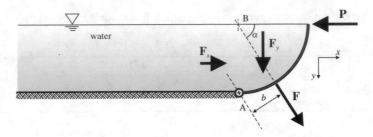

**Fig. 2.36** Schematic for calculating the forces

The arm with respect to the pivot in A has the following expression:

$$b = R \sin \left( \frac{\pi}{2} - \alpha \right).$$

The magnitude of the force **P** is calculated by imposing the balance of moment about the pivot A:

$$PR + Fb = 0 \rightarrow P = -\frac{Fb}{R}.$$

If the gate was pivoted in B, the force required to keep the gate in the position is zero, since the force exerted by the water on the sluice gate has no arm with respect to the new centre of rotation B. We can add that any value of **P** is admissible, because it has no moment about B.

For $C_u = C_{pu} = 0$ it results $L = 2\,\text{m}$, $R = 4\,\text{m}$,

$$F_x = \frac{1}{2} \gamma_w R^2 L = \frac{1}{2} \times 9806 \times 4^2 \times 2 = \mathbf{156\,896\,N},$$

$$F_y = \gamma_w \frac{\pi R^2}{4} L = 9806 \times \frac{\pi \times 4^2}{4} \times 2 = \mathbf{246\,452\,N},$$

$$|\mathbf{F}| = \sqrt{F_x^2 + F_y^2} = \sqrt{156\,896^2 + 246\,452^2} = \mathbf{292\,156\,N},$$

$$\alpha = \tan^{-1} \left( \frac{F_y}{F_x} \right) = \tan^{-1} \left( \frac{246\,452}{156\,896} \right) = \mathbf{57° \, 31'},$$

$$b = R \sin \left( \frac{\pi}{2} - \alpha \right) = 4 \times \sin \left( \frac{\pi}{2} - \alpha \right) = \mathbf{2.15\,m},$$

$$\therefore \qquad P = -\frac{Fb}{R} = -\frac{292\,156 \times 2.15}{4} = \mathbf{-157\,kN}.$$

The force **P** must be pointing to the left.

---

**Exercise 2.13**  The pressurized tank in Fig. 2.37 contains water and air in its upper part; on the wall of the tank there is an indentation of conical shape, with height $b$ and diameter of the base $D$. The depth of the cone tip with respect to the air-water interface is $h$. The air above has a pressure measured by the pressure gauge connected to the upper part of the tank.

– Determine the magnitude and the horizontal inclination of the hydrostatic force on the conical surface.

Numerical data: $\gamma_w = 9806\,\mathrm{N\,m^{-3}}$, $p_{air} = (3 + 0.1 \times C_u) \times 10^5\,\mathrm{Pa}$ (gage), $h = (1 + C_{pu} \times 0.5)\,\mathrm{m}$, $b = (0.5 + C_{pu} \times 0.5)\,\mathrm{m}$, $D = 0.30\,\mathrm{m}$.

**Solution**  We select the coordinate system shown in Fig. 2.38. Applying the global equation of static equilibrium to the conical volume ideally filled with liquid (global equation method), yields

$$\boldsymbol{\Pi}_0 + \boldsymbol{\Pi}_1 + \mathbf{G} = \mathbf{0},$$

where $\boldsymbol{\Pi}_1$ is the force on the flat surface of circular shape, $\mathbf{G}$ is the weight of the water that ideally fills the conical volume, $\boldsymbol{\Pi}_0$ is the force on the lateral surface of the cone. The required hydrostatic force $\mathbf{F}$ coincides with $\boldsymbol{\Pi}_0$, hence

$$\mathbf{F} \equiv \boldsymbol{\Pi}_0 = -\boldsymbol{\Pi}_1 - \mathbf{G}.$$

Since the vectors $\mathbf{G}$ and $\boldsymbol{\Pi}_1$ are vertical and horizontal, respectively, it yields that the hydrostatic force on the cone has component $F_x$ equal to the magnitude of $\boldsymbol{\Pi}_1$, i.e.

$$F_x = (\gamma_w h + p_{air})\,\frac{\pi D^2}{4},$$

**Fig. 2.37** Pressurized tank with a conical indentation

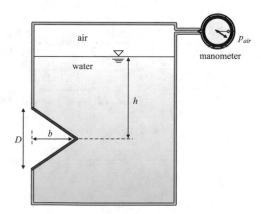

**Fig. 2.38** Schematic for the computation of the forces acting on the conical surface

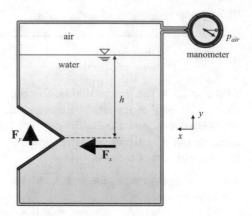

and it is pointing to the left.

The vertical component $F_y$ is pointing upwards and is equal to the magnitude of **G**:

$$F_y = \gamma_w \frac{\pi D^2}{4} \frac{b}{3}.$$

The magnitude of the total force is equal to

$$|\mathbf{F}| = \sqrt{F_x^2 + F_y^2},$$

and the force is at an angle to the horizontal equal to

$$\alpha = \tan^{-1}\left(\frac{F_y}{F_x}\right).$$

For $C_u = C_{pu} = 0$ it results $p_{air} = 3 \times 10^5$ Pa, $h = 1$ m, $b = 0.5$ m, $D = 0.30$ m,

$$F_x = (\gamma_w h + p_{air})\frac{\pi D^2}{4} = \left(9806 \times 1.0 + 3 \times 10^5\right) \times \frac{\pi \times 0.3^2}{4} = \mathbf{21.90\,kN},$$

$$F_y = \gamma_w \frac{\pi D^2}{4}\frac{b}{3} = 9806 \times \frac{\pi \times 0.3^2}{4} \times \frac{0.5}{3} = \mathbf{0.11\,kN},$$

$$\therefore \qquad |\mathbf{F}| = \sqrt{F_x^2 + F_y^2} = \sqrt{21.90^2 + 0.11^2} = \mathbf{21.90\,kN},$$

$$\therefore \qquad \alpha = \tan^{-1}\left(\frac{F_y}{F_x}\right) = \tan^{-1}\left(\frac{0.11}{21.90}\right) \approx \mathbf{17'}.$$

**Exercise 2.14** The cylindrical gate in Fig. 2.39 is $b = 2$ m long and is pivoted in O.

- Calculate the total force exerted by the water on the gate.
- Calculate the action line of the force.
- Calculate the force **S** required to open the gate.

Neglect the weight of the gate and friction. Assume $\gamma_w = 9806\,\mathrm{N\,m^{-3}}$.

**Solution** In the coordinate system shown in Fig. 2.40, the horizontal component of the force is equal to

$$F_x = \gamma_w \left(h + \frac{r}{2}\right) rb = 9806 \times \left(3 + \frac{2}{2}\right) \times 2 \times 2 = \mathbf{156.9\,kN},$$

and the vertical component is equal to the opposite of the weight of the hatched volume filled with water:

$$F_y = \gamma_w hrb + \gamma_w \frac{\pi}{4}r^2 b = \gamma_w b \left(hr + \frac{\pi}{4}r^2\right) =$$
$$9806 \times 2 \times \left(3 \times 2 + \frac{\pi}{4} \times 2^2\right) = \mathbf{179.3\,kN}.$$

Total force has magnitude

$$\therefore \qquad |\mathbf{F}| = \sqrt{F_x^2 + F_y^2} = \sqrt{156.9^2 + 179.3^2} = \mathbf{238.3\,kN},$$

passes through the axis O at an angle to the horizontal

**Fig. 2.39** Cylindrical gate closing the tank

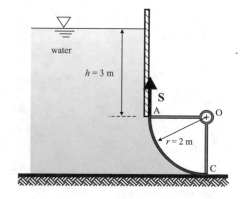

**Fig. 2.40** Schematic for the calculation of the forces acting on the gate

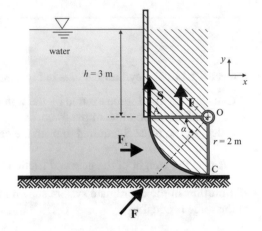

$$\therefore \qquad \alpha = \tan^{-1}\left(\frac{F_y}{F_x}\right) = \tan^{-1}\left(\frac{179.3}{156.9}\right) \approx \mathbf{49°}.$$

The force **S** necessary to open the gate is null, if one neglects friction and weight of the gate, because the total force **F** has zero moment about the pivot of the gate.

---

**Exercise 2.15** The gate in Fig. 2.41 is 3/8 of a cylindrical shell with a circular base; it is $l = 3$ m long in a direction perpendicular to the sheet, it is pivoted in B and rests without friction in A.

– Calculate the constraining reactions in A and B, if the fluid is seawater with a specific weight of $\gamma = 10\,050\,\mathrm{N\,m^{-3}}$.

**Fig. 2.41** Pivoted cylindrical shell

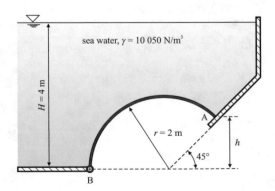

**Fig. 2.42** Schematic for the calculation of the forces

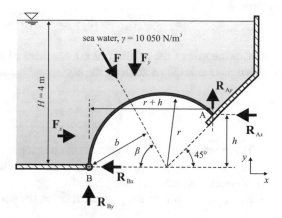

**Solution** We select the coordinate system shown in Fig. 2.42. The force of the fluid on the curved surface $\overarc{AB}$ can be calculated as the composition of the vertical and of the horizontal forces.

The vertical force is equal to the weight of the volume above the gate, and is equal to

$$F_y = -\gamma V = -\gamma l \left[ (r+h) H - \frac{3}{8} \pi r^2 - \frac{h^2}{2} \right],$$

where $V$ is the volume delimited by vertical planes through A and B, by the horizontal free surface and by the cylindrical shell, with $l$ being the length of the shell. Since it results

$$h = r \frac{\sqrt{2}}{2} = 2 \times \frac{\sqrt{2}}{2} = \sqrt{2}\,\mathrm{m},$$

by substituting the numerical values, the vertical force is equal to

$$F_y = -\gamma V =$$

$$- 10\,050 \times 3 \times \left[ \left(2 + \sqrt{2}\right) \times 4 - \frac{3}{8} \times \pi \times 2^2 - \frac{\left(\sqrt{2}\right)^2}{2} \right]$$

$$= -\mathbf{240\,kN}.$$

The horizontal force is pointing to the right and is equal to

$$F_x = \gamma \left( H - \frac{h}{2} \right) hl = 10\,050 \times \left( 4 - \frac{\sqrt{2}}{2} \right) \times \sqrt{2} \times 3 = \mathbf{140\,kN}.$$

Total force has magnitude

$$|\mathbf{F}| = \sqrt{|\mathbf{F}_x|^2 + |\mathbf{F}_y|^2} = \sqrt{240^2 + 140^2} = \mathbf{278\,kN},$$

crosses the axis of the cylinder (it is a resultant of elementary vectors all concurrent towards the axis) at an angle to the horizontal equal to

$$\beta = \tan^{-1}\left(\frac{F_y}{F_x}\right) = \tan^{-1}\left(\frac{-240}{140}\right) = \mathbf{-59°\ 44'}.$$

The reaction in A is orthogonal to the bearing and, for the geometry of the system, has $x$- and $y$-components with the same magnitude. The reaction in B has two independent components along $x$ and $y$.

Imposing the moment equilibrium about the pivot in B, yields

$$-|\mathbf{F}|\,b + R_{Ay}\,(r+h) - R_{Ax}h = 0.$$

The arm $b$ is

$$b = r\sin\beta = 2 \times \sin 59°\ 44' = \mathbf{1.73\,m},$$

hence

$$\therefore \qquad R_{Ay} = -R_{Ax} = \frac{|\mathbf{F}|\,b}{(r+2h)} = \frac{278 \times 1.73}{2 + 2\sqrt{2}} = \mathbf{99\,kN}.$$

Imposing the equilibrium of the forces along $x$ yields

$$R_{Bx} + F_x + R_{Ax} = 0,$$

hence

$$\therefore \qquad R_{Bx} = -F_x - R_{Ax} = -140 + 99 = \mathbf{-41\,kN}.$$

Imposing the equilibrium of the forces along $y$ yields

$$R_{By} + F_y + R_{Ay} = 0,$$

hence

$$\therefore \qquad R_{By} = -F_y - R_{Ay} = 240 - 99 = \mathbf{141\,kN}.$$

---

**Exercise 2.16** Water fills the cylindrical tank in Fig. 2.43 up to the top end of the inlet pipe. The tank is 1.0 m long and is obtained by means of two semi-cylindrical

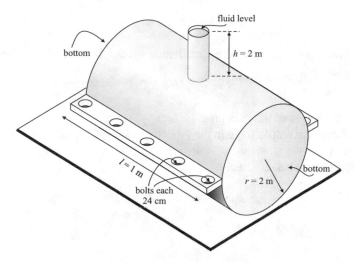

**Fig. 2.43** Bolted cylindrical tank

**Fig. 2.44** Control volume
and schematic of the forces

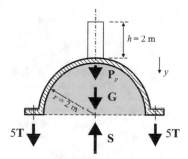

shells each weighing 4.5 kN, with a radius $r = 2$ m and bolted along two generatrices
with bolts every 24 cm. If the tank is resting on a horizontal plane, calculate:

– the stress in the bolts while neglecting the contribution of the two end walls;
– the force orthogonal to the two end walls.

Assume a specific weight of water $\gamma_w = 9806 \, \mathrm{N \, m^{-3}}$.

**Solution** We consider the balance of the control volume of length $l$ in Fig. 2.44.
The acting forces are (1) the weight of the fluid, downwards, equal to

$$G = \gamma_w \frac{\pi r^2}{2} l.$$

(We have neglected the weight of the volume of fluid contained in the feed tube.)
(2) The own weight of the semi-cylindrical shell, downwards, equal to

$$P_p = 4500\,\text{N}.$$

(3) The force due to the pressure distribution on the horizontal meridian plane, upwards, equal to

$$S = -\gamma_w(r + h)2rl,$$

where $h$ is the height of the inlet pipe.

(4) The resultant tensile force in the 10 bolts, unknown, positive downwards and equal to $10\,|\mathbf{T}|$.

By imposing the balance, yields

$$P_p + G + 10T + S = 0,$$

and substituting, yields

$$P_p + \gamma_w \frac{\pi r^2}{2}l + 10T - \gamma_w(r + h)2rl = 0,$$

from which it is derived

$$\therefore\quad T = \frac{\gamma_w(r + h)2rl - P_p - \gamma_w\dfrac{\pi r^2}{2}l}{10} =$$

$$\frac{9806 \times (2 + 2) \times 2 \times 2 \times 1 - 4500 - 9806 \times \dfrac{\pi \times 2^2}{2} \times 1}{10}$$

$$= \mathbf{9078\,N}.$$

**Fig. 2.45** Pascal's barrel experiment, from *The forces of nature* by Amédée Guillemin, 1872

We notice that the traction in the bolts increases linearly with $h$, independent of the diameter of the feed tube. A sufficiently long vertical tube, even with a very small diameter, could break the bolts and the tank. Figure 2.45 shows the socalled Pascal's barrel experiment in 1646 (although it is not documented and the experiment could have been performed by another scientist).

The force orthogonal to each of the two end walls has a magnitude equal to

$$\therefore \qquad S_o = \gamma_w \, (r + h) \, \pi r^2 = 9806 \times (2 + 2) \times \pi \times 2^2 = \mathbf{493\,kN},$$

and it is pointing outwards.

---

**Exercise 2.17** A wooden ball with specific gravity equal to 0.6 separates the two tanks in Fig. 2.46. The two tanks are pressurized and the difference in pressure is indicated by a column of 150 mm of mercury in the U-tube (specific gravity of mercury equal to 13.6). In the tank on the left there is water and in the tank on the right there is oil, with a specific weight relative to water of 0.8.

– Calculate the forces acting on the sphere.

Assume a specific weight of water $\gamma_w = 9806\,\mathrm{N\,m^{-3}}$.

**Solution** In the coordinate system shown in Fig. 2.46, the horizontal force exerted by the water is equal to

$$F_{xw} = \left[ \gamma_w \left( h_1 + \frac{D}{2} \right) + p_1 \right] \frac{\pi D^2}{4},$$

**Fig. 2.46** Tanks with spherical plug in the separating vertical wall

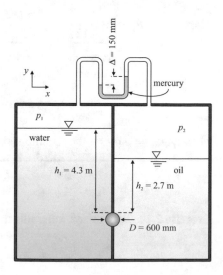

where $\gamma_w$ is the specific weight of water, and the horizontal force exerted by the oil is equal to

$$F_{xo} = -\left[\gamma_o\left(h_2 + \frac{D}{2}\right) + p_2\right]\frac{\pi D^2}{4},$$

where $\gamma_o = 0.8\gamma_w$ is the specific weight of oil.

The resulting horizontal force is equal to

$$F_x = F_{xw} + F_{xo} = \left[\gamma_w\left(h_1 + \frac{D}{2}\right) - \gamma_o\left(h_2 + \frac{D}{2}\right) + p_1 - p_2\right]\frac{\pi D^2}{4}.$$

The difference in pressure between the two tanks, ignoring the specific weight of the air, is equal to

$$p_1 - p_2 = \gamma_{Hg}\Delta,$$

where $\gamma_{Hg} = 13.6\gamma_w$ is the specific weight of mercury.

Substituting the numerical values, yields

$$\therefore \quad F_x = \left[\gamma_w\left(h_1 + \frac{D}{2}\right) - \gamma_o\left(h_2 + \frac{D}{2}\right) + \gamma_{Hg}\Delta\right]\frac{\pi D^2}{4} =$$

$$\left[9806 \times \left(4.3 + \frac{0.6}{2}\right) - 0.8 \times 9806 \times \left(2.7 + \frac{0.6}{2}\right)\right.$$

$$\left. + 13.6 \times 9806 \times 0.15\right] \times \frac{\pi \times 0.6^2}{4} = \mathbf{11.76\,kN}.$$

The vertical component of the force acting on the sphere, including the sphere own weight, is equal to

$$\therefore \quad F_y = \underbrace{\gamma_w\frac{\pi D^3}{12}}_{\substack{\text{buoyancy} \\ \text{in water}}} + \underbrace{\gamma_o\frac{\pi D^3}{12}}_{\substack{\text{buoyancy} \\ \text{in oil}}} - \underbrace{\gamma_{wood}\frac{\pi D^3}{6}}_{\text{own weight}} =$$

$$(9806 + 0.8 \times 9806 - 2 \times 0.6 \times 9806) \times \frac{\pi \times 0.6^3}{12} = \mathbf{332\,N},$$

where $\gamma_{wood} = 0.6\gamma_w$ is the specific weight of wood.

---

**Exercise 2.18**  Pressurized water fills the container in Fig. 2.47.

**Fig. 2.47** Pressurized tank
with conical indented surface

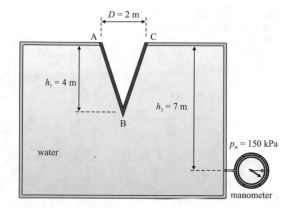

**Fig. 2.48** Schematic for the
calculation of the forces
acting on the conical surface

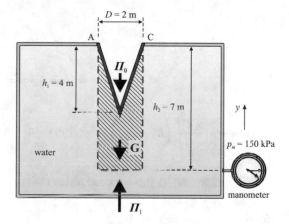

- Calculate the total force on the conical surface if the pressure gauge indicates a
relative pressure of $p_m = 150$ kPa.

Assume a specific weight of water $\gamma_w = 9806$ N m$^{-3}$.

**Solution** After selecting the dashed control volume in Fig. 2.48, the vertical force
component is computed by imposing the balance (along $y$ only):

$$\mathbf{G} + \mathbf{\Pi}_1 + \mathbf{\Pi}_0 = \mathbf{0}.$$

$\mathbf{\Pi}_0$ is the force exerted by the conical surface on the control volume, $\mathbf{F} = -\mathbf{\Pi}_0$ is
the force exerted by the control volume on the conical surface, $\mathbf{G}$ is the weight of the
fluid contained in the volume, $\mathbf{\Pi}_1$ is the force exerted by the pressure acting on the
base of the cylinder of height $h_2$, which is equal to the base of the cone.
The forces have the following expressions:

**Fig. 2.49**  Alternative schematic for the calculation of the forces acting on the conical surface

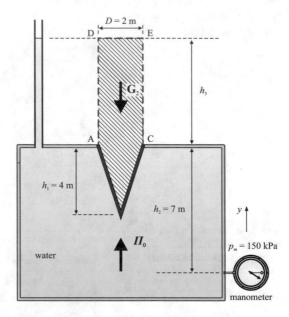

$$G = -\gamma_w \frac{\pi D^2}{4}\left(h_2 - \frac{h_1}{3}\right) = -9806 \times \frac{\pi \times 2^2}{4} \times \left(7 - \frac{4}{3}\right) = -174.57\,\text{kN},$$

$$\Pi_1 = p_m \frac{\pi D^2}{4} = 150\,000 \times \frac{\pi \times 2^2}{4} = 471.24\,\text{kN}.$$

Hence,

$$\therefore \qquad \Pi_0 = -G - \Pi_1 = 174.57 - 471.24 = -296.67\,\text{kN} \rightarrow$$
$$F \equiv -\Pi_0 = 296.67\,\text{kN}.$$

The force on the conical surface is pointing upwards and is only vertical, because the horizontal force is null by symmetry.

An alternative choice of the control volume is shown in Fig. 2.49, where the dashed line represents the trace of the plane where pressure equals the atmospheric pressure, i.e. the rising level of a piezometer with one end at contact with the atmosphere. The control volume, a cylinder plus a cone, is dashed.

The force acting on the base of the cylinder DE is null, since it is at contact with the atmospheric pressure; the vertical force exerted by the conical surface on the control volume balances the weight of the fluid:

$$G_2 + \Pi_0 = 0.$$

The value of $h_3$ is

$$\gamma_w(h_2 + h_3) = p_m \rightarrow h_3 = \frac{p_m}{\gamma_w} - h_2 = \frac{150\,000}{9806} - 7 = \mathbf{8.30\,m},$$

and

$$G_2 = -\gamma_w \frac{\pi D^2}{4}\left(h_3 + \frac{h_1}{3}\right) = -9806 \times \frac{\pi \times 2^2}{4} \times \left(8.30 + \frac{4}{3}\right) = \mathbf{-296.67\,kN},$$

hence,

$$\therefore \qquad \Pi_0 = -G = \mathbf{296.67\,kN} \rightarrow F \equiv \Pi_0 = \mathbf{296.67\,kN}.$$

**Fig. 2.50** Cylindrical floodgate

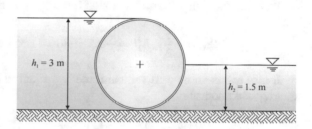

$h_1 = 3$ m

$h_2 = 1.5$ m

**Exercise 2.19** A cylindrical gate has a diameter $D = 3$ m and a length, measured in the direction orthogonal to the drawing in Fig. 2.50, equal to $l = 6$ m.

– Calculate the magnitude and direction of the total force exerted by the water on the gate.

Assume a specific weight of water $\gamma_w = 9806\,\mathrm{N\,m^{-3}}$.

**Solution** Observing the schematic in Fig. 2.51, water on the left exerts a horizontal force $S_{1x}$ pointing to the right and equal to

$$S_{1x} = \frac{1}{2}\gamma_w h_1^2 l = 0.5 \times 9806 \times 3^2 \times 6 = \mathbf{265\,kN}.$$

It also exerts an upward vertical force $S_{1y}$ equal to the Archimedes' force of half the floodgate:

$$S_{1y} = \frac{1}{2}\gamma_w \pi \frac{D^2}{4}l = \frac{1}{8} \times 9806 \times \pi \times 3^2 \times 6 = \mathbf{208\,kN}.$$

**Fig. 2.51** System of forces
acting on the cylindrical
floodgate

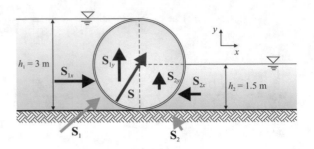

The resulting force $S_1$ will be radial and inclined at an angle to the horizontal equal to

$$\alpha_1 = \tan^{-1}\left(\frac{S_{1y}}{S_{1x}}\right) = \tan^{-1}\left(\frac{208}{265}\right) = \mathbf{38° \ 6'}.$$

Water on the right exerts a horizontal force $S_{2x}$ pointing to the left and equal to

$$S_{2x} = -\frac{1}{2}\gamma_w h_2^2 l = -0.5 \times 9806 \times 1.5^2 \times 6 = -\mathbf{66\,kN}.$$

It also exerts an upward vertical force $S_{2y}$ equal to the Archimedes' force of a quarter of a floodgate:

$$S_{2y} = \frac{1}{4}\gamma_w \pi \frac{D^2}{4} l = \frac{1}{16} \times 9806 \times \pi \times 3^2 \times 6 = \mathbf{104\,kN}.$$

The resulting force $S_2$ will be radial and inclined at an angle to the horizontal equal to

$$\alpha_2 = \tan^{-1}\left(\frac{S_{2y}}{S_{2x}}\right) = \tan^{-1}\left(\frac{104}{66}\right) = \mathbf{57° \ 36'}.$$

The sum of the horizontal components results in a positive force $S_x$ equal to

$$S_x = S_{1x} + S_{2x} = \mathbf{199\,kN},$$

and the sum of the vertical components results in a positive (upward) $S_y$ force, with magnitude

$$S_y = S_{1y} + S_{2y} = \mathbf{312\,kN}.$$

Total force has magnitude

$$\therefore \qquad |\mathbf{S}| = \sqrt{S_x^2 + S_y^2} = \sqrt{199^2 + 312^2} = \mathbf{370\,kN},$$

crosses the axis of the cylinder at an angle to the horizontal equal to

$$\therefore \qquad \alpha = \tan^{-1}\left(\frac{S_y}{S_x}\right) = \tan^{-1}\left(\frac{312}{199}\right) = 57°\ 28',$$

and it is directed as shown in the schematic in Fig. 2.51.

---

**Fig. 2.52** Sealing sphere at the bottom of a tank

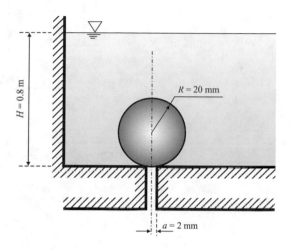

**Exercise 2.20** A sphere of radius $R$, made of material of specific gravity $s$, is immersed in a water tank, see Fig. 2.52. The sphere occludes a hole of radius $a$ at the bottom of the tank.

– Calculate the general expression of the specific gravity $s$, as a function of $H$, $R$ and $a$, necessary for the sphere to float.
– Calculate the minimum value of $s$ necessary for the sphere to remain in position occluding the hole.

Assume a specific weight of water $\gamma_w = 9806\,\mathrm{N\,m^{-3}}$.

**Solution** The sphere is subject to the weight of the fluid cylinder $V_1$ shown in Fig. 2.53, to the own weight and to the buoyancy force acting to the portion of volume immersed in the fluid.

To simplify the calculation, the volume $V_2$ is added both to the volume of fluid above the sphere and to the volume for the calculation of the buoyancy (the overall effect is null, since the weight of the fluid above the sphere and the buoyancy have the same direction, have centre of pressure aligned along the vertical but are pointing in the opposite side). Hence,

$$\mathbf{S}_y + \mathbf{G}_{(V_1+V_2)} + \mathbf{P} + \mathbf{S}_A = \mathbf{0},$$

**Fig. 2.53** Schematic for computing the forces acting on the sphere

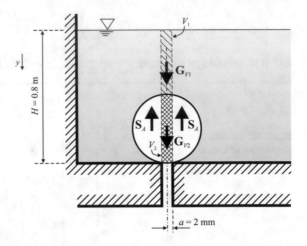

$$G_{(V_1+V_2)} = \gamma_w \pi a^2 H,$$

$$P = \gamma_m \frac{4\pi R^3}{3},$$

$$S_A = -\gamma_w \frac{4\pi R^3}{3},$$

where $\gamma_m = s\gamma_w$. We have assumed that the radius $a$ of the hole is much smaller than $R$.

In the selected coordinate system, it results

$$S_y = \gamma_w \pi a^2 H + \gamma_m \frac{4\pi R^3}{3} - \gamma_w \frac{4\pi R^3}{3}.$$

For buoyancy, it is necessary that $S_y < 0$, i.e.,

$$\pi a^2 H + \frac{\gamma_m}{\gamma_w} \frac{4\pi R^3}{3} - \frac{4\pi R^3}{3} < 0 \rightarrow$$

$$\frac{\gamma_m}{\gamma_w} \equiv s < \frac{\dfrac{4\pi R^3}{3} - \pi a^2 H}{\dfrac{4\pi R^3}{3}} \rightarrow s < s_{min} \equiv 1 - \frac{3a^2 H}{4R^3}.$$

Substituting the numerical data shown in Fig. 2.53, the limit equilibrium condition yields

$$\therefore \qquad s_{min} = 1 - \frac{3 \times (2 \times 10^{-3})^2 \times 0.8}{4 \times (20 \times 10^{-3})^3} = \mathbf{0.7}.$$

**Fig. 2.54** Cork at the bottom of a tank

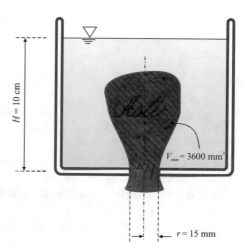

**Exercise 2.21** The cork in Fig. 2.54, of irregular shape and specific gravity $s = 0.45$, closes a circular hole of radius $r = 15$ mm at the bottom of a water-filled tank. The volume of the cork immersed in water, dashed in the figure, is equal to $V_{imm} = 3600$ mm$^3$, the total volume of the cork is equal to $V = 4200$ mm$^3$.

– Calculate the total force acting on the cork.

Assume a specific weight of water $\gamma_w = 9806$ N m$^{-3}$.

**Solution** The horizontal forces acting on the cork are balanced, while in the vertical direction the cork is subject to buoyancy (upward) for the volume $V'$ dashed in Fig. 2.55, to the weight of the overlying water contained in the volume $V''$ (downward) and to its own weight (downward).

Total force acting is equal to

$$F_z = s\gamma_w V - \gamma_w V' + \gamma_w V''.$$

Adding and subtracting $V'''$, yields

$$F_z = s\gamma_w V - \gamma_w \underbrace{(V' + V''')}_{V_{imm}} + \gamma_w (V'' + V''').$$

**Fig. 2.55** Schematic for the
calculation of the forces
acting on the cork

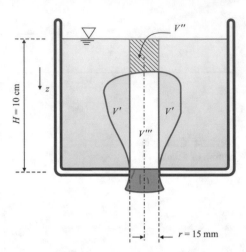

The second contribution is buoyancy relative to the entire volume of the immersed
cork, while the third contribution coincides with the weight of the cylindrical volume
of height $H$ and circular base of radius $r$ filled with water, and is calculated as follows:

$$\gamma_w \left( V'' + V''' \right) = \gamma_w H \pi r^2.$$

Hence,

$$F_z = s\gamma_w V - \gamma_w V_{imm} + \gamma_w H \pi r^2.$$

Notice that for the calculation of the magnitude of the hydrostatic force, it is
sufficient to know only the volume of the immersed body and the area of the cross-
section of the occluded hole (which does not necessarily has to be circular), regardless
of the shape of the cork and its possible asymmetry. The information on the shape
of the immersed body is, however, necessary if you also want to calculate the point
of application of the force.

Substituting the numerical values, yields

$$\therefore \quad F_z = s\gamma_w V - \gamma_w V_{imm} + \gamma_w H \pi r^2 = \gamma_w (sV - V_{imm} + H \pi r^2) =$$
$$9806 \times \left( 0.45 \times 4200 \times 10^{-9} - 3600 \times 10^{-9} + 0.1 \times \pi \times 15^2 \times 10^{-6} \right)$$
$$= \mathbf{0.68\,N}.$$

This force is pointing downwards.

# Chapter 3
# Immersed and Floating Bodies

Looking at an iron ship one wonders how it is possible that it floats, possibly in a stable configuration for the safety of cargo and people, even if the iron has a mass density much greater than that of water. We also wonder about the principle behind the rising motion of hot air balloons, or behind the vertical movement of submarines or fishes. All these are called floating bodies and the analysis of their equilibrium requires only a small addition to the analysis of the hydrostatic forces acting on a curved surface. Some basic principles were discovered by Archimedes more than two thousand years ago and are perfectly suited to the modern theory of Fluid Statics. In essence, a body immersed (partially or totally) in a fluid is subject to its own weight, and receives an additional force, called Archimedes' thrust, equal to the weight of the volume of fluid displaced by the body and directed against gravity.

In this chapter we consider the condition of equilibrium of floating bodies with some examples of analysis of the stability of the equilibrium.

**Exercise 3.1** A wooden beam floating in water is pivoted along a hedge as shown in Fig. 3.1. The beam, of unitary length, is in equilibrium in the geometric configuration shown in the drawing.

– Calculate the relative specific weight of the wood neglecting friction at the pivot.

Assume $\gamma_w = 9800 \, \text{N} \, \text{m}^{-3}$, $L = (1.00 + C_u/10) \, \text{m}$, $D = (0.50 + C_{pu}/20) \, \text{m}$.

**Solution** The beam is in rotational equilibrium if the resulting moment of all the forces about the pivot is zero. Observing the schematic in Fig. 3.2, the forces acting are the hydrostatic force on the base of the beam (per unit length), equal to

---

$C_u$ and $C_{pu}$, that are two integer numbers between 0 and 9, for example, the last and second-last digit of the registration number.

© Springer Nature Switzerland AG 2021
S. Longo et al., *Problems in Hydraulics and Fluid Mechanics*, Springer Tracts in Civil Engineering, https://doi.org/10.1007/978-3-030-51387-0_3

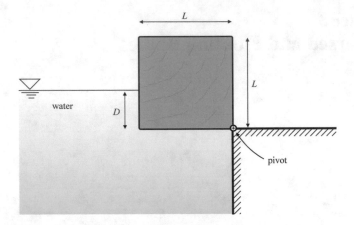

**Fig. 3.1**   Floating wooden beam pivoted in one corner

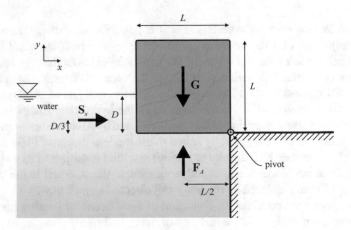

**Fig. 3.2**   Schematic of the forces acting on the beam

$$F_{Ay} = \gamma_w L D,$$

which has arm $L/2$ with respect to the pivot; the own weight of the beam (per unit length),

$$G_y = -\gamma_w s L^2,$$

where $s$ is the relative specific weight of the wood, which has arm $L/2$ with respect to the pivot; the horizontal force of the water on the vertical left wall of the beam (per unit length),

$$S_x = \frac{1}{2}\gamma_w D^2,$$

which has arm $D/3$ with respect to the pivot.

By imposing balance at rotation about the pivot, yields

$$\left(F_{Ay} + G_y\right)\frac{L}{2} + S_x\frac{D}{3} = 0 \rightarrow \gamma_w\frac{L^2 D}{2} - s\gamma_w\frac{L^3}{2} + \gamma_w\frac{D^3}{6} = 0.$$

Hence

$$s = \frac{D}{L} + \frac{D^3}{3L^3}.$$

For $C_u = C_{pu} = 0$ it results $L = 1.0\,\mathrm{m}$, $D = 0.5\,\mathrm{m}$,

$$\therefore \qquad s = \frac{D}{L} + \frac{D^3}{3L^3} = \frac{0.5}{1.0} + \frac{(0.5)^3}{3 \times (1.0)^3} = \mathbf{0.54}.$$

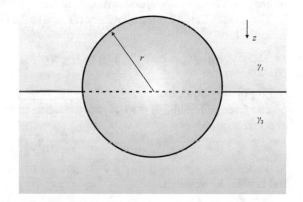

**Fig. 3.3** Floating sphere at the interface between two liquids

**Exercise 3.2** The sphere in Fig. 3.3 is in equilibrium at the interface between two liquids of specific weight $\gamma_1 = 7\,\mathrm{kN\,m^{-3}}$ and $\gamma_2 = 9\,\mathrm{kN\,m^{-3}}$, respectively. The interface passes through the centroid of the sphere.

– Calculate the specific weight of the material of the sphere.

**Solution** The sphere is in vertical equilibrium under the action of its own weight and of the forces exerted by the two liquids, see Fig. 3.4. The equation of equilibrium in the vertical direction is

$$G_z + \Pi_{1z} + \Pi_{2z} = 0.$$

The own weight is equal to

**Fig. 3.4** Forces acting on
the sphere

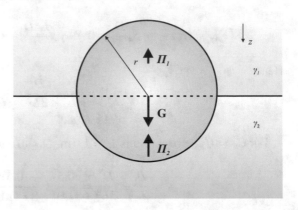

$$G_z = \frac{4}{3}\pi r^3 \gamma_s,$$

where $\gamma_s$ is the unknown specific weight of the material of the sphere.

The force of liquid 1 is the buoyancy relative to the volume of the sphere immersed in liquid 1,

$$\Pi_{1z} = -\frac{2}{3}\pi r^3 \gamma_1,$$

and is pointing upwards.

The force of liquid 2 is the buoyancy relative to the volume of sphere immersed in liquid 2,

$$\Pi_{2z} = -\frac{2}{3}\pi r^3 \gamma_2,$$

and is pointing upwards.

In equilibrium condition, yields

$$\frac{4}{3}\pi r^3 \gamma_s - \frac{2}{3}\pi r^3 \gamma_1 - \frac{2}{3}\pi r^3 \gamma_2 = 0.$$

Substituting the numerical values, it results

$$\therefore \qquad \gamma_s = \frac{\gamma_1 + \gamma_2}{2} = \frac{7+9}{2} = 8\,\mathrm{kN\,m}^{-3}.$$

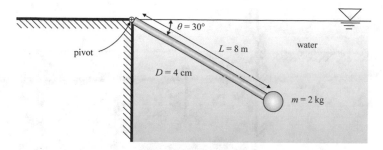

**Fig. 3.5** Pivoted cylinder with lead mass at the free end

**Exercise 3.3** A cylinder with a circular cross-section is pivoted and is in equilibrium in water with a sphere of lead with a mass of $m = 2$ kg hanging at the free end, see Fig. 3.5.

- Calculate the specific weight of the cylinder material.
- Calculate the reaction of the pivot.
- Study the stability of the equilibrium to variations of the free surface level.

The ambient fluid is water with $\gamma_w = 9806$ N m$^{-3}$, and the specific weight of lead is $\gamma_{Pb} = 11.4\gamma_w$.

**Solution** In the coordinate system shown in Fig. 3.6, the following forces are applied to the cylinder and the sphere:

(1) the weight of the cylinder applied in the gravity centre, positive downwards, equal to

$$P_t = \gamma_m V_{cyl},$$

where $\gamma_m$ is the specific weight of the material of the cylinder.

(2) The buoyancy force applied in the centroid of the immersed volume of the cylinder, upwards, equal to

$$S_{At} = -\gamma_w V_{cyl}.$$

(3) The weight of the lead sphere applied in the centre of gravity of the sphere, downwards, equal to

$$P_s = mg.$$

(4) The buoyancy force applied in the centroid of the immersed volume of the sphere, upwards, equal to

$$S_{As} = -\gamma_w V_{sphere} = -\frac{\gamma_w mg}{\gamma_{Pb}}.$$

(5) The reaction of the pivot, directed along the vertical since there are no horizontal forces to balance, with unknown magnitude and direction.

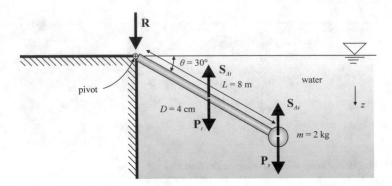

**Fig. 3.6** Schematic of the forces

The equilibrium to the translation in the vertical is

$$\gamma_m V_{cyl} + mg - \gamma_w V_{cyl} - \frac{\gamma_w mg}{\gamma_{Pb}} + R = 0.$$

The equation of equilibrium at rotation about the pivot, neglecting the size of the lead sphere (i.e., considering the forces applied at the end of the cylinder), is as follows:

$$(\gamma_m - \gamma_w) V_{cyl} \frac{L}{2} \cos\theta + mg \left(1 - \frac{\gamma_w}{\gamma_{Pb}}\right) L \cos\theta = 0.$$

The reaction of the pivot is equal to

$$R = (\gamma_w - \gamma_m) V_{cyl} + mg \left(\frac{\gamma_w}{\gamma_{Pb}} - 1\right).$$

Substituting the numerical values, yields

$$\therefore \quad \gamma_m = \gamma_w - 2 \frac{mg}{\frac{\pi D^2}{4} L} \left(1 - \frac{\gamma_w}{\gamma_{Pb}}\right) =$$

$$9806 - 2 \times \frac{2 \times 9.806}{\frac{\pi \times 0.04^2}{4} \times 8} \times \left(1 - \frac{1}{11.4}\right) = \mathbf{6245\,N\,m^{-3}},$$

$$\therefore \quad R = (9806 - 6245) \times \frac{\pi \times 0.04^2}{4} \times 8 + 2 \times 9.806 \times \left(\frac{1}{11.4} - 1\right) = \mathbf{17.9\,N}.$$

The reaction of the pivot is pointing downwards. Notice that the value of the angle, 30°, is not necessary to solve the problem.

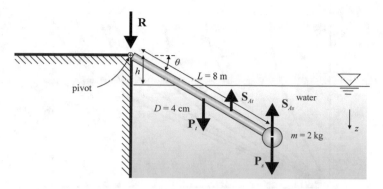

**Fig. 3.7** Schematic of the forces acting when the cylinder is partially immersed in water

*Stability of the Equilibrium*

(*a*) If the free surface level is above the pivot, nothing changes with respect to the case already studied. Equilibrium is indifferent.

(*b*) If the free surface level decreases, but with the lead sphere entirely immersed in water, the reaction of the pivot decreases. The angle of inclination of the bar is uniquely defined since the buoyancy force of the immersed part of the cylinder is a function of $\theta$.

We indicate with $h$ the distance between the free surface level and the pivot, as shown in Fig. 3.7. The equation of equilibrium in the vertical is

$$\gamma_m V_{cyl} + mg - \gamma_w V_{cyl}\left(1 - \frac{h}{L\sin\theta}\right) - \frac{\gamma_w mg}{\gamma_{Pb}} + R = 0,$$

and the equation of rotational equilibrium is

$$\gamma_m V_{cyl}\frac{L}{2}\cos\theta - \gamma_w V_{cyl}\left(1 - \frac{h}{L\sin\theta}\right)\left(\frac{L}{2}\cos\theta + \frac{h}{2\tan\theta}\right)$$
$$+ mg\left(1 - \frac{\gamma_w}{\gamma_{Pb}}\right)L\cos\theta = 0,$$

which is simplified as

$$\frac{h}{L\sin\theta} = \sqrt{\left(1 - \frac{\gamma_m}{\gamma_w}\right) - \frac{2mg}{\gamma_w V_{cyl}}\left(1 - \frac{\gamma_w}{\gamma_{Pb}}\right)}.$$

The root argument must be positive, i.e. it must result

$$\frac{\gamma_m}{\gamma_w} < 1 - \frac{2mg}{\gamma_w V_{cyl}}\left(1 - \frac{\gamma_w}{\gamma_{Pb}}\right).$$

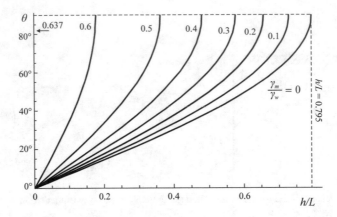

**Fig. 3.8** Angle of equilibrium as a function of the water level with respect to the pivot and for different values of the relative specific weight of the cylinder material

In addition, the solution is coherent with the hypothesis providing that

$$\frac{h}{L \sin \theta} < 1.$$

In fact, if $h > L \sin \theta$ the sphere is floating at the interface between water and air and the buoyancy force must be calculated with reference to the partial volume of the sphere immersed in water. The angle of equilibrium is shown in Fig. 3.8 as a function of $h/L$ and the relative specific weight of the cylinder material.

The clockwise torque due to the weight and to the buoyancy force acting on the sphere decreases when $h$ increases, because the arm is reduced. The counter-clockwise torque due to the weight and to the buoyancy force acting on the cylinder is reduced when $h$ increases, both by the reduction of the arm and by the reduction of the buoyancy force. Hence, it is reasonable to expect that there is a value of $h$ beyond which the angle of equilibrium is equal to 90°. This limit value is shown in Fig. 3.9 as a function of the relative specific weight of the material. Beyond this limit condition, the only stable and possible equilibrium is for $\theta = 90°$.

(c) If the free surface level further decreases, leaving the whole sphere out of the water, the angle is 90°, the buoyancy force is null, the reaction in the pivot is maximum, positive upwards, with magnitude equal to the sum of the weight of the cylinder and the lead sphere. Equilibrium is stable.

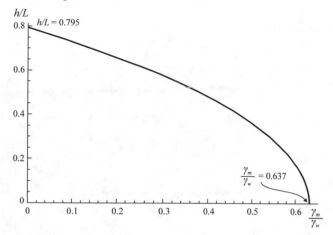

Fig. 3.9 Limit value of the water level distance from the pivot for different values of the specific weight of the cylinder material relative to water

**Exercise 3.4** The wooden tree trunk in Fig. 3.10 is immersed in water and is anchored to the bottom with a cable.

- Calculate the tensile force in the cable.
- Calculate the specific weight of wood.

In addition, analyze the equilibrium as the water depth $H$ varies.

**Fig. 3.10** Floating tree trunk with anchoring cable

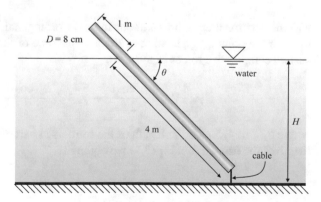

**Solution** Referring to the schematic shown in Fig. 3.11, the forces acting on the tree trunk are:

(1) the force of buoyancy, applied in the centre of gravity of the immersed volume (centre of buoyancy), pointing upwards and equal to

$$S_A = -\gamma_w V_i = -\gamma_w \frac{4}{5} V_{tot},$$

**Fig. 3.11** Schematic of the forces

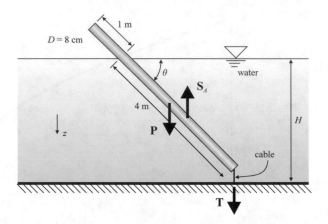

where $V_i$ is the immersed volume, $V_{tot}$ is the total volume of the trunk, $\gamma_w = 9806\,\mathrm{N\,m^{-3}}$ is the specific weight of the water.

(2) The weight of the trunk, applied in the centre of gravity of the body, pointing downwards and equal to

$$P = \gamma_m V_{tot},$$

where $\gamma_m$ is the specific weight of the wood.

(3) The force exerted by the cable, necessarily of traction, pointing downwards and equal to $T$.

We can write the following equation of equilibrium to the vertical translation:

$$-\gamma_w \frac{4}{5} V_{tot} + \gamma_m V_{tot} + T = 0,$$

and the rotational equilibrium about the axis orthogonal to the sheet and passing through the point of application of the tensile force of the cable (the cable cannot sustain compressive forces or torques):

$$\gamma_w \frac{4}{5} V_{tot} \frac{l_{imm} \cos\theta}{2} - \gamma_m V_{tot} \frac{l_{tot} \cos\theta}{2} = 0, \tag{3.1}$$

where $l_{imm} = 4/5 l_{tot}$ is the length of the immersed part of the trunk, $l_{tot}$ is the total length of the trunk. From Eq. (3.1) we obtain

$$\frac{\gamma_m}{\gamma_w} = \frac{16}{25},$$

hence

$$T = \gamma_w \left( \frac{4}{5} V_{tot} - \frac{16}{25} V_{tot} \right) = \frac{4}{25} \gamma_w V_{tot} = \frac{1}{4} \gamma_m V_{tot} = \frac{P}{4}.$$

By substituting the numerical values, yields

$$\therefore \qquad \gamma_m = \frac{16}{25}\gamma_w = \frac{16}{25} \times 9806 = \mathbf{6275\,N\,m^{-3}},$$

and

$$\therefore \qquad T = \frac{1}{4}\gamma_m V_{tot} = \frac{1}{4} \times 6275 \times \frac{\pi \times 0.08^2}{4} \times 5 = \mathbf{39.4\,N}.$$

The angle of inclination $\theta$ will depend only on the water depth $H$ and on the length of the trunk.

*Equilibrium for varying water depth*

We neglect the length of the cable.

(a) If $H < 4l_{tot}/5$ (but $H$ enough to prevent the trunk from touching the bottom), the angle $\theta$ depends only on the geometry. Since if $\gamma_m/\gamma_w = 16/25$ it is also $l_{imm}/l_{tot} = 4/5$, for increasing $H$ the cylinder rotates at an angle $\theta = \sin^{-1}[5H/(4l_{tot})]$. The equilibrium is stable, the tensile force in the cable is constant and equal to $P/4$.

(b) If $4l_{tot}/5 < H < l_{tot}$, the angle is $\theta = 90°$ and the tension in the cable increases linearly until it reaches the maximum value of $T = 9P/16$. Equilibrium is stable.

(c) If $H > l_{tot}$, the angle is still $\theta = 90°$ and the tensile force in the cable is constant and equal to the maximum value. Equilibrium is stable.

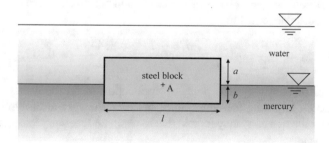

**Fig. 3.12** Parallelepiped block floating at the interface between two liquids

**Exercise 3.5** The steel block in Fig. 3.12 floats at the interface between mercury and water. The specific weight of steel is equal to $\gamma_{Fe}/\gamma_w = 7.85$, the mercury specific weight is equal to $\gamma_{Hg}/\gamma_w = 13.56$.

– Calculate the ratio $a/b$ in equilibrium condition.
– Calculate the $l/b$ ratio required for rotational stability around an axis orthogonal to the sheet plane of trace A.

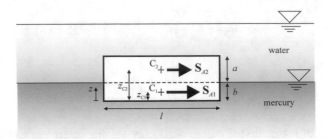

**Fig. 3.13** Schematic for the calculation of the centre of buoyancy

Assume a unit depth of the block.

**Solution** The condition of equilibrium along the vertical is obtained by imposing that the buoyancy force equals the weight of the block, i.e.:

$$\gamma_{Hg} lb + \gamma_w la = \gamma_{Fe} l (a + b) \rightarrow$$

$$\therefore \qquad \frac{a}{b} = \frac{\gamma_{Hg} - \gamma_{Fe}}{\gamma_{Fe} - \gamma_w} = \mathbf{0.834}.$$

To check the linear rotational stability of equilibrium for (small) rotations about A, we first calculate the position of the centre of gravity. Selecting a coordinate system with the origin on the lowest basis of the block and positive upwards, the centre of gravity has coordinate equal to

$$z_G = \frac{a+b}{2} \rightarrow \frac{z_G}{b} = \frac{a/b+1}{2} = \frac{0.834+1}{2} = \mathbf{0.917}.$$

The centre of buoyancy (the point of application of the buoyancy force) can be calculated by imposing the equivalence between the system of applied vectors represented by the two buoyancy forces (due to water and mercury, respectively) and the resulting vector.

This equivalence requires that the moment of the resulting vector with respect to any axis be equal to the vector sum of the moments of the two buoyancy forces with respect to the same axis. It can be demonstrated that the result does not change if the two vectors are rotated at the same angle, and for simplicity we rotate $S_{A1}$ and $S_{A2}$ in order to make them horizontal, see Fig. 3.13. Actually, they are collinear and pointing downwards. Considering the horizontal axis through the origin of the coordinate system, yields

$$S_{A2} z_{C2} + S_{A1} z_{C1} = S_A z_C \rightarrow \gamma_w al \left( \frac{a}{2} + b \right) + \gamma_{Hg} bl \frac{b}{2} = \left( \gamma_w al + \gamma_{Hg} bl \right) z_C,$$

hence,

$$\frac{z_C}{b} = \frac{1}{2} + \frac{\dfrac{a}{b}\left(\dfrac{a}{2b} + \dfrac{1}{2}\right)}{\dfrac{a}{b} + \dfrac{\gamma_{Hg}}{\gamma_w}}.$$

By substituting the numerical values, it yields

$$\frac{z_C}{b} = \frac{1}{2} + \frac{0.834 \times \left(\dfrac{0.834}{2} + \dfrac{1}{2}\right)}{0.834 + \dfrac{13.56}{1}} = \mathbf{0.553}.$$

The relative distance between the centre of gravity and the centre of buoyancy is equal to

$$\frac{z_G - z_C}{b} = 0.917 - 0.553 = \mathbf{0.364},$$

with the buoyancy centre below the centre of gravity. In this condition stability is not guaranteed, and it is necessary to verify the position of the metacentre. In the hypothesis of small rolling rotations (rotation around an axis orthogonal to the sheet), the distance between the centre of buoyancy and the metacentre is equal to

$$z_M - z_C = \frac{(\gamma_{Hg} - \gamma_w)\, I_{Gxx}}{\gamma_{Hg} bl + \gamma_w al},$$

where $I_{Gxx} = l^3/12$ is the second moment of the cross-section of the block at the interface with respect to the barycentric axis of rotation. Substituting, it yields

$$\frac{z_M - z_C}{b} = \frac{\left(\dfrac{\gamma_{Hg}}{\gamma_w} - 1\right)\dfrac{1}{12}\dfrac{l^2}{b^2}}{\dfrac{\gamma_{Hg}}{\gamma_w} + \dfrac{a}{b}} = \frac{\left(\dfrac{13.56}{1} - 1\right) \times \dfrac{1}{12} \times \dfrac{l^2}{b^2}}{\dfrac{13.56}{1} + 0.834} = 0.0727\dfrac{l^2}{b^2}.$$

For the stability of the equilibrium it is required that

$$\frac{z_M - z_C}{b} > \frac{z_G - z_C}{b},$$

hence

$$\therefore \qquad 0.0727\frac{l^2}{b^2} > 0.364 \rightarrow \frac{l}{b} > \mathbf{2.237}.$$

Let us consider the condition with the specific weight of the upper fluid (water) negligible if compared to that of the lower fluid (mercury). In this case the vertical equilibrium requires that

$$\gamma_{Hg} l b = \gamma_{Fe} l \left( a + b \right),$$

or

$$\frac{a}{b} = \frac{\gamma_{Hg}}{\gamma_{Fe}} - 1 = 0.727.$$

The centre of gravity has still the coordinate

$$z_G = \frac{a + b}{2},$$

hence

$$z_G = \frac{a + b}{2} \rightarrow \frac{z_G}{b} = \frac{a/b + 1}{2} = \frac{0.727 + 1}{2} = 0.864.$$

The centre of buoyancy has the coordinate

$$\frac{z_C}{b} = \frac{1}{2}.$$

The distance between the centre of gravity and the centre of buoyancy is equal to

$$\frac{z_G - z_C}{b} = 0.864 - 0.5 = 0.364.$$

with the centre of gravity below the buoyancy centre.

The distance between the buoyancy centre and the metacentre is equal to

$$z_M - z_C = \frac{I_{Gxx}}{bl} \rightarrow \frac{z_M - z_C}{b} = \frac{1}{12} \frac{l^2}{b^2}.$$

For the stability of the equilibrium it is required that

$$\frac{z_M - z_C}{b} > \frac{z_G - z_C}{b},$$

or

$$\frac{1}{12} \frac{l^2}{b^2} > 0.364 \rightarrow \frac{l}{b} > 2.09.$$

This result is almost equal to the one obtained considering also the action of the overlying fluid. If the overlying fluid had been air, the results on the stability condition of the equilibrium analyzed by including or neglecting the overlying fluid would have been practically coincident. This also applies if the underlying fluid is water with air above. It is for this reason that, as a rule, the action of the air is always neglected when studying the stability of the equilibrium of boats and of floating bodies in water.

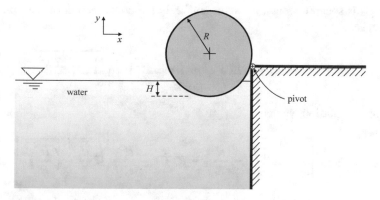

**Fig. 3.14** Floating cylinder pivoted along the generatrix

**Exercise 3.6** The cylinder in Fig. 3.14 floats in water and is pivoted to its full length along the generatrix.

– Calculate the relative specific weight of the cylinder material as a function of the ratio $\alpha = H/R$ for $0 \leq \alpha < 1$.

Friction at the pivot is negligible.

**Solution** The forces acting on the cylinder are:

– the weight force, applied in the centre of gravity and pointing downwards;
– the buoyancy force, applied in the centre of buoyancy and pointing upwards;
– the reaction of the constraint, necessarily passing through the pivot.

The first two forces are always vertically aligned due to the symmetry of the body. Since they must have zero torque about the pivot, they must have the same magnitude. Therefore, the reaction of the pivot is null, since there are no horizontal acting forces.

In equilibrium condition, it results

$$- |\mathbf{P}| + |\mathbf{S}_A| = 0.$$

The weight of the cylinder per unit of depth has magnitude

$$|\mathbf{P}| = \gamma_m \pi R^2,$$

where $\gamma_m$ is specific weight of the cylinder.

Buoyancy force per unit depth has magnitude

$$|\mathbf{S}_A| = \gamma_w A,$$

where $\gamma_w$ is specific weight of water and $A$ is the cross-section of the immersed volume, equal to

$$A = R^2\cos^{-1}\left(1 - \frac{H}{R}\right) - R^2\left(1 - \frac{H}{R}\right)\sqrt{2\frac{H}{R} - \left(\frac{H}{R}\right)^2}.$$

Upon substitution, yields

$$-\gamma_m\pi R^2 + \gamma_w\left[R^2\cos^{-1}\left(1 - \frac{H}{R}\right) - R^2\left(1 - \frac{H}{R}\right)\sqrt{2\frac{H}{R} - \left(\frac{H}{R}\right)^2}\right] = 0,$$

and, as function of $\alpha$:

$$\therefore \qquad \frac{\gamma_m}{\gamma_w} = \frac{\cos^{-1}(1 - \alpha) - (1 - \alpha)\sqrt{2\alpha - \alpha^2}}{\pi}.$$

---

**Exercise 3.7** The barge in Fig. 3.15 carries oil floating on a layer of water. The weight of the hull per unit length is $P_{hull} = (3000 + 100 \times C_{pu})\,\mathrm{N\,m^{-1}}$. The width is $b = (10 + C_u)\,\mathrm{m}$, the depth of the water layer is $y_1 = 1.5\,\mathrm{m}$ and the depth of the oil layer is $y_2 = 2.0\,\mathrm{m}$.

– Calculate the draught $y_3$ of the boat if it is immersed in water.

Assume $\gamma_w = 9806\,\mathrm{N\,m^{-3}}$, $\gamma_o = 0.8\gamma_w$.

**Solution** Vertical equilibrium is satisfied if buoyancy force equals the weight of the hull and its contents. The buoyancy force per unit length of the hull (orthogonal to the sheet) is equal to the product of the immersed volume of the hull and the specific weight of the water:

$$F_A = \gamma_w\,(b + y_3)\,y_3,$$

and is pointing upwards.

The weight of the hull and its contents, per unit of length, is equal to

$$P = P_{hull} + \gamma_w\,(b + y_1)\,y_1 + \gamma_o\,(b + 2y_1 + y_2)\,y_2,$$

and is pointing downwards.

By equating and solving the following second-order equation in $y_3$,

$$\underbrace{\gamma_w\,(b + y_3)\,y_3}_{V_{displaced}} = \underbrace{P_{hull} + \gamma_w\,(b + y_1)\,y_1 + \gamma_o\,(b + 2y_1 + y_2)\,y_2}_{P},$$

**Fig. 3.15** Floating barge
containing stratified liquids

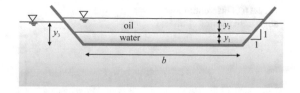

the draught of the barge is calculated.

For $C_u = C_{pu} = 0$ it results $P_{hull} = 3\,\text{kN}\,\text{m}^{-1}$, $b = 10.0\,\text{m}$, $y_1 = 1.5\,\text{m}$, $y_2 = 2.0\,\text{m}$,

$$\therefore \quad P = P_{hull} + \gamma_w\,(b + y_1)\,y_1 + \gamma_o\,(b + 2y_1 + y_2)\,y_2 =$$
$$3000 + 9806 \times (10 + 1.5) \times 1.5$$
$$+ 0.8 \times 9806 \times (10 + 2 \times 1.5 + 2.0) \times 2.0 = \mathbf{407.5\,kN\,m^{-1}},$$

$$\gamma_w\,(b + y_3)\,y_3 = P \rightarrow 9806 \times (10.0 + y_3)\,y_3 = 407\,500,$$

which admits the physically acceptable solution

$$\therefore \qquad y_3 = \mathbf{3.16\,m}.$$

**Exercise 3.8** The device in Fig. 3.16 is a differential level gauge. The two cylinders, of specific weight $\gamma_m$, are connected by an inextensible cable suspended with a pulley of diameter $D_p = 200\,\text{mm}$. The cylinders have diameter $D_1 = D_2 = 150\,\text{mm}$ and height $h_1 = h_2 = h = 400\,\text{mm}$, and the length of the cable is $L = 1000\,\text{mm}$. In the hypothesis that the friction torque at the axis of the pulley is negligible:

– analyze the behavior of the system as a function of the absolute level and the difference in level in the two measuring wells, if $\gamma_m < \gamma_w$, where $\gamma_w$ is specific weight of water.
– Perform the same analysis if $\gamma_m > \gamma_w$.

In addition, analyze the effects of a friction torque at the pulley axis equal to $M = 5 \times 10^{-3}\,\text{Nm}$.

**Fig. 3.16** Differential level
gauge with floating cylinders

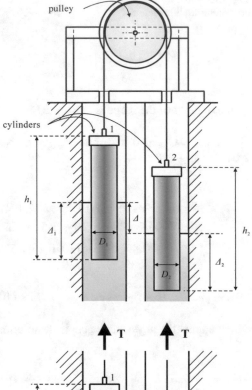

**Fig. 3.17** Schematic of the
forces

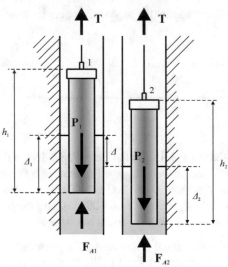

**Solution** We consider the equilibrium of the two cylinders, subject to their own
weight, to the buoyancy and to the cable traction, see Fig. 3.17.

In the absence of friction torque at the pulley axis, the tensile force in the cable is
constant. The draught of the two cylinders is calculated on the basis of the following
two equilibrium equations:

$$
\begin{cases}
\gamma_m \dfrac{\pi D_1^2}{4} h_1 - \gamma_w \dfrac{\pi D_1^2}{4} \Delta_1 - T = 0, \\[4mm]
\gamma_m \dfrac{\pi D_2^2}{4} h_2 - \gamma_w \dfrac{\pi D_2^2}{4} \Delta_2 - T = 0.
\end{cases}
$$

By subtracting the corresponding sides of the two equations, yields

$$
\gamma_m \left( \frac{\pi D_1^2}{4} h_1 - \frac{\pi D_2^2}{4} h_2 \right) - \gamma_w \left( \frac{\pi D_1^2}{4} \Delta_1 - \frac{\pi D_2^2}{4} \Delta_2 \right) = 0.
$$

If the diameters of the two cylinders are equal ($D_1 = D_2 = D$), it results

$$
\gamma_m \frac{\pi D^2}{4} (h_1 - h_2) - \gamma_w \frac{\pi D^2}{4} (\Delta_1 - \Delta_2) = 0 \rightarrow (\Delta_1 - \Delta_2) = \frac{\gamma_m (h_1 - h_2)}{\gamma_w}.
$$

This means that the difference between the draught of the two cylinders is independent of the level of the fluid in the two wells. Therefore, if the level in one well remains fixed and the level in the other well varies, the cable will run by a value exactly equal to the differential level variation, and the angle of rotation of the pulley will be linearly proportional to the difference in level between the two wells:

$$
\alpha \propto \frac{\Delta}{D_p},
$$

and the device works as a differential level gauge with a linear characteristic.

If $\gamma_m < \gamma_w$, the device no longer works as a differential level gauge when the traction in the cable is null, i.e. when the following condition occurs:

$$
\begin{cases}
\Delta_1 = \dfrac{\gamma_m}{\gamma_w} h_1, \\[4mm]
\Delta_2 = \dfrac{\gamma_m}{\gamma_w} h_2.
\end{cases}
\tag{3.2}
$$

This is equivalent to the condition of floating cylinders with buoyancy force able to support them without traction in the cable.

The analysis of the operating range can be carried out by referring (i) to the average level of free surface level in the two wells, and (ii) to the excursion of free surface level in the two wells about this average level.

After selecting a system of coordinates $z$ with the origin at the level of the axis of the pulley, see Fig. 3.18, assuming $h_1 = h_2 = h$ the limit condition (3.2) allows to calculate one of the two limit positions of the average level:

$$
max \; z_{av} = \left( L - \frac{\pi D_p}{2} \right) \frac{1}{2} + h \left( 1 - \frac{\gamma_m}{\gamma_w} \right),
$$

**Fig. 3.18** Schematic for the
analysis of the measurement
range

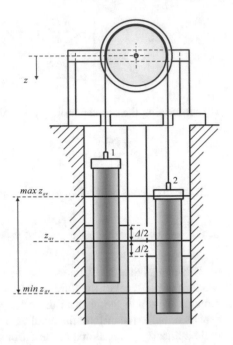

which becomes:

$$max\ z_{av} = \left(L - \frac{\pi D_p}{2}\right)\frac{1}{2} + h\left(1 - \frac{\gamma_m}{\gamma_w}\right) \rightarrow$$

$$\left(1000 - \frac{\pi \times 200}{2}\right) \times \frac{1}{2} + 400 \times \left(1 - \frac{\gamma_m}{\gamma_w}\right) = 742.9 - 400\left(\frac{\gamma_m}{\gamma_w}\right)\ \text{mm}.$$

The other limit position of the average level corresponds to the condition of
cylinders suspended out of the water (buoyancy force is null and the cylinders are
supported only by the traction in the cable), and depends on the geometry of the
system:

$$min\ z_{av} = \left(L - \frac{\pi D_p}{2}\right)\frac{1}{2} + h,$$

which becomes

$$min\ z_{av} = \left(L - \frac{\pi D_p}{2}\right)\frac{1}{2} + h \rightarrow \left(1000 - \frac{\pi \times 200}{2}\right) \times \frac{1}{2} + 400 = 742.9\ \text{mm}.$$

On the basis of the previous results, the range of excursion of the average level is
equal to

$$\therefore \qquad\qquad max\ z_{av} - min\ z_{av} = h\frac{\gamma_m}{\gamma_w}.$$

The range of excursion of the differential level in the wells, equal to the average level plus $\pm \Delta/2$, depends only on the geometry of the system, i.e. the length of the cable and the height of the cylinders.

If $\gamma_m > \gamma_w$ the device no longer works as a differential level gauge if the level in the wells is such that the two cylinders sink or emerge completely. Unlike the case where $\gamma_m < \gamma_w$, the traction in the cable never cancels.

The condition of fully sunk cylinders results in a limit position of the average level of:

$$max\ z_{av} = \left( L - \frac{\pi D_p}{2} \right) \frac{1}{2} \rightarrow \left( 1000 - \frac{\pi \times 200}{2} \right) \times \frac{1}{2} = 342.9\ \text{mm}.$$

The condition of cylinders completely out of water leads to calculate the other limit position of the mean level, which depends on the geometry of the system and is equal to:

$$min\ z_{av} = \left( L - \frac{\pi D_p}{2} \right) \frac{1}{2} + h,$$

and which becomes:

$$min\ z_{av} = \left( L - \frac{\pi D_p}{2} \right) \frac{1}{2} + h \rightarrow \left( 1000 - \frac{\pi \times 200}{2} \right) \times \frac{1}{2} + 400 = 742.9\ \text{mm}.$$

The average level excursion range is equal to:

$$\therefore \qquad max\ z_{av} - min\ z_{av} = \textbf{h}.$$

Notice that this value is greater than the value calculated if $\gamma_m < \gamma_w$.

Also in this case, the range of excursion of the differential level in the wells, equal to the average level $\pm \Delta/2$, depends only on the geometry of the system, i.e. the length of the cable and the height of the cylinders.

If we include the effect of the friction torque, the traction in the cable is not necessarily constant and the two equilibrium equations for the two cylinders are rewritten as follows:

$$\begin{cases} \gamma_m \dfrac{\pi D_1^2}{4} h_1 - \gamma_w \dfrac{\pi D_1^2}{4} \Delta_1 - T = 0, \\[4mm] \gamma_m \dfrac{\pi D_2^2}{4} h_2 - \gamma_w \dfrac{\pi D_2^2}{4} \Delta_2 - T \pm \dfrac{2M}{D_p} = 0. \end{cases}$$

By subtracting the corresponding sides of the two equations, yields

$$\gamma_m \left( \frac{\pi D_1^2}{4} h_1 - \frac{\pi D_2^2}{4} h_2 \right) - \gamma_w \left( \frac{\pi D_1^2}{4} \Delta_1 - \frac{\pi D_2^2}{4} \Delta_2 \right) \mp \frac{2M}{D_p} = 0.$$

If the diameters of the two cylinders are equal ($D_1 = D_2 = D$), it results

$$\gamma_m \frac{\pi D^2}{4} (h_1 - h_2) - \gamma_w \frac{\pi D^2}{4} (\Delta_1 - \Delta_2) \mp \frac{2M}{D_p} = 0 \rightarrow$$

$$(\Delta_1 - \Delta_2) = \frac{\gamma_m (h_1 - h_2)}{\gamma_w} \mp \frac{8M}{\gamma_w \pi D^2 D_p}.$$

$$(3.3)$$

Compared to the analysis in the absence of friction torque, there is an uncertainty in the estimate which is systematic and leads to an underestimation of the differential level if the difference in level is increasing, to an overestimation if the difference in level is decreasing. It can be demonstrated that the uncertainty in the estimation of $(\Delta_1 - \Delta_2)$ (only the contribution due to the friction torque) is equal to the uncertainty in the differential level:

$$\delta (\Delta_1 - \Delta_2) \equiv \delta (\Delta) .$$

In the present condition, this uncertainty is equal to

$$\delta (\Delta_1 - \Delta_2) \equiv \delta (\Delta) = \pm \frac{8M}{\gamma_w \pi D^2 D_p} =$$

$$\pm \frac{8 \times 5 \times 10^{-3}}{9806 \times \pi \times 0.15^2 \times 0.2} = \pm 0.3 \, \text{mm},$$

$$(3.4)$$

and it can be reduced by increasing the diameter of the pulley and of the cylinders.

---

**Exercise 3.9** The concrete caisson in Fig. 3.19 has a horizontal section with walls $d = 15$ cm thick and a bottom $2d$ thick. The external side has length $l = (2.0 + 0.01 \times C_u)$ m, and the height is $H = (3.2 + 0.1 \times C_{pu})$ m.

- Check the buoyancy equilibrium.
- Check the stability of the equilibrium for small rotations about a horizontal axis.

Assume the specific weight of concrete $\gamma_c = 24 \, \text{kN m}^{-3}$ and the specific weight of seawater $\gamma_w = 10.15 \, \text{kN m}^{-3}$.

**Solution** Consider the schematic shown in Fig. 3.20. For buoyancy, it is necessary that the force of buoyancy balances the weight of the caisson:

$$\gamma_c V_c = \gamma_w V_b,$$

where $V_c$ is the net volume of concrete and $V_b$ is the volume of buoyancy.

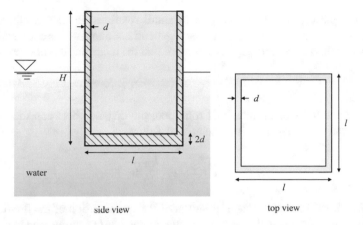

**Fig. 3.19** Concrete floating caisson

**Fig. 3.20** Schematic for calculation of the centre of gravity and of centre of buoyancy

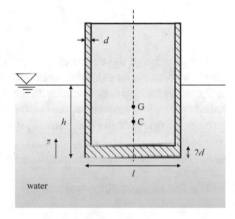

The net volume of concrete is equal to

$$V_c = l^2 H - (l - 2d)^2 (H - 2d),$$

and the volume of buoyancy is equal to

$$V_b = l^2 h,$$

where $h$ refers to the immersed part of the caisson.

Hence,

$$\gamma_c \left[ l^2 H - (l - 2d)^2 (H - 2d) \right] = \gamma_w l^2 h \rightarrow$$

$$h = \frac{\gamma_c \left[ l^2 H - (l - 2d)^2 (H - 2d) \right]}{\gamma_w l^2}.$$

For the stability of the equilibrium to small rotations about a horizontal axis orthogonal to the sheet, it is necessary that the centre of buoyancy is above the centre of gravity. If this is not verified, it is sufficient that the metacentre is above the centre of gravity, that is:

$$(z_M - z_C) > (z_G - z_C).$$

In the hypothesis of infinitesimal rotations, the distance between the centre of buoyancy and the metacentre is calculated as follows:

$$(z_M - z_C) = \frac{I_{Gxx}}{V_b},$$

where $I_{Gxx}$ is the second moment of inertia of the cross-section of the floating body at the waterline with respect to an axis parallel to the axis of rotation and barycentric. For a square cross-section, results:

$$I_{Gxx} = \frac{1}{12} l^4.$$

The position of the centre of gravity is calculated by dividing the box into elementary volumes, for example the bottom and the walls. Therefore, the first moment of the whole caisson is equalised with the sum of the first moments of the elementary volumes. The first moment of the bottom, with respect to a horizontal plane passing through the origin of the coordinate system, is

$$S_{z-b} = 2d^2 l^2.$$

The first moment of the walls is

$$S_{z-w} = [2 (l - 2d) d + 2ld] (H - 2d) \left( \frac{H - 2d}{2} + 2d \right).$$

The first moment of the caisson (net volume filled with concrete) is equal to

$$S_{z-c} = \left[ l^2 H - (l - 2d)^2 (H - 2d) \right] z_G.$$

Hence

$$S_{z-c} = S_{z-b} + S_{z-w} \rightarrow \left[ l^2 H - (l - 2d)^2 (H - 2d) \right] z_G =$$
$$2d^2 l^2 + [2 (l - 2d) d + 2ld] (H - 2d) \left( \frac{H - 2d}{2} + 2d \right) \rightarrow$$
$$z_G = \frac{2d^2 l^2 + [2 (l - 2d) d + 2ld] (H - 2d) \left( \dfrac{H - 2d}{2} + 2d \right)}{l^2 H - (l - 2d)^2 (H - 2d)}.$$

The distance between the centre of buoyancy and the centre of gravity is equal to $z_G - z_C$ and the equilibrium is stable if $(z_M - z_C) > (z_G - z_C)$.

For $C_u = C_{pu} = 0$ it results $d = 15\,\text{cm}$, $l = 2.00\,\text{m}$, $H = 3.2\,\text{m}$, $\gamma_c = 24\,\text{kN}\,\text{m}^{-3}$, $\gamma_w = 10.15\,\text{kN}\,\text{m}^{-3}$,

$$\therefore \quad h = \frac{\gamma_c \left[ l^2 H - (l - 2d)^2 (H - 2d) \right]}{\gamma_w l^2} =$$

$$\frac{24\,000 \times \left[ 2.0^2 \times 3.2 - (2.0 - 2 \times 0.15)^2 \times (3.2 - 2 \times 0.15) \right]}{10\,150 \times 2.0^2} = 2.61\,\text{m},$$

$$(z_M - z_C) = \frac{I_{Gxx}}{V_b} = \frac{\frac{1}{12} l^4}{l^2 h} = \frac{\frac{1}{12} \times 2.0^4}{2.0^2 \times 2.61} = 0.13\,\text{m},$$

$$z_G = \frac{\left[ 2d^2 l^2 + [2(l - 2d)d + 2ld](H - 2d)\left( \frac{H - 2d}{2} + 2d \right) \right]}{l^2 H - (l - 2d)^2 (H - 2d)} =$$

$$\frac{\left[ 2 \times 0.15^2 \times 2.0^2 + [2 \times (2.0 - 2 \times 0.15) \times 0.15 + 2 \times 2.0 \times 0.15] \times \right.}{}$$

$$\frac{(3.2 - 2 \times 0.15) \times \left( \frac{3.2 - 2 \times 0.15}{2} + 2 \times 0.15 \right) \left. \right]}{2.0^2 \times 3.2 - (2.0 - 2 \times 0.15)^2 (3.2 - 2 \times 0.15)}$$

$$= 1.315\,\text{m},$$

$$(z_G - z_C) = 1.315 - \frac{2.61}{2} = 0.01\,\text{m}.$$

Hence, the centre of buoyancy is below the centre of gravity. However

$$(z_M - z_C) > (z_G - z_C) \rightarrow 0.13 > 0.01,$$

and the equilibrium is **stable**.

---

**Exercise 3.10** The breakwater of a marina in a lake shown in Fig. 3.21, is made of floating caissons with dimensions $l = 5260\,\text{mm}$, $b = 3000\,\text{mm}$, $h_c = 1800\,\text{mm}$, free to slide vertically on pairs of vertical cylindrical circular piers, equipped with a screen to reduce wave penetration, having height $h_s = 3000\,\text{mm}$. To ensure the verticality of the caisson, a counterweight equal to the weight of the screen is placed inside. The sliding takes place through circular pipes with a diameter of 1400 mm,

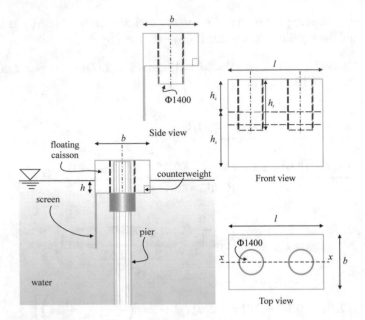

**Fig. 3.21**  Floating breakwater caisson free to slide in the vertical

**Fig. 3.22**  Schematic for the
calculation of the centre of
mass, centre of buoyancy
and metacentre

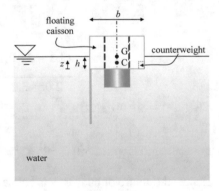

extended beyond the bottom of the box. The caissons, screen and pipes are made of
steel sheet with a thickness of $s = 10\,\text{mm}$.

- Calculate the submersion $h$ of the caissons.
- Check the stability of the equilibrium for small rotations about the two horizontal
  axes.
- Calculate the period of vertical oscillation of the caissons.

Assume the specific weight of steel $\gamma_{Fe} = 77\,\text{kN}\,\text{m}^{-3}$ and the specific weight of
fresh water $\gamma_w = 9.80\,\text{kN}\,\text{m}^{-3}$. The added mass coefficient is equal to $C_m = 1.7$.

**Solution**  Consider the diagram in Fig. 3.22.

For buoyancy (equilibrium in the vertical direction), the force of buoyancy must balance the weight of the caisson:

$$P_c = \gamma_w V_b,$$

where $V_b$ is the volume of buoyancy.

The weight of the caisson is equal to

$$P_c = \gamma_{Fe}\left[\underbrace{2h_cls + 2h_cbs}_{\text{volume of sides}} + \underbrace{2\left(bl - 2\frac{\pi D^2}{4}\right)s}_{\text{volume horizontal plates}} + \underbrace{2\pi Dh_ts}_{\substack{\text{volume}\\\text{of tubes}}} + \underbrace{h_sls}_{\text{volume screen}}\right] + P_{cw},$$

where $P_{cw}$ is the counterweight, equal to the weight of the screen. Hence

$$P_c = \gamma_{Fe}\left[\underbrace{2h_cls + 2h_cbs}_{\text{volume of sides}} + \underbrace{2\left(bl - 2\frac{\pi D^2}{4}\right)s}_{\text{volume horizontal plates}} + \underbrace{2\pi Dh_ts}_{\substack{\text{volume}\\\text{of tubes}}} + \underbrace{2h_sls}_{\substack{\text{volume screen +}\\\text{counterweight}}}\right].$$

The volume of buoyancy is equal to

$$V_b = \underbrace{\left(bl - 2\frac{\pi D^2}{4}\right)h}_{\substack{\text{volume displaced by}\\\text{the parallelepiped}}} + \underbrace{2\pi D\,(h_t - h_c)\,s}_{\substack{\text{volume of the tubes}\\\text{out of the caisson}}} + \underbrace{h_sls}_{\text{volume screen}},$$

where $h$ is the depth of the submerged part of the caisson.

In equilibrium condition, it results

$$\gamma_w\left[\underbrace{\left(bl - 2\frac{\pi D^2}{4}\right)h}_{\substack{\text{volume displaced by}\\\text{the parallelepiped}}} + \underbrace{2\pi D\,(h_t - h_c)\,s}_{\substack{\text{volume of the tubes}\\\text{out of the caisson}}} + \underbrace{h_sls}_{\text{volume screen}}\right] =$$

$$\gamma_{Fe}\left[\underbrace{2h_cls + 2h_cbs}_{\text{volume of sides}} + \underbrace{2\left(bl - 2\frac{\pi D^2}{4}\right)s}_{\text{volume horizontal plates}} + \underbrace{2\pi Dh_ts}_{\substack{\text{volume}\\\text{of tubes}}} + \underbrace{2h_sls}_{\substack{\text{volume screen +}\\\text{counterweight}}}\right].$$

By substituting the numerical values, yields

$$P_c = \gamma_{Fe} \left[ \underbrace{2h_c ls + 2h_c bs}_{\text{volume of sides}} + \underbrace{2\left(bl - 2\frac{\pi D^2}{4}\right)s}_{\text{volume horizontal plates}} + \underbrace{2\pi D h_t s}_{\substack{\text{volume} \\ \text{of tubes}}} + \underbrace{2h_s ls}_{\substack{\text{volume screen} \\ \text{plus counterweight}}} \right] =$$

$$77 \times \left[ \underbrace{2 \times 1.80 \times 5.26 \times 0.01 + 2 \times 1.80 \times 3 \times 0.01 +}_{\text{volume of sides}} \right.$$
$$\underbrace{2 \times \left(3 \times 5.26 - 2 \times \frac{\pi \times 1.4^2}{4}\right) \times 0.01}_{\text{volume horizontal plates}} + \underbrace{2 \times \pi \times 1.4 \times 2.8 \times 0.01 +}_{\substack{\text{volume} \\ \text{of tubes}}}$$
$$\left. \underbrace{2 \times 3 \times 5.26 \times 0.01}_{\substack{\text{volume screen +} \\ \text{counterweight}}} \right] =$$

$$77 \times 1.1132 = \mathbf{85.72\,kN},$$

where the volume of iron equals $1.1132$ m$^3$,

$$V_b = \underbrace{\left(bl - 2\frac{\pi D^2}{4}\right)h}_{\substack{\text{volume displaced by} \\ \text{the parallelepiped}}} + \underbrace{2\pi D \,(h_t - h_c)\,s}_{\substack{\text{volume of the tubes} \\ \text{out of the caisson}}} + \underbrace{h_s ls}_{\text{volume screen}} =$$

$$\left(3 \times 5.26 - 2 \times \frac{\pi \times 1.4^2}{4}\right) \times h + 2 \times \pi \times 1.4 \times (2.80 - 1.80) \times 0.01$$

$$+ 3 \times 5.26 \times 0.01 = 12.70 \times h + 0.24576$$

In equilibrium condition, it results:

$$\therefore \quad P_c = \gamma_w V_b \rightarrow 85\,720 = 9800 \times (12.70h + 0.24576) \rightarrow h = \mathbf{0.67\,m}.$$

The volume of buoyancy is equal to

$$V_b = \frac{P_{cb}}{\gamma_w} = \frac{85\,720}{9800} = \mathbf{8.747\,m^3}.$$

For the stability of the equilibrium to the small rotations around a horizontal axis orthogonal to the sheet, it is necessary to consider the level of the centre of mass, of the centre of buoyancy and possibly of the metacentre. If the centre of buoyancy is above the centre of mass, the equilibrium is stable and no further verification is necessary. Otherwise it is sufficient that the metacentre is above the centre of mass, i.e.

$$(z_M - z_C) > (z_G - z_C).$$

The position of the centre of mass is calculated by dividing the caisson into elementary volumes. Then, the first moment of the whole caisson is equalized with the sum of the first moments of the elementary volumes.

The first moment of the parallelepiped portion of the caisson, with respect to a horizontal plane passing through the origin of the coordinate system, is as follows:

$$
S_{z1} = \left[ \underbrace{2h_c ls + 2h_c bs}_{\text{volume of sides}} + \underbrace{2\left( bl - 2\frac{\pi D^2}{4} \right)s}_{\text{volume horizontal plates}} + \underbrace{2\pi D h_c s}_{\substack{\text{volume of tubes} \\ \text{within the caisson}}} \right] \frac{h_c}{2}.
$$

The first moment of the screen is negative and is equal to

$$
S_{z2} = - \left( \underbrace{h_s ls}_{\text{volume screen}} \right) \frac{h_s}{2}.
$$

The first moment of the out-of-caisson portion of the tubes is negative and is equal to

$$
S_{z3} = - \underbrace{2\pi D\,(h_t - h_c)\,s}_{\substack{\text{volume of tubes} \\ \text{out of the caisson}}} \frac{(h_t - h_c)}{2}.
$$

We neglect the first moment of the counterweight, assuming that the counterweight is at a short distance from the origin of the coordinate system. The total first moment is equal to the product of the total volume of steel (including the equivalent volume of the counterweight) multiplied by the unknown distance of the centre of mass

$$
\left[ \underbrace{2h_c ls + 2h_c bs}_{\text{volume of sides}} + \underbrace{2\left( bl - 2\frac{\pi D^2}{4} \right)s}_{\text{volume horizontal plates}} + \underbrace{2\pi D h_t s}_{\substack{\text{volume} \\ \text{of tubes}}} + \underbrace{2h_s ls}_{\substack{\text{volume screen +} \\ \text{counterweight}}} \right] z_G =
$$

$$
S_{z1} + S_{z2} + S_{z3}.
$$

By substituting the numerical values, it results

$$
S_{z1} = \left[ \underbrace{2h_c ls + 2h_c bs}_{\text{volume of sides}} + \underbrace{2\left( bl - 2\frac{\pi D^2}{4} \right)s}_{\text{volume horizontal plates}} + \underbrace{2\pi D h_c s}_{\substack{\text{volume of tubes} \\ \text{in the caisson}}} \right] \frac{h_c}{2} =
$$

$$
\left[ \underbrace{2 \times 1.80 \times 5.26 \times 0.01 + 2 \times 1.80 \times 3 \times 0.01}_{\text{volume of sides}} \right.
$$

$$+ 2 \times \underbrace{\left( 3 \times 5.26 - 2 \times \frac{\pi \times 1.4^2}{4} \right) \times 0.01}_{\text{volume horizontal plates}} + \underbrace{2 \times \pi \times 1.4 \times 1.8 \times 0.01}_{\substack{\text{volume of tubes} \\ \text{in the caisson}}} \Bigg] \times \frac{1.8}{2}$$

$$= 0.639 \, \text{m}^4,$$

$$S_{z2} = - \left( \underbrace{h_s l s}_{\text{volume screen}} \right) \frac{h_s}{2} = -3 \times 5.26 \times 0.01 \times \frac{3}{2} = -0.2367 \, \text{m}^4,$$

$$S_{z3} = - \underbrace{2 \pi D \, (h_t - h_c) \, s}_{\substack{\text{volume of tubes} \\ \text{out of caisson}}} \frac{(h_t - h_c)}{2} =$$

$$- 2 \times \pi \times 1.4 \times (2.8 - 1.8) \times 0.01 \times \frac{(2.8 - 1.8)}{2} = - 0.04398 \, \text{m}^4,$$

$$z_G = \frac{S_{z1} + S_{z2} + S_{z3}}{\underbrace{2 h_c l s + 2 h_c b s}_{\text{volume of sides}} + \underbrace{2 \left( b l - 2 \frac{\pi D^2}{4} \right) s}_{\text{volume horizontal plates}} + \underbrace{2 \pi D h_t s}_{\substack{\text{volume} \\ \text{of tubes}}} + \underbrace{2 h_s l s}_{\substack{\text{volume screen +} \\ \text{counterweight}}}} =$$

$$\frac{0.639 - 0.2367 - 0.04398}{1.1132} = 0.32 \, \text{m}.$$

In the same way, it is possible to calculate the centre of buoyancy with reference to the first moment of the displaced volumes of water. $S_{z2}$ and $S_{z3}$ are unchanged. The first moment of the volume of water displaced by the caisson is

$$S_{z4} = \left[ b l h - 2 \frac{\pi D^2 h}{4} \right] \frac{h}{2} = \underbrace{\left[ 3 \times 5.26 \times 0.67 - 2 \times \frac{\pi \times 1.4^2}{4} \times 0.67 \right]}_{V_4} \times \frac{0.67}{2} =$$

$$\underbrace{8.509}_{V_4} \times \frac{0.67}{2} = 2.850 \, \text{m}^4,$$

$$z_C = \frac{S_{z4} + S_{z2} + S_{z3}}{V_b} = \frac{2.850 - 0.2367 - 0.04398}{8.747} = 0.29 \, \text{m}.$$

As already mentioned in the previous paragraphs, if the centre of buoyancy were above the centre of mass, the floating body would be stable for small rotations, since the incipient torque associated with a small rotation would tend to reverse the rotation itself. Since the centre of buoyancy is below the centre of mass, it is necessary to check the position of the metacentre.

In the hypothesis of small roll rotations (rotation around an axis orthogonal to the figure), the distance between the centre of buoyancy and the metacentre is calculated as follows:

$$(z_M - z_C) = \frac{I_{Gxx}}{V_b},$$

where $I_{Gxx}$ is the second moment of the cross-section of the floating body at the waterline level, with respect to an axis parallel to the axis of rotation and barycentric. In the present case, it results

$$I_{Gxx} = \frac{1}{12}lb^3 - 2\frac{\pi D^4}{64}.$$

By substituting the numerical values, yields

$$I_{Gxx} = \frac{1}{12}lb^3 - 2\frac{\pi D^4}{64} = \frac{1}{12} \times 5.26 \times 3^3 - 2 \times \frac{\pi \times 1.4^4}{64} = 11.458\,\text{m}^4.$$

Hence

$$(z_M - z_C) = \frac{I_{Gxx}}{V_b} = \frac{11.458}{8.747} = 1.31\,\text{m}.$$

The condition of stability of equilibrium at small rotations about a horizontal axis orthogonal to the sheet requires that

$$(z_M - z_C) > (z_G - z_C) \rightarrow 1.31 > 0.32 - 0.29 = 0.03\,\text{m}.$$

Equilibrium is **stable**. The stability is guaranteed even more for the rotation about an axis orthogonal to the previous one and contained in the horizontal plane, since the moment of inertia $I_{Gxx}$ previously calculated is the smallest.

For the calculation of the resonance period for vertical oscillations, we write the equation of the dynamics in which the force acting is buoyancy as a consequence of a vertical displacement $z$ with respect to the position of equilibrium:

$$(m + m_a)\frac{d^2z}{dt^2} + \gamma_w Az = 0,$$

where $m$ is the mass of the floating body, $m_a$ is the added mass (due to the surrounding fluid involved in the movement of the floating body, Fig. 3.23), $A$ is the area of the cross-section of the buoyancy volume at the waterline.

The added mass is obtained by multiplying the mass of water displaced by the volume of buoyancy (also equal to the mass of the floating body) by the added mass coefficient $C_m$ (which is experimentally evaluated):

$$m_a = C_m \rho_w V_b,$$

**Fig. 3.23** Volume of the added mass (hatched) for vertical oscillations analysis. The shape is purely conceptual

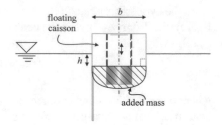

where $\rho_w$ is the mass density of water.

The homogeneous 2nd order differential equation can be rewritten in a compact form as

$$\frac{d^2 z}{dt^2} + \underbrace{\frac{\gamma_w A}{m + m_a}}_{\omega^2} z = 0.$$

The solution is

$$z(t) = c_1 \sin \omega t + c_2 \cos \omega t,$$

where $c_1$ and $c_2$ are the two integration constants, which depend on the initial conditions. The period of oscillation is equal to

$$T = \frac{2\pi}{\omega} = 2\pi \sqrt{\frac{m + m_a}{\gamma_w A}}.$$

By substituting the numerical values, yields

$$m = \frac{P_c}{g} = \frac{85\,720}{9.806} = \textbf{8742}\,\textbf{kg},$$

$$m_a = C_m \rho_w V_b = 1.7 \times 1000 \times 8.747 = \textbf{14\,870}\,\textbf{kg},$$

$$A = 3 \times 5.26 - 2 \times \frac{\pi \times 1.4^2}{4} = \textbf{12.70}\,\textbf{m}^2,$$

$$\therefore \qquad T = 2\pi \sqrt{\frac{m + m_a}{\gamma_w A}} = 2 \times \pi \times \sqrt{\frac{8742 + 14\,870}{9800 \times 12.70}} = \textbf{2.74}\,\textbf{s}.$$

# Chapter 4
# Balances of Linear and Angular Momentum

Several practical problems of Fluid Dynamics can be solved with the use of the linear momentum balance equation, mostly in integral form. We select an appropriate control volume and evaluate all the forces acting on it, separating the surface forces (acting on the outer surface of the control volume), and the volume forces (acting on the fluid particles contained in the volume). For ideal fluids (zero viscosity), only pressure is present and the surface forces are normal to the external surface. The volume forces are the weight, the local inertia (in unsteady flows) and the apparent forces (e.g., centrifugal, Coriolis, Euler), defined in non-inertial control volumes.

Other equations are mass conservation and, for ideal fluids, energy conservation. Energy conservation is often expressed as Bernoulli's theorem: the total energy of each particle of an ideal fluid body is invariant along a path provided no energy enters or leaves the system. Energy is defined as "total head", has a length dimension (energy per unit of weight), and is shared between (i) elevation head (represents the gravitational potential energy arising from elevation), (ii) pressure head (represents the energy due to fluid pressure) and (iii) velocity head (represents the kinetic energy). The magnitude of each of the three terms may vary but their sum is invariant. The real fluid dissipates energy during the flow, but in many cases the dissipation is negligible or can be parametrised with coefficients. Bernoulli's theorem has different formulations if the fluid is compressible or not, is unsteady, and is specialized for non inertial frames.

A careful selection of the control volume is essential to derive solutions also in complex systems. In some cases, the equation of conservation of the angular momentum is also invoked.

---

$C_u$ and $C_{pu}$, that are two integer numbers between 0 and 9, for example, the last and second-last digit of the registration number.

© Springer Nature Switzerland AG 2021

S. Longo et al., *Problems in Hydraulics and Fluid Mechanics*, Springer Tracts in Civil Engineering, https://doi.org/10.1007/978-3-030-51387-0_4

Although the equations used here are specialized for fluids, they apply to any continuum in a broader definition.

**Exercise 4.1** The reducing elbow curve shown in Fig. 4.1 is contained in a vertical plane and deflects a water current by an angle of 135°. The inlet and outlet diameters are $D_1$ and $D_2$. The water flow rate $Q$ and the pressures $p_1$ and $p_2$, barycentric in sections 1 and 2, are known; the water volume $W$ between sections 1 and 2 and the weight of the curve $P$ are also known.

– Determine the forces in the horizontal and vertical directions needed to keep the curve in equilibrium.

Numerical data: $D_1 = (400 + C_u)$ mm, $D_2 = (200 + C_u)$ mm, $Q = (400 + C_{pu})$ l s$^{-1}$, $p_1 = (150 + C_u)$ kPa, $p_2 = (90 + C_{pu})$ kPa, $W = (0.2 + 0.01 \times C_u)$ m$^3$, $P = (120 + C_{pu})$ N.

**Fig. 4.1** Reducing elbow curve contained in a vertical plane

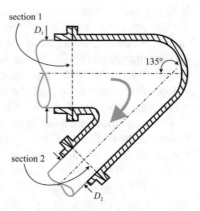

**Solution** We apply the momentum balance in integral form to the control volume delimited by sections 1 and 2 and by the walls of the curve, see Fig. 4.2:

$$\mathbf{G} + \mathbf{\Pi} + \mathbf{I} + \mathbf{M}_1 - \mathbf{M}_2 = 0 \rightarrow \mathbf{G} + \mathbf{F} + \mathbf{\Pi}_1 + \mathbf{\Pi}_2 + \mathbf{M}_1 - \mathbf{M}_2 = 0,$$

where $\mathbf{F}$ is the force exerted by the curve on the control volume and local inertia $\mathbf{I}$ is cancelled because the flow is stationary. The balance equation of the momentum in the $x$-direction is

$$F_x = -p_1 \frac{\pi D_1^2}{4} - \rho Q V_1 - p_2 \frac{\pi D_2^2}{4} \cos\theta - \rho Q V_2 \cos\theta,$$

where $F_x$ represents the $x$-component of the force exerted by the curve on the control volume. The balance equation of the momentum in the $y$-direction is

**Fig. 4.2** Schematic for calculation of the forces

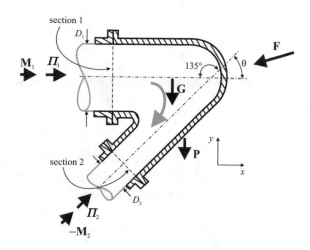

$$F_y = -p_2 \frac{\pi D_2^2}{4} \sin\theta - \rho Q V_2 \sin\theta + \gamma_w W,$$

where $F_y$ represents the y-component of the action exerted by the curve on the control volume. The total vertical force to be exerted on the curve will be equal to the sum of $F_y$ and the weight of the curve, $F_{toty} = F_y - P$.

For $C_u = C_{pu} = 0$ it results $D_1 = 400$ mm, $D_2 = 200$ mm, $Q = 400$ l s$^{-1}$, $p_1 = 150$ kPa, $p_2 = 90$ kPa, $W = 0.2$ m$^3$, $P = 120$ N, $\theta = 45°$, $\rho = 1000$ kg m$^{-3}$,

$$\therefore \quad F_x = -p_1 \Omega_1 - \rho \frac{Q^2}{\Omega_1} - p_2 \Omega_2 \cos\theta - \rho \frac{Q^2}{\Omega_2} \cos\theta =$$

$$- 150\,000 \times \frac{\pi \times 0.4^2}{4} - 1000 \times \frac{0.4^2}{\frac{\pi \times 0.4^2}{4}} - 90\,000 \times \frac{\pi \times 0.2^2}{4} \times \cos 45°$$

$$- 1000 \times \frac{0.4^2}{\frac{\pi \times 0.2^2}{4}} \times \cos 45° = \mathbf{-25.70\ kN},$$

$$\therefore \quad F_y = -p_2 \Omega_2 \sin\theta - \rho \frac{Q^2}{\Omega_2} \sin\theta + \gamma_w W = -90\,000 \times \frac{\pi \times 0.2^2}{4} \times \sin 45°$$

$$- 1000 \times \frac{0.4^2}{\frac{\pi \times 0.2^2}{4}} \times \sin 45° + 9806 \times 0.2 = \mathbf{-3.64\ kN},$$

$$\therefore \qquad F_{toty} = F_y - P = -3639 + 120 = \mathbf{-3.52\ kN}.$$

**Fig. 4.3** Skater with nozzle

**Exercise 4.2** The skater in Fig. 4.3, of mass $M = (60 + C_u)$ kg, holds a nozzle of diameter $D = 20$ mm and negligible mass. The water outflow speed is $10$ m s$^{-1}$. Starting from rest:

– calculate the speed of the skater after 5 s.
– Calculate the distance travelled by the skater in the same time interval.

    **Solution** In a control volume attached to the skater (in general, this frame is non-inertial), the momentum balance allows the calculation of a force acting on the skater equal to the out-flowing momentum:

$$F = \rho V^2 \frac{\pi D^2}{4}.$$

    This force is parallel to the outflow velocity $V$ and is pointing in the opposite direction. The velocity must always be calculated in relation to the control volume and, if the nozzle supply is not affected by the skater's motion, the force $F$ is constant. Therefore, the dynamic equation for the skater can be written as

$$M \frac{d^2 s}{dt^2} = F,$$

where $s$ is the space coordinate. Double integrating, yields

$$s = \frac{F}{M} \frac{t^2}{2} + c_1 t + c_2.$$

    The initial conditions of zero speed and position in the origin of the coordinate system $s$, render null the two constants of integration $c_1$ and $c_2$. Hence,

$$s = \frac{F}{M}\frac{t^2}{2}, \quad U \equiv \frac{ds}{dt} = \frac{F}{M}t.$$

For $C_u = C_{pu} = 0$ it results $M = 60$ kg, $D = 20$ mm, $V = 10$ m s$^{-1}$,

$$\therefore \qquad F = \rho V^2 \frac{\pi D^2}{4} = 1000 \times 10^2 \times \frac{\pi \times 0.02^2}{4} = \mathbf{31.4\ N},$$

$$\therefore \qquad s = \frac{F}{M}\frac{t^2}{2} = \frac{31.4}{60} \times \frac{5^2}{2} = \mathbf{6.54\ m},$$

$$\therefore \qquad U = \frac{F}{M}t = \frac{31.4}{60} \times 5 = \mathbf{2.62\ m\,s^{-1}}.$$

---

**Exercise 4.3** The water jet in Fig. 4.4, having diameter $D = (100 + 10 \times C_{pu})$ mm and speed $V = (30 + C_u)$ m s$^{-1}$, axially strikes a conical surface.

- Calculate the thickness of the water sheet at a distance of $R = 200$ mm from the axis of the cone.
- Calculate the force that must be applied to the cone to move it to the left with constant speed $V_c = 15$ m s$^{-1}$.

**Solution** We consider the dashed trajectory in Fig. 4.5, between section 1 and section 2. Neglecting the energy losses, from Bernoulli's theorem results

**Fig. 4.4** Water jet striking a conical surface

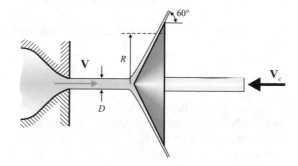

**Fig. 4.5** Trajectory for the application of Bernoulli's theorem and pressure diagram

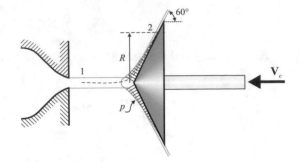

$$z_1 + \frac{\cancel{p_1}}{\cancel{\gamma_w}} + \frac{V_1^2}{2g} = z_2 + \frac{p_2}{\gamma_w} + \frac{V_2^2}{2g}.$$

Figure 4.5 shows the qualitative diagram of the pressure distribution, hatched area. The pressure is maximum in the axis and is zero in the section of detachment of the current from the cone. In the sections of the current at contact with the conical surface, the pressure distribution along the normal wall is almost hydrostatic (however, the trajectories are slowly converging towards the wall and therefore the distribution is not rigorously hydrostatic) and decreases from the axis towards the trailing edge, where it assumes a value equal to the atmospheric pressure. Since the calculation of $p_2$ requires the integration of the motion equations, the application of Bernoulli's theorem is in general not advantageous for the calculation of the $V_2$ speed. But if we neglect the variation of elevation head, and if section 2 is external to the cone, where the current is again all in air (and the relative pressure $p_2$ is zero), it results that the velocity of the current is equal to the velocity of the incident jet. This also applies approximately at a distance sufficiently large from the axis. At a distance of $R$ from the axis, the current section is equal to:

$$\Omega = 2\pi R\delta + \pi\delta^2 \sin 30° \approx 2\pi R\delta,$$

where $\delta \ll R$ is the thickness of the jet. Applying the mass conservation equation, yields

$$\frac{\pi D^2}{4} V = 2\pi R\delta V \rightarrow \delta = \frac{D^2}{8R}.$$

To calculate the force required to move the cone at a constant speed, we choose the mobile control volume attached to the cone in Fig. 4.6. In the application of momentum balance, all flows must be calculated with reference to the surface delimiting the control volume, with quantities that have been transformed into the mobile reference system attached to the control volume. The speed of the input flow in the mobile inertial frame is equal to $V + V_c$. The pressure is equal to the atmospheric pressure (it can be demonstrated that the pressure inside a current of a liquid in air, must be

**Fig. 4.6** Mobile control
volume for forces calculation

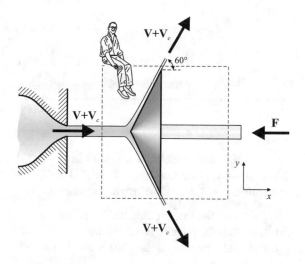

uniform and equal to the pressure at the boundary). The speed of the outlet fluid is
still equal to $V + V_c$. Hence, yields

$$\underbrace{\rho(V + V_c)^2 \frac{\pi D^2}{4}}_{\text{input momentum flux along } x} - \underbrace{\rho(V + V_c)^2 \Omega_u \cos 60°}_{\text{output momentum flux along } x} + F_x = 0,$$

where $\Omega_u$ is the area of the cross-section of the current flowing out of the control
volume. For the conservation of the mass, it results:

$$(V + V_c) \frac{\pi D^2}{4} = (V + V_c) \Omega_u,$$

hence

$$F_x = -\rho(V + V_c)^2 \frac{\pi D^2}{4} (1 - \cos 60°).$$

Neglecting the weight force, any other component of the force is null by symmetry.
The force applied to the cone must be pointing to the left.

For $C_u = C_{pu} = 0$ it results $D = 100$ mm, $V = 30$ m s$^{-1}$, $R = 200$ mm, $V_c =$
15 m s$^{-1}$,

$$\therefore \qquad \delta = \frac{D^2}{8R} = \frac{0.1^2}{8 \times 0.2} = \textbf{6.25 mm},$$

$$\therefore \qquad F_x = -\rho(V + V_c)^2 \frac{\pi D^2}{4} (1 - \cos 60°) =$$

$$- 1000 \times (30 + 15)^2 \times \frac{\pi \times 0.1^2}{4} \times (1 - \cos 60°) = -7950 \text{ N}.$$

---

**Exercise 4.4** The horizontal water jet in Fig. 4.7, having cross-sectional area $a = (50 + 10 \times C_{pu})$ cm$^2$ and speed $V = (5.0 + C_u)$ m s$^{-1}$, strikes a flat plate inclined at an angle $\theta = (20 + C_{pu})°$ to the horizontal.

(a)  Calculate the force if the plate is stationary.
(b)  Calculate the force if the plate is moving with horizontal velocity $U = 1.0$ m s$^{-1}$.
(c)  Calculate the force if the plate is moving with velocity $U = 1.0$ m s$^{-1}$ along its normal.
(d)  Calculate the power of the force and the efficiency for the three cases.

**Solution** *Case* (a). The fluid is ideal and the forces can only be normal to the plate. The momentum balance in integral form reduces to

$$\cancel{\mathcal{G}} + \Pi + \cancel{\mathbf{I}} + \mathbf{M}_1 - \mathbf{M}_2 = 0 \rightarrow -\mathbf{F} + \mathbf{M}_1 - \mathbf{M}_2 = 0,$$

where we have neglected the weight, local inertia is zero, and where $\mathbf{F} \equiv -\Pi$ is the force of the control volume on the flat plate. If we choose a coordinate system with $x$ normal to the plate and $y$ parallel to the plate, results

$$F_x = M_{1x}, \quad F_y = 0. \tag{4.1}$$

Hence,

$$F_x = \rho a V^2 \cos \theta.$$

The power and the efficiency are null since the force has a fixed application point.

*Case* (b). If the plate is moving with horizontal velocity $U$, we choose an inertial control volume translating with the same velocity, see Fig. 4.8, and the linear momentum balance equations are still coincident with Eq. (4.1), with velocities computed in the mobile control volume. The result is

$$F_x = \rho a (V - U)^2 \cos \theta,$$

since the velocity in the inertial control volume is $\mathbf{V} - \mathbf{U}$.

The power is

$$P = \mathbf{F} \cdot \mathbf{U} = F_x U \cos \theta = \rho a U (V - U)^2 \cos^2 \theta,$$

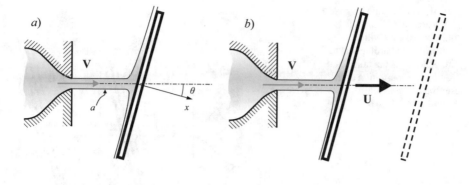

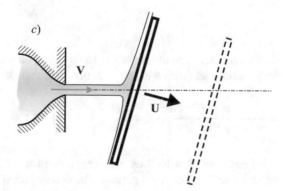

**Fig. 4.7**   Water jet striking a flat plate, case (a), (b) and (c)

**Fig. 4.8**   Schematic for case
(b), water jet striking a flat
plate translating with
uniform horizontal speed

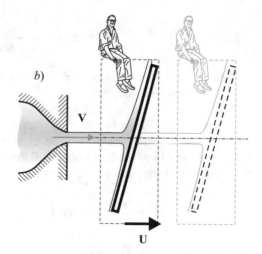

**Fig. 4.9** Schematic for case
(c), water jet striking a flat
plate translating with
uniform speed orthogonal to
the plate. $\mathbf{U'}$ is the horizontal
velocity of translation of the
inertial control volume, with
magnitude $U / \cos\theta$

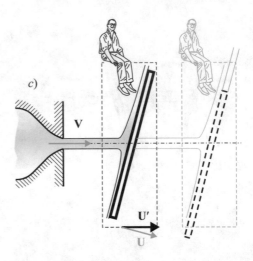

and the efficiency is the ratio between the power of force acting on the plate and the
kinetic power of the incoming jet (with a unitary energy coefficient):

$$\eta = \frac{\rho a U \, (V - U)^2 \cos^2\theta}{\rho V^3 a / 2} = 2\frac{U}{V}\left(1 - \frac{U}{V}\right)^2 \cos^2\theta.$$

*Case* (c). The third case can be solved by selecting a control volume which, in
order to guarantee a zero inertial term, must contain a time invariant volume of fluid.
This result is achieved if the control volume translates horizontally with a velocity
$U / \cos\theta$, see Fig. 4.9. Hence,

$$F_x = \rho a \left(V - \frac{U}{\cos\theta}\right)^2 \cos\theta. \tag{4.2}$$

This result can be obtained as follows, by reasoning without separating flux of
momentum and local inertia.

The resultant of the forces balances the variation of momentum in unit time. The
mass leaving the nozzle in unit time is $\rho a V$ and the mass used to extend the jet
is $\rho a (U / \cos\theta)$, hence the mass striking the plate is $\rho a (V - U / \cos\theta)$. The initial
velocity normal to the plate is $V\cos\theta$ and the final velocity normal to the plate is $U$,
with a velocity variation equal to $V\cos\theta - U$. Hence,

$$F_x = \frac{m}{\Delta t}\Delta V \equiv \rho a \left(V - \frac{U}{\cos\theta}\right)(V\cos\theta - U),$$

which is equivalent to Eq. (4.2).

**Fig. 4.10** *Case* (c), water jet striking a flat plate translating with uniform speed orthogonal to the plate. Disadvantageous choice of the control volume

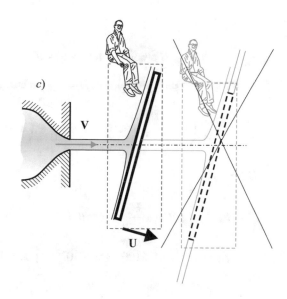

The power is

$$P = \mathbf{F} \cdot \mathbf{U} = F_x U = \rho a U \left( V - \frac{U}{\cos\theta} \right) (V \cos\theta - U),$$

and the efficiency is the ratio between the power of force acting on the plate and the kinetic power of the incoming jet (by assuming again a unitary energy coefficient):

$$\eta = \frac{\rho a U \left( V - \dfrac{U}{\cos\theta} \right)(V\cos\theta - U)}{\rho V^3 a/2} = \frac{2}{\cos\theta}\frac{U}{V}\left( \cos\theta - \frac{U}{V} \right)^2.$$

The control volume in a reference system translating with the velocity $\mathbf{U}$, shown in Fig. 4.10, should be avoided, since it does not follow the fluid and requires extra terms for the correct balance.

For $C_u = C_{pu} = 0$ it results $a = 50$ cm$^2$, $V = 5.0$ m s$^{-1}$, $U = 1.0$ m s$^{-1}$, $\theta = 20°$,

*Case* (a)

$$\therefore \qquad F_x = \rho a V^2 \cos\theta = 1000 \times 50 \times 10^{-4} \times 5.0^2 \cos 20° = \mathbf{117.5\ N},$$

$$\therefore \qquad\qquad P = 0,\ \eta = 0.$$

*Case* (b)

$$\therefore \quad F_x = \rho a \, (V - U)^2 \cos \theta = 1000 \times 50 \times 10^{-4} \times (5.0 - 1.0)^2 \cos 20° = \mathbf{75.2 \ N},$$

$$\therefore \quad P = F_x U \cos \theta = 75.2 \times 1.0 \times \cos 20° = \mathbf{70.7 \ W},$$

$$\therefore \quad \eta = 2 \frac{U}{V} \left(1 - \frac{U}{V}\right)^2 \cos^2 \theta = 2 \times \frac{1.0}{5.0} \times \left(1 - \frac{1.0}{5.0}\right)^2 \times \cos^2 20° = \mathbf{22.6 \ \%}.$$

*Case* (c)

$$\therefore \quad F_x = \rho a \left(V - \frac{U}{\cos \theta}\right)^2 \cos \theta =$$

$$1000 \times 50 \times 10^{-4} \times \left(5.0 - \frac{1.0}{\cos 20°}\right)^2 \cos 20° = \mathbf{72.8 \ N},$$

$$\therefore \quad P = F_x U = 72.8 \times 1.0 = \mathbf{72.8 \ W},$$

$$\therefore \quad \eta = \frac{2}{\cos \theta} \frac{U}{V} \left(\cos \theta - \frac{U}{V}\right)^2 = \frac{2}{\cos 20°} \times \frac{1.0}{5.0} \times \left(\cos 20° - \frac{1.0}{5.0}\right)^2 = \mathbf{23.3 \ \%}.$$

---

**Exercise 4.5** The fuel tank of the rocket shown in Fig. 4.11 is cylindrical, with diameter $D = 3$ m, coaxial with the rocket and pressurized to an absolute pressure $p = (3 + C_{pu}/10) \times 10^5$ Pa. The mass density of the liquid fuel is equal to $\rho_f = 0.8\rho_w$, and the initial level is equal to $h = (3 + C_u/10)$ m.

- Calculate the force at the bottom of the tank at take-off, if the initial acceleration is $a = 10 \ \mathrm{m \, s^{-2}}$ and the launch ramp is at sea level.
- Calculate the same force when 40% of the fuel is burned and the total rocket mass is 70% of the initial mass, assuming the engine thrust is the same as the take-off thrust, the tank pressurization is unchanged, the gravity acceleration is equal to 50% of the standard gravity acceleration and the atmospheric pressure is reduced to 20% of the sea level atmospheric pressure.

Assume $\rho_w = 1000 \ \mathrm{kg \, m^{-3}}$, $p_{atm} = 10^5$ Pa at sea level.

**Solution** Integrating the indefinite equation of statics in the non-inertial reference attached to the rocket:

$$\rho_f \, (g + a) = \frac{\partial p}{\partial z},$$

and requiring the condition that the absolute pressure at the fuel free interface is equal to $p$, the absolute pressure at the bottom of the tank is equal to

**Fig. 4.11** Rocket with cylindrical circular tank

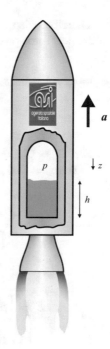

$$p_{fint} = p + \rho_f \left( g + a \right) h.$$

Atmospheric pressure acts from the outside. The net force acting on the bottom, on take-off, is equal to

$$F = \left( p_{fint} - p_{atm} \right) \frac{\pi D^2}{4} = \left[ p + \rho_f \left( g + a \right) h - p_{atm} \right] \frac{\pi D^2}{4}.$$

If after a certain interval of time in flight, during which the mass reduces from $m$ to $m' = 0.7m$ (due to fuel and comburent consumption), the thrust of the vector is unchanged, there will be an increase in acceleration, with a new value equal to

$$a' = a \frac{m}{m'} = \frac{a}{0.7}.$$

In addition, a 40% reduction in fuel leads to a reduction of the fuel level in the tank, the new fuel level being equal to $h' = 0.6h$.

Therefore, also considering the reduction of the acceleration of gravity, the value of the new force on the bottom will be equal to:

$$F' = \left(p'_{fint} - 0.2 p_{atm}\right) \frac{\pi D^2}{4} \qquad = \left[p + \rho_f \left(0.5g + a'\right) h' - 0.2 p_{atm}\right] \frac{\pi D^2}{4} =$$
$$\left[p + \rho_f \left(0.5g + \frac{a}{0.7}\right) 0.6h - 0.2 p_{atm}\right] \frac{\pi D^2}{4}.$$

For $C_u = C_{pu} = 0$ it results $D = 3$ m, $p = 3 \times 10^5$ Pa, $\rho_w = 1000$ kg m$^{-3}$, $\rho_f = 0.8 \rho_w = 800$ kg m$^{-3}$, $h = 3$ m, $a = 10$ m s$^{-2}$,

$$\therefore \quad F = \left[p + \rho_f \left(g + a\right) h - p_{atm}\right] \frac{\pi D^2}{4} =$$
$$\left[3 \times 10^5 + 800 \times (9.806 + 10) \times 3 - 1 \times 10^5\right]$$
$$\times \frac{\pi \times 3^2}{4} = \textbf{1750 kN},$$

$$\therefore \quad F' = \left[p + \rho_f \left(0.5g + \frac{a}{0.7}\right) 0.6h - 0.2 p_{atm}\right] \frac{\pi D^2}{4} =$$
$$\left[3 \times 10^5 + 800 \times \left(0.5 \times 9.806 + \frac{10}{0.7}\right) \times 0.6 \times 3 - 0.2 \times 10^5\right]$$
$$\times \frac{\pi \times 3^2}{4} = \textbf{2174 kN}.$$

---

**Exercise 4.6** The helicopter in Fig. 4.12, of mass $M = (10\,000 + 100 \times C_u)$ kg, has blades of diameter $D = (14 + C_{pu}/10)$ m rotating at 400 rpm. Calculate:

- the volumetric air flow rate required for lifting, if the helicopter is at sea level.
- The power required.
- The minimum blade rotation rate required, if the helicopter is at an altitude of 3000 m, with air density equal to $\rho'_{air} = 0.79$ kg m$^{-3}$.
- The same values if the helicopter moves vertically with a speed $U = 20$ m s$^{-1}$.
- The same values if the helicopter has attached a mass load $M_1 = 12\,000$ kg.

Assume $\rho_{air} = 1.18$ kg m$^{-3}$ at sea level. Assume the volumetric flow rate linearly variable with the rotation rate of the blades.

**Solution** Experimental results indicate that the streamlines of the flow generated by the blades converge from the upstream region, and are arranged in such a way as to define the streamtube shown in Fig. 4.13.

The pressure varies slightly along the single pathline (coinciding with the streamline in stationary condition) except across the blades, where a significant gradient

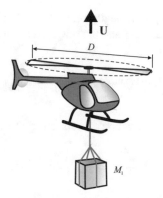

**Fig. 4.12**  Helicopter for transporting loads

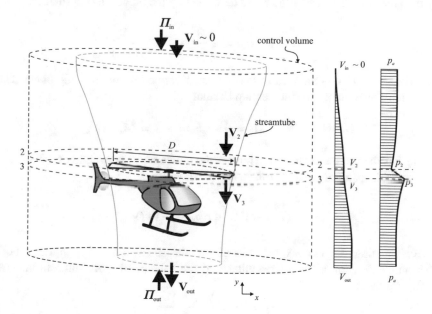

**Fig. 4.13**  Schematic and control volume for calculating forces

of pressure is located. The speed of the current is asymptotically zero upstream, increases progressively through the propeller, reaches a uniform value $V_{out}$ downstream. We choose a control volume with a lateral surface outside the flow tube and with inlet and outlet sections where the trajectories are straight and parallel. Applying Bernoulli's theorem for a streamline between the asymptotic upstream section and section 2, yields

$$p_a + \rho_{air}\frac{V_{in}^2}{2} = p_2 + \rho_{air}\frac{V_2^2}{2}.$$

The change in level head gives a negligible contribution. The same theorem applied for a streamline between section 3 and the output section downstream, results in the following balance equation

$$p_3 + \rho_{air} \frac{V_3^2}{2} = p_a + \rho_{air} \frac{V_{out}^2}{2}.$$

Since $V_2 = V_3$ (mass conservation), subtracting the corresponding sides of the two equations results in

$$p_3 - p_2 = \rho_{air} \frac{V_{out}^2 - V_{in}^2}{2}.$$

The thrust is obtained by multiplying the difference of pressure across the blades by the area of the surface of the circle described by the blades during rotation,

$$F = \Omega \, (p_3 - p_2) = \frac{\pi D^2}{4} \rho_{air} \frac{V_{out}^2 - V_{in}^2}{2},$$

where $\Omega = \pi D^2/4$ is the area of the surface described by the rotating blade. The thrust must also satisfy the momentum balance,

$$\cancel{\mathbf{G}} + \mathbf{\Pi} + \mathbf{M}_1 - \mathbf{M}_2 = 0 \rightarrow \mathbf{F} = -\mathbf{\Pi} = \mathbf{M}_1 - \mathbf{M}_2, \qquad (4.3)$$

where $\mathbf{F}$ is the force exerted by the fluid to the helicopter. The weight of the air $\mathbf{G}$ is negligible. Expanding all the terms in Eq. (4.3), yields

$$F = \rho_{air} Q_1 \, (V_{out} - V_{in}) \equiv \rho_{air} \frac{\pi D^2}{4} V_2 \, (V_{out} - V_{in}) \, .$$

By equating the two expressions of the thrust (obtained by applying both Bernoulli's theorem along a streamline, and the balance of momentum in integral form), results in

$$V_2 = V_3 = \frac{V_{out} + V_{in}}{2}.$$

If the helicopter is at rest, the input speed of the fluid is zero, thus

$$V_2 = V_3 = \frac{V_{out}}{2},$$

$$F = \frac{\pi D^2}{4} \rho_{air} \frac{V_{out}^2}{2} = \frac{\pi D^2}{4} \rho_{air} V_2 V_{out} = 2 \rho_{air} \frac{Q_1^2}{\Omega},$$

hence,

$$Q_1 = \sqrt{\frac{\Omega}{2 \rho_{air}} M g},$$

where $Mg$ is the weight of the helicopter.

The power required to take off is equal to

$$P_1 = \gamma_{air} Q_1 H = \gamma_{air} Q_1 \left( \cancel{z} + \frac{\cancel{p_a}}{\cancel{\gamma_{air}}} + \frac{V_{out}^2}{2g} \right) - \rho_{air} Q_1 2 V_2^2 = 2\rho_{air} \frac{Q_1^3}{\Omega^2}.$$

If the helicopter ascends vertically with uniform velocity $U$, having chosen a control volume attached to the moving helicopter, we can use all the previous equations by imposing $V_{in} = U$,

$$V_2 = V_3 = \frac{V_{out} + U}{2} \rightarrow V_{out} = 2V_2 - U,$$

$$F = \rho_{air} Q_2 (V_{out} - V_{in}) = \rho_{air} \Omega V_2 (V_{out} - U) = \rho_{air} \Omega V_2 (2V_2 - 2U) =$$

$$2\rho_{air} Q_2 \left( \frac{Q_2}{\Omega} - U \right)$$

and then, solving the 2nd order equation, yields

$$Q_2 = \frac{U\Omega}{2} + \sqrt{\frac{U^2\Omega^2}{4} + \frac{Mg\Omega}{2\rho_{air}}}.$$

The power is equal to

$$P_2 = \gamma_{air} Q_2 \Delta H = \gamma_{air} Q_2 \left( \cancel{z} + \frac{p_a}{\gamma_{air}} + \frac{V_{out}^2}{2g} - \cancel{z} - \frac{p_a}{\gamma_{air}} - \frac{U^2}{2g} \right) =$$

$$\rho_{air} Q_2 \frac{1}{2} \left( V_{out}^2 - U^2 \right) = \rho_{air} Q_2 2 V_2 (V_2 - U) = 2\rho_{air} \frac{Q_2^2}{\Omega} \left( \frac{Q_2}{\Omega} - U \right).$$

If the mass density of the air changes, the new volumetric flow rate will be equal to

$$Q_1' = \sqrt{\frac{\Omega}{2\rho_{air}'} Mg} \rightarrow \frac{Q_1'}{Q_1} = \sqrt{\frac{\rho_{air}}{\rho_{air}'}}.$$

To calculate the rotation speed of the blades, consider that the volumetric flow rate varies linearly with the rotation speed. Therefore,

$$\frac{n_1'}{n_1} = \sqrt{\frac{\rho_{air}}{\rho_{air}'}}.$$

For $C_u = C_{pu} = 0$ it results $M = 10\,000$ kg, $D = 14$ m, $n_1 = 400$ rpm, $\Omega = 153.93$ m$^2$.

(i) If $U = 0$:

$$\therefore \quad Q_1 = \sqrt{\frac{\Omega}{2\rho_{air}} Mg} = \sqrt{\frac{153.93}{2 \times 1.18} \times 10\,000 \times 9.806} = \mathbf{2529 \ m^3 \ s^{-1}},$$

$$\therefore \quad P_1 = 2\rho_{air} \frac{Q_1^3}{\Omega^2} = 2 \times 1.18 \times \frac{2529^3}{153.93^2} = \mathbf{1.61 \ MW},$$

$$\therefore \quad \frac{n_1'}{n_1} = \sqrt{\frac{\rho_{air}}{\rho_{air}'}} = \sqrt{\frac{1.18}{0.79}} = 1.22 \to n_1' = \mathbf{489 \ rpm}.$$

(ii) If $U = 20 \ \mathrm{m\,s^{-1}}$:

$$\therefore \quad Q_2 = \frac{U\Omega}{2} + \sqrt{\frac{U^2 \Omega^2}{4} + \frac{Mg\Omega}{2\rho_{air}}} = \frac{20 \times 153.93}{2}$$

$$+ \sqrt{\frac{20^2 \times 153.93^2}{4} + \frac{10\,000 \times 9.806 \times 153.93}{2 \times 1.18}} = \mathbf{4500 \ m^3 \ s^{-1}},$$

$$\therefore \quad P_2 = 2\rho_{air} \frac{Q_2^2}{\Omega} \left( \frac{Q_2}{\Omega} - U \right) =$$

$$2 \times 1.18 \times \frac{4500^2}{153.93} \times \left( \frac{4500}{153.93} - 20 \right) = \mathbf{2.87 \ MW}.$$

The case of a payload in addition to the own weight of the helicopter, has an immediate solution.

---

**Exercise 4.7** The system in Fig. 4.14 is a helicopter turbine of total mass $M = (1500 + 20 \times C_{pu})$ kg. The inlet has a circular cross-section with diameter $D = (3 + C_u/10)$ m, the outlet section is a circular crown with internal diameter $d = (2.7 + C_u/10)$ m and external diameter $D$. Assuming in the inlet and outlet sections a uniform pressure distribution equal to the atmospheric pressure, considering the take-off conditions, calculate:

– the velocity of the air at the inlet and outlet sections.
– The required power of the turbine.

**Fig. 4.14** Helicopter turbine

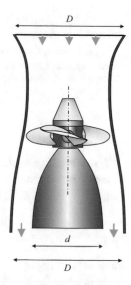

**Fig. 4.15** Control volume

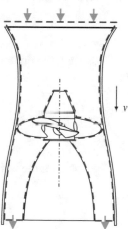

Assume an isochoric flow for the air, with $\rho_{air} = 1.2 \ \mathrm{kg \, m^{-3}}$.

**Solution** After choosing the control volume in Fig. 4.15, we project the balance of momentum equation along $y$:

$$\cancel{G_y} + \Pi_y + \cancel{V_y} + M_{1y} - M_{2y} = 0. \tag{4.4}$$

The surface forces are only the action exerted by the propeller and the turbine casing on the control volume,

$$\Pi_y = -F,$$

because, by hypothesis, the pressure in the inlet and outlet sections is equal to the atmospheric pressure.

In particular, the pressure acting on the lower side of the propeller is greater than the pressure acting on the upper side. Expanding all terms in Eq. (4.4) results in

$$-F + \rho_{air} \frac{Q^2}{\frac{\pi D^2}{4}} - \rho_{air} \frac{Q^2}{\frac{\pi (D^2 - d^2)}{4}} = 0.$$

Therefore, the action of the fluid in the control volume on the propeller-carter system (i.e. the thrust of the turbine) is equal to:

$$F = -\frac{4\rho_{air} Q^2}{\pi} \left[ \frac{1}{(D^2 - d^2)} - \frac{1}{D^2} \right],$$

and it is pointing upwards. At the take-off, the thrust must be equal to the weight of the helicopter (in incipient motion, we are neglecting the acceleration):

$$F + Mg = 0.$$

Therefore, the minimum volume discharge rate required for the take-off is equal to

$$Q = \sqrt{\frac{Mg}{\frac{4\rho_{air}}{\pi} \left[ \frac{1}{(D^2 - d^2)} - \frac{1}{D^2} \right]}}.$$

The power of the turbine shall be equal to

$$P = \gamma_{air} Q \Delta H = \rho_{air} Q \left( \frac{V_o^2}{2} - \frac{V_i^2}{2} \right),$$

since the pressure contribution is null and the gravitational term is negligible. $V_i$ and $V_o$ are the average air velocity in the inlet and outlet section calculated in the control volume reference, respectively, using the mass conservation equation:

$$V_i = \frac{4Q}{\pi D^2}, \quad V_o = \frac{4Q}{\pi (D^2 - d^2)}.$$

For $C_u = C_{pu} = 0$ it results $M = 1500$ kg, $D = 3$ m, $d = 2.7$ m, $\rho_{air} = 1.2$ kg m$^{-3}$,

$$Q = \sqrt{\frac{Mg}{\dfrac{4\rho_{air}}{\pi}\left[\dfrac{1}{(D^2-d^2)}-\dfrac{1}{D^2}\right]}} =$$

$$= \sqrt{\frac{1500 \times 9.806}{\dfrac{4 \times 1.2}{\pi} \times \left[\dfrac{1}{(3^2-2.7^2)}-\dfrac{1}{3^2}\right]}} = 142.6 \text{ m}^3 \text{ s}^{-1},$$

$$\therefore \quad V_i = \frac{4Q}{\pi D^2} = \frac{4 \times 142.6}{\pi \times 3^2} = 20.2 \text{ m s}^{-1},$$

$$\therefore \quad V_o = \frac{4Q}{\pi\left(D^2-d^2\right)} = \frac{4 \times 142.6}{\pi \times \left(3^2-2.7^2\right)} = 106.2 \text{ m s}^{-1},$$

$$\therefore \quad P = \rho_{air}\,Q\left(\frac{V_o^2}{2}-\frac{V_i^2}{2}\right) = 1.2 \times 142.6 \times \left(\frac{106.2^2}{2}-\frac{20.2^2}{2}\right) = 930 \text{ kW}.$$

**Exercise 4.8** Water flows through the pipeline system in Fig. 4.16, with outflow in atmosphere through pipelines 2 and 3.

– Calculate the outflow velocity $V_2$.
– Calculate the force exerted on the flange.

Neglect weight and losses. Numerical data: $V_1 = (6 + C_u/10) \text{ m s}^{-1}$, $V_3 = (9 + C_u/10) \text{ m s}^{-1}$, $D_1 = D_2 = (300 + C_{pu}) \text{ mm}$, $D_3 = (200 + C_u) \text{ mm}$, $\alpha = (40 + C_u)°$, $p_m = (0.7 + C_{pu} \times 0.1) \times 10^5 \text{ Pa}$.

**Solution** The velocity $V_2$ is calculated using mass conservation,

$$V_1\frac{\pi D_1^2}{4} = V_2\frac{\pi D_2^2}{4} + V_3\frac{\pi D_3^2}{4} \rightarrow V_2 = V_1\frac{D_1^2}{D_2^2} - V_3\frac{D_3^2}{D_2^2}.$$

The forces along the $x$ and $y$ axes are calculated by applying the momentum balance,

$$\cancel{G_x} + \Pi_{0x} + \Pi_{1x} + \Pi_{2x} + \cancel{K_x} + M_{1x} - M_{2x} = 0,$$

$$\cancel{G_y} + \Pi_{0y} + \Pi_{1y} + \Pi_{2y} + \cancel{K_y} + M_{1y} - M_{2y} = 0,$$

equivalent to

$$\Pi_{0x} + \left(\rho V_1^2 + p_m\right) \frac{\pi D_1^2}{4} + \rho V_3^2 \frac{\pi D_3^2}{4} - \rho V_2^2 \frac{\pi D_2^2}{4} \cos \alpha = 0,$$

$$\Pi_{0y} + \rho V_2^2 \frac{\pi D_2^2}{4} \sin \alpha = 0,$$

where $\Pi_{0x}$ and $\Pi_{0y}$ are the two components of the force exerted by the flange, through the pipe sections connected to it, on the control volume, and $p_m$ is the relative pressure indicated by the manometer. According to the principle of action and reaction, the actions on the flange are equal and opposite:

$$S_x = -\Pi_{0x} = \left(\rho V_1^2 + p_m\right) \frac{\pi D_1^2}{4} + \rho V_3^2 \frac{\pi D_3^2}{4} - \rho V_2^2 \frac{\pi D_2^2}{4} \cos \alpha,$$

$$S_y = -\Pi_{0y} = \rho V_2^2 \frac{\pi D_2^2}{4} \sin \alpha.$$

For $C_u = C_{pu} = 0$ it results $V_1 = 6.0 \text{ m s}^{-1}$, $V_3 = 9.0 \text{ m s}^{-1}$, $D_1 = D_2 = 300$ mm, $D_3 = 200$ mm, $\alpha = 40°$, $p_m = 0.7 \times 10^5$ Pa,

$$\therefore \qquad V_2 = V_1 \frac{D_1^2}{D_2^2} - V_3 \frac{D_3^2}{D_2^2} = 6.0 - 9.0 \times \frac{200^2}{300^2} = \mathbf{2.0 \text{ m s}^{-1}},$$

$$\therefore \quad S_x = \left(\rho V_1^2 + p_m\right) \frac{\pi D_1^2}{4} + \rho V_3^2 \frac{\pi D_3^2}{4} - \rho V_2^2 \frac{\pi D_2^2}{4} \cos \alpha =$$

$$\left(1000 \times 6.0^2 + 0.7 \times 10^5\right) \times \frac{\pi \times 0.3^2}{4} + 1000 \times 9.0^2 \times \frac{\pi \times 0.2^2}{4}$$

$$- 1000 \times 2.0^2 \times \frac{\pi \times 0.3^2}{4} \times \cos 40° = \mathbf{9821 \text{ N}},$$

**Fig. 4.16** Pipeline system

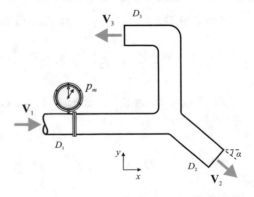

$$\therefore \quad S_y = \rho V_2^2 \frac{\pi D_2^2}{4} \sin \alpha = 1000 \times 2.0^2 \times \frac{\pi \times 0.3^2}{4} \times \sin 40° = \mathbf{182 \ N}.$$

**Exercise 4.9** The fighter aircraft in Fig. 4.17 flights in a straight line with uniform speed $V = (100 + 10 \times C_u)$ m s$^{-1}$. The direction control system is supported by a deflector for changing the direction of the reactor jet. The input mass flow rate is equal to the output mass flow rate and is equal to $Q_m = (230 + 10 \times C_{pu})$ kg s$^{-1}$. $V_{out} = (300 + 10 \times C_u)$ m s$^{-1}$ is measured with respect to the aircraft. Calculate:

– the thrust of the engine if the jet is coaxial to the fuselage.
– The thrust of the engine if the jet is inclined by $\theta = (4 + C_{pu})°$ in the vertical plane with respect to the axis of the fuselage.
– The pitch torque (rotation around the longitudinal axis of the wings) for the two cases, calculated with respect to the axis with trace $C_g$.

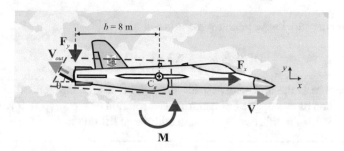

**Fig. 4.17** Fighter aircraft, the dashed line delimits the control volume

**Solution** We select the dashed control volume in Fig. 4.17. If the motion is rectilinear and uniform, the reference system attached to the control volume is inertial, and we choose a coordinate system with the $x$-axis coaxial to the fuselage and the $y$-axis positive upwards. In the mobile reference, the inflow has a velocity equal to $V$, and applying the momentum balance to the control volume, the forces acting on the fuselage are equal and opposite to the forces exerted by the fuselage (via the carter of the engine) on the control volume:

$$\begin{cases} F_x = -\Pi_{0x} = M_{1x} - M_{2x} = -Q_m \left( V - V_{out} \cos \theta \right), \\ F_y = -\Pi_{0y} = M_{1y} - M_{2y} = -Q_m V_{out} \sin \theta. \end{cases}$$

In particular, the $F_y$ component is due to the deflector and is assumed to be applied in the pivot of the deflector. The pitch torque is equal to

$$|\mathbf{M}| = F_y b.$$

For $C_u = C_{pu} = 0$ it results $Q_m = 230$ kg s$^{-1}$, $V = 100$ m s$^{-1}$, $V_{out} = 300$ m s$^{-1}$, $\theta = 4°$.

If the jet is coaxial to the fuselage, it results

$$\therefore \qquad F_x = Q_m (V_{out} - V) = 230 \times (300 - 100) = \textbf{46 000 N},$$

$$\therefore \qquad F_y = 0,$$

and the pitching torque is null. Notice that the airplane weight is balanced by the lift of the wings.

If the jet is inclined of $\theta = 4°$, it results

$$\therefore \qquad F_x = Q_m (V_{out} \cos \theta - V) = 230 (300 \times \cos 4° - 100) = \textbf{45 832 N},$$

$$\therefore \qquad F_y = -Q_m V_{out} \sin \theta = -230 \times 300 \times \sin 4° = \textbf{-4813 N},$$

and the counter-clockwise pitch torque is equal to

$$\therefore \qquad |\mathbf{M}| = F_y b = 4813 \times 8 = \textbf{38 504 N m},$$

where $b = 8$ m is the arm with respect to $C_g$. This torque is intended as incipient torque: as soon as it is applied, it determines a rotation of the aircraft corresponding to a nose-up. Also in this case there is an additional lift due to the wings.

---

**Exercise 4.10** A dredger and a barge are linked by a chain, as shown in Fig. 4.18. The dredger draws the mud through a vertical pipe from the bottom and discharges into the barge a mixture of water and slurry with a specific gravity equal to 1.9. The horizontal outflow velocity is equal to $V = (5 + C_u/10)$ m s$^{-1}$ and the nozzle diameter is equal to $D = (30 + C_{pu})$ cm.

– Calculate the traction force in the chain.

**Solution** Consider the control volume in Fig. 4.19. By applying the momentum balance, the horizontal component of the input momentum flow, equal to $M_{1x} = \rho Q V$, tends to push the barge to the right and is counteracted by the tractive force in the chain. The vertical component depends on the vertical speed reached by the current of the jet due to gravity, tends to increase the sinking of the barge and is balanced by the greater buoyancy force.

In the momentum balance applied to the dredger, see Fig. 4.20, the input momentum flow has no horizontal component (it has a vertical component, neglected in the

**Fig. 4.18** Dredger
unloading water and slurry
into a barge

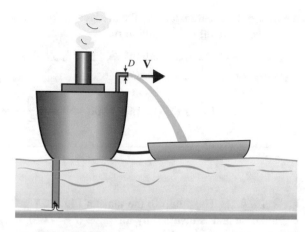

**Fig. 4.19** Control volume
for calculating forces acting
on the barge

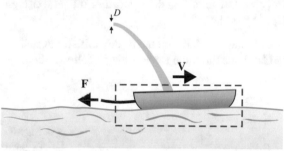

**Fig. 4.20** Control volume
for calculating forces acting
on the dredger

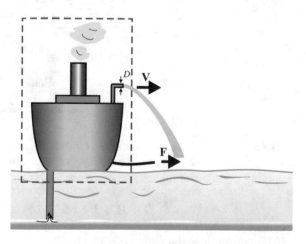

present analysis) and there is only the output momentum flow equal to $M_{2x} = \rho Q V$,
balanced by the chain traction force. Applying the momentum balance to the dredger-
barge system, the outgoing and incoming momentum flows in the horizontal direction
are zero and the system remains at rest. Therefore, the chain is in traction with a force
equal to $F = \rho Q V$ that balances both the flow of momentum entering the barge and

the flow of momentum leaving the dredger, and the dredger-chain-barge system does not move, since all forces are internal.

For $C_u = C_{pu} = 0$ it results $V = 5 \, \mathrm{m\,s^{-1}}$, $D = 30 \, \mathrm{cm}$, and

$$\therefore \qquad F = \rho Q V = 1.9 \times 1000 \times 5 \times \frac{\pi \times 0.3^2}{4} \times 5 = \textbf{3.35 kN}.$$

**Exercise 4.11** The pump in the tank in Fig. 4.21 generates a jet of water that hits a crankcase and is diverted by an angle $\alpha = (120 + C_{pu})°$. The flow rate is equal to $Q = 0.01 \, \mathrm{m^3\,s^{-1}}$ and the speed of the jet exiting the pump is equal to $V = (2 + C_u/3) \, \mathrm{m\,s^{-1}}$.

– Calculate the force on the trolley initially at rest if the jet follows the trajectory A.
– Calculate the same force if the jet follows the trajectory B.

**Fig. 4.21** Trolley with pump

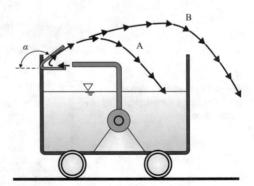

**Solution** We first choose the dashed control volume in Fig. 4.22. If the jet follows the trajectory A, the force on the trolley is the weight of the volume of fluid only. Applying the momentum balance equation, yields

$$\mathbf{G} + \mathbf{\Pi} + \mathbf{I} + \mathbf{M_1} - \mathbf{M_2} = \mathbf{0}.$$

The projection in the $x$-direction yields

$$\cancel{G_x} + \Pi_x + \cancel{I_x} + M_{1x} - M_{2x} = 0.$$

It can be shown that $M_{1x} = M_{2x}$ and, therefore, $F_x = -\Pi_x = 0$. $F_x$ is the action on the trolley, equal and opposite to the action $\Pi_x$ exerted by the trolley on the control volume.

The projection in the $y$-direction yields

**Fig. 4.22** Mobile control volume

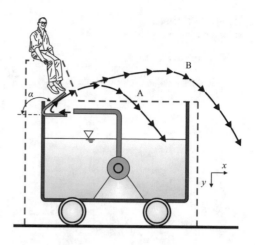

$$G_y + \Pi_y + \cancel{V_y} + M_{1y} - M_{2y} = 0.$$

It can be shown that $M_{1y} = M_{2y}$ and, therefore, $F_y = -\Pi_y = G_y$.

If the jet follows the trajectory B, applying the momentum balance and projecting it in the $x$-direction yields

$$\cancel{G_x} + \Pi_x + \cancel{V_x} + \cancel{M_{1x}} - M_{2x} = 0,$$

hence $F_x = -\Pi_x - -M_{2x}$. The trolley is subject to a force equal to

$$F_x = -\rho Q V \cos (\pi - \alpha)$$

which, for $\alpha > \pi/2$, is pointing to the left. In the vertical direction it results $F_y = -\Pi_y = G_y - M_{2y}$. The trolley is subject to a force in excess of the weight of the fluid and equal to

$$F_y - G_y = \rho Q V \sin (\pi - \alpha).$$

For $C_u = C_{pu} = 0$ it results $\alpha = 120°$, $Q = 0.01$ m$^3$ s$^{-1}$, $V = 2$ m s$^{-1}$,

$$\therefore \quad F_x = -\rho Q V \cos (\pi - \alpha) = -1000 \times 0.01 \times 2 \times \cos (180° - 120°) = -\mathbf{10\ N},$$

pointing left, and

$$\therefore \quad F_y - G_y = \rho Q V \sin (\pi - \alpha) = 1000 \times 0.01 \times 2 \times \sin (180° - 120°) = \mathbf{17.3\ N},$$

pointing downwards.

**Fig. 4.23** Three-arm
splinker

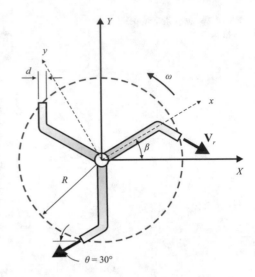

**Exercise 4.12** The sprinkler in Fig. 4.23 is operated with a flow rate $Q = (50 + 10 \times C_u)\,l\,min^{-1}$. The diameter of the nozzles is $d = 6$ mm and the radius of the pinwheel is $R = (80 + 10 \times C_{pu})$ mm.

– Calculate the outflow rate.
– Calculate the uniform angular speed at full speed by neglecting the friction torque on the axis of rotation.

**Solution** The exercise can be approached by choosing (i) a fixed inertial frame, and (ii) a rotating non-inertial frame attached to the pinwheel. The choice of the coordinate system is arbitrary: in both cases a fixed or mobile coordinate system can be selected.

We first choose a coaxial cylindrical control volume *in a fixed inertial frame*, with dimensions such as to intersect the three outgoing water jets with the lateral surface, and the incoming fluid current with the lower base. We also choose a fixed $XYZ$ coordinate system with the $Z$-axis coincident with the axis of rotation, see Fig. 4.24.

Velocities and accelerations must be evaluated in the control volume frame, i.e. in the fixed inertial frame. The equation of balance of the angular momentum is as follows:

$$\mathbf{G}_m + \mathbf{I}_m + \mathbf{\Pi}_m + \mathbf{M}_{1m} - \mathbf{M}_{2m} = \mathbf{0}.$$

Since the only degree of freedom of the system is the rotation around the vertical axis, we project the equation in the direction of the $Z$-axis coincident with $z$-axis. We analyze the different contributions on moments of the acting forces.

The moment of the mass forces is equal to

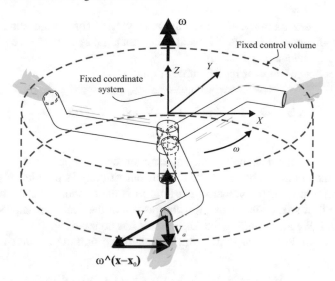

**Fig. 4.24** Inertial fixed control volume

$$G_{mZ} = \int_{CV} [(\mathbf{X} - \mathbf{X}_0) \wedge \rho \mathbf{f}]_Z \, dW = \int_{CV} [(\mathbf{X} - \mathbf{X}_0) \wedge \rho \mathbf{g}]_Z \, dW = 0,$$

where $\mathbf{X}_0$ is a point belonging to the $Z$-axis (for example, the origin of the coordinate system). The acceleration vector of gravity $\mathbf{g}$ is parallel to $Z$ and the cross product $(\mathbf{X} - \mathbf{X}_0) \wedge \rho \mathbf{g}$ is always orthogonal to a plane that contains the $Z$-axis. As a consequence, the component of the vector cross product along the $Z$-axis direction is null.

The rotational inertial term is equal to

$$I_{mZ} = -\int_{CV} \left[ (\mathbf{X} - \mathbf{X}_0) \wedge \frac{\partial \rho \mathbf{V}_a}{\partial t} \right]_Z dW = 0,$$

because the analysis is in steady state and all the variables and characteristic quantities of the fluid are independent of time. $\mathbf{V}_a$ is the absolute velocity of the fluid, as observed in the fixed inertial frame.

The moment of the surface forces is equal to

$$\Pi_{mZ} = \int_{S} [(\mathbf{X} - \mathbf{X}_0) \wedge (\mathbf{T} \cdot \mathbf{n})]_Z \, dS = 0,$$

($\mathbf{T}$ is the stress tensor) since the pressure of the fluid in the outlet sections is equal to the atmospheric pressure (the relative pressure is, therefore, zero). To the fluid pressure in the supply line (parallel to $Z$) is associated (i) a force with a zero arm, if the pressure distribution has an axial symmetry; (ii) a force generating a moment with null component along the $Z$-axis, if it has not axial symmetry. In both cases its moment component along $Z$ is null.

The inlet flow of angular momentum is equal to

$$M_{1mZ} = \int\limits_{S_{in}} [(\mathbf{X} - \mathbf{X}_0) \wedge \rho \mathbf{V}_a (\mathbf{V}_a \cdot \mathbf{n})]_Z \, dS = 0,$$

because the only input section is the intersection of the supply line with the base surface of the control volume. There the current velocity is parallel to the $Z$-axis and the arm is null, if the velocity distribution is axial-symmetric. If the velocity distribution is not axial-symmetric, the component of the vector cross product along $Z$ is still null. We are assuming that the current has no swirling.

The only non-zero term is the component of the outflow angular momentum $M_{2mZ}$:

$$M_{2mZ} = \int\limits_{S_{out}} [(\mathbf{X} - \mathbf{X}_0) \wedge \rho \mathbf{V}_a (\mathbf{V}_a \cdot \mathbf{n})]_Z \, dS.$$

In the outlet section of each of the three small pipes the velocity of the fluid has components, in a rotating coordinate system attached to the pinwheel, equal to

$$\mathbf{V}_r = |\mathbf{V}_r| \sin \theta \mathbf{i}' - |\mathbf{V}_r| \cos \theta \mathbf{j}',$$

where $\mathbf{i}'$ and $\mathbf{j}'$ are the versors of the rotating coordinate system. If $\beta(t)$ is the angle (function of time) between the $x$-axis of the rotating coordinate system and the $X$-axis of the fixed coordinate system, the instantaneous components of the relative velocity vector in the outlet section, read in the fixed coordinate system, are

$$\mathbf{V}_r = |\mathbf{V}_r| (\sin \theta \cos \beta + \cos \theta \sin \beta) \mathbf{i} + |\mathbf{V}_r| (\sin \theta \sin \beta - \cos \theta \cos \beta) \mathbf{j},$$

where $\mathbf{i}$ and $\mathbf{j}$ are the $X$ and $Y$ versors in the fixed coordinate system. The relative velocity is composed with the transport velocity of the relative frame of reference:

$$\mathbf{V}_a = \mathbf{V}_r + \mathbf{V}_0 + \boldsymbol{\omega} \wedge (\mathbf{x} - \mathbf{x}_0).$$

$\mathbf{V}_a$ is the velocity in the absolute reference, $\mathbf{V}_0$ is the translation velocity of the origin of the relative reference with respect to the absolute reference, $\boldsymbol{\omega}$ is the rotation velocity of the relative reference, $(\mathbf{x} - \mathbf{x}_0)$ is the position vector in the relative coordinate system. In the present case $\mathbf{V}_0 = \mathbf{0}$ and the angular velocity vector is expressed as $\boldsymbol{\omega} = |\boldsymbol{\omega}| \, \mathbf{k}$; the vector cross product is equal to

$$\boldsymbol{\omega} \wedge (\mathbf{x} - \mathbf{x}_0) = \begin{vmatrix} \mathbf{i} & \mathbf{j} & \mathbf{k} \\ 0 & 0 & |\omega| \\ R\cos\beta & R\sin\beta & 0 \end{vmatrix} = -|\omega|\,R\sin\beta\,\mathbf{i} + |\omega|\,R\cos\beta\,\mathbf{j}.$$

In summary, the absolute velocity of the current in the exit section, read in the fixed coordinate system, is equal to

$$\mathbf{V}_a = [|\mathbf{V}_r|\,(\sin\theta\cos\beta + \cos\theta\sin\beta) - |\omega|\,R\sin\beta]\,\mathbf{i} +$$
$$[|\mathbf{V}_r|\,(\sin\theta\sin\beta - \cos\theta\cos\beta) + |\omega|\,R\cos\beta]\,\mathbf{j} \equiv V_{aX}\mathbf{i} + V_{aY}\mathbf{j}.$$

For a stationary observer who observes the sprinkler during its operation, the jet of nozzles will not appear coaxial to the tubes in the outflow section, due to the dragging component, see Fig. 4.24. The term $(\mathbf{X} - \mathbf{X}_0) \wedge \rho\mathbf{V}_a$ is uniform on the outlet surface (the outlet section of the small pipe) and results

$$(\mathbf{X} - \mathbf{X}_0) \wedge \rho\mathbf{V}_a = \rho \begin{vmatrix} \mathbf{i} & \mathbf{j} & \mathbf{k} \\ R\cos\beta & R\sin\beta & 0 \\ V_{aX} & V_{aY} & 0 \end{vmatrix} = \rho\,(V_{aY}\,R\cos\beta - V_{aX}\,R\sin\beta)\,\mathbf{k} =$$

$$\rho\left(-|\mathbf{V}_r|\,R\cos\theta + |\omega|\,R^2\right)\mathbf{k}.$$

The relative output velocity magnitude can be calculated by applying the mass conservation to the control volume, by dividing the input flow rate into three equal flow rates (for symmetry),

$$|\mathbf{V}_r| = \frac{4Q}{3\pi d^2}.$$

Ultimately, it results

$$M_{2mZ} = \int_{S_{out}} [(\mathbf{X} - \mathbf{X}_0) \wedge \rho\mathbf{V}_a(\mathbf{V}_a \cdot \mathbf{n})]_Z\,dS =$$

$$\rho\left(-|\mathbf{V}_r|\,R\cos\theta + |\omega|\,R^2\right) \int_{S_{out}} \mathbf{V}_a \cdot \mathbf{n}\,dS = -\rho QR\,(|\mathbf{V}_r|\cos\theta - |\omega|\,R),$$

already extended to all three pipes. In fact, due to mass conservation, it results

$$\int_{S_{out}} \mathbf{V}_a \cdot \mathbf{n}\,dS = Q.$$

The angular speed in steady state regime, in the absence of friction, assumes the value that cancels the outflow of angular momentum, since no other torque can balance it:

$$-\rho Q R \left(|\mathbf{V}_r| \cos\theta - |\boldsymbol{\omega}| R\right) = 0 \rightarrow |\boldsymbol{\omega}| = \frac{|\mathbf{V}_r| \cos\theta}{R}.$$

In this condition, results

$$\mathbf{V}_a = |\mathbf{V}_r| (\sin\theta \cos\beta)\, \mathbf{i} + |\mathbf{V}_r| (\sin\theta \sin\beta)\, \mathbf{j} \equiv |\mathbf{V}_r| \sin\theta \, (\cos\beta \mathbf{i} + \sin\beta \mathbf{j}),$$

i.e. the velocity of the outlet current is only radial since $\cos\beta \mathbf{i} + \sin\beta \mathbf{j} \equiv \mathbf{i}'$, where $\mathbf{i}'$ is the radial versor rotating with the sprinkler and read in the fixed coordinate system $XYZ$.

We perform the same calculations by fixing an $xyz$ rotating coordinate system, but with a fixed (inertial) reference system. The only non-zero contribution is still the outflow of angular momentum,

$$M_{2mz} = \int_{S_{out}} [(\mathbf{X} - \mathbf{X}_0) \wedge \rho \mathbf{V}_a (\mathbf{V}_a \cdot \mathbf{n})]_z \, dS.$$

Suppose, for simplicity, that one of the three tubes is aligned with the $x$-axis. The absolute velocity in the outflow section, read in the $xyz$ rotating coordinate system, is equal to

$$\mathbf{V}_a = |\mathbf{V}_r| \sin\theta \, \mathbf{i}' - (|\mathbf{V}_r| \cos\theta - |\boldsymbol{\omega}| R)\, \mathbf{j}'.$$

The arm has component

$$(\mathbf{X} - \mathbf{X}_0) = R \mathbf{i}',$$

hence

$$(\mathbf{X} - \mathbf{X}_0) \wedge \rho \mathbf{V}_a = \rho \begin{vmatrix} \mathbf{i}' & \mathbf{j}' & \mathbf{k}' \\ R & 0 & 0 \\ |\mathbf{V}_r| \sin\theta & -(|\mathbf{V}_r| \cos\theta - |\boldsymbol{\omega}| R) & 0 \end{vmatrix} =$$

$$\rho \left(-|\mathbf{V}_r| R \cos\theta + |\boldsymbol{\omega}| R^2\right) \mathbf{k}'.$$

By integrating in the outlet section of the single small pipe, the contribution to the outflow momentum, equal to 1/3 of the total outflow, is computed. For symmetry, the total outflow of momentum flux will be equal to

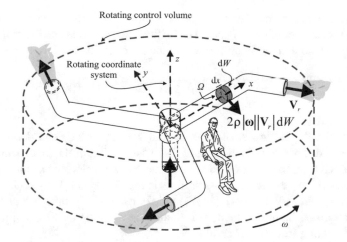

**Fig. 4.25** Non-inertial rotating control volume

$$M_{2mz} = \int\limits_{S_{out}} [(\mathbf{X} - \mathbf{X}_0) \wedge \rho \mathbf{V}_a (\mathbf{V}_a \cdot \mathbf{n})]_z \, dS =$$

$$\rho \left( -|\mathbf{V}_r| R \cos\theta + |\boldsymbol{\omega}| R^2 \right) \int\limits_{S_{out}} \mathbf{V}_a \cdot \mathbf{n} \, dS = -\rho Q R \left( |\mathbf{V}_r| \cos\theta - |\boldsymbol{\omega}| R \right).$$

This last expression coincides with the one previously calculated using the fixed coordinate system $XYZ$. The final result, of course, does not depend on the choice of the coordinate system.

If we choose a control volume in the *non-inertial frame of reference* as shown in Fig. 4.25, geometrically defined as the previous one but rotating together with the pinwheel, the balance equation of the angular momentum remains unchanged, but the velocities and accelerations must be evaluated in the relative non-inertial rotating frame of reference. Again, we are free to choose a fixed or a rotating coordinate system. We choose a rotating coordinate system $xyz$. In addition to the mass force of gravity, there are also fictitious mass forces deriving from the evaluation of accelerations in the non-inertial frame. The relative acceleration to be introduced in the calculation of $\mathbf{G}_m$ is obtained from the following vectorial composition:

$$\mathbf{a}_a = \mathbf{a}_r + \mathbf{a}_0 + 2\boldsymbol{\omega} \wedge \mathbf{V}_r + \boldsymbol{\omega} \wedge (\boldsymbol{\omega} \wedge (\mathbf{x} - \mathbf{x}_0)) + \frac{d\boldsymbol{\omega}}{dt} \wedge (\mathbf{x} - \mathbf{x}_0),$$

with obvious meaning of the terms. Since in the present case $\mathbf{a}_0 = 0$, $\mathbf{a}_a = \mathbf{g}$, $d\boldsymbol{\omega}/dt = 0$ (in steady state regime), it results

$$\mathbf{a}_r = \mathbf{g} - 2\boldsymbol{\omega} \wedge \mathbf{V}_r - \boldsymbol{\omega} \wedge (\boldsymbol{\omega} \wedge (\mathbf{x} - \mathbf{x}_0)).$$

The moment of gravity acceleration has no component along $z$; the moment of the centrifugal acceleration is zero because centrifugal acceleration and the arm $(\mathbf{x} - \mathbf{x}_0)$ are parallel. The only contribution is due to the acceleration of Coriolis,

$$G_{mz} = \int\limits_{CV} [(\mathbf{x} - \mathbf{x}_0) \wedge \rho(-2\boldsymbol{\omega} \wedge \mathbf{V}_r)]_z \, dW.$$

We consider the outflow small pipe coaxial with the $x$-axis and define the volume $dW = \Omega dx$, where $\Omega$ is the area of the cross section of the pipe. The position vector will have component $(\mathbf{x} - \mathbf{x}_0) = x\mathbf{i}'$; the relative velocity has the only component $\mathbf{V}_r = |\mathbf{V}_r|\mathbf{i}'$; the vector cross product is $\boldsymbol{\omega} \wedge \mathbf{V}_r = |\boldsymbol{\omega}| |\mathbf{V}_r|\mathbf{j}'$ and, therefore, $(\mathbf{x} - \mathbf{x}_0) \wedge \rho(-2\boldsymbol{\omega} \wedge \mathbf{V}_r) = -2\rho |\boldsymbol{\omega}| |\mathbf{V}_r| x\mathbf{k}'$. For simplicity sake, we assume that the end portion of the pipe, which leads to an axis variation of $30°$, is of negligible length. Integrating from 0 to $R$ results in

$$\frac{G_{mz}}{3} = \int\limits_0^R -2\rho |\boldsymbol{\omega}| |\mathbf{V}_r| \Omega x \, dx = -\rho |\boldsymbol{\omega}| \frac{Q}{3} R^2.$$

It can be demonstrated that this result is valid regardless of the geometry of the small pipe. The contribution of girotoric inertia and of surface forces are null, and the inflowing angular momentum at the inlet is zero. The outflowing angular momentum is equal to

$$M_{2mz} = \int\limits_{S_{out}} \left[(\mathbf{x} - \mathbf{x}_0) \wedge \rho \mathbf{V}_r(\mathbf{V}_r \cdot \mathbf{n}')\right]_z dS =$$

$$\rho(-|\mathbf{V}_r| R \cos\theta) \int\limits_{S_{out}} \mathbf{V} \cdot \mathbf{n}' \, dS = -\rho Q R |\mathbf{V}_r| \cos\theta,$$

already extended to all three small pipes.

In equilibrium conditions, it results

$$G_{mz} - M_{2mz} = \rho Q R |\mathbf{V}_r| \cos\theta - \rho |\boldsymbol{\omega}| Q R^2 = \rho Q R (|\mathbf{V}_r| \cos\theta - |\boldsymbol{\omega}| R) = 0,$$

which coincides with the expression found by choosing a control volume in an inertial frame.

In the non-inertial rotating frame, in equilibrium conditions the torque generated by the outflowing currents is balanced by the torque generated by the distribution of surface forces (mainly pressure forces) on the internal walls of the pipes. This distribution of surface forces is such as to divert the current from the uniform rectilinear motion, forcing it to follow also the circular motion of the pipe, with pathlines that, for an external observer in a fixed inertial frame, are spirals.

The results would be identical, making the appropriate transformations, if the coordinate system had been chosen fixed and coincident, for example, with the $XYZ$ system.

Ultimately, the choice of frame of reference and coordinate system does not change the results and must be conveniently operated in such a way as to simplify the calculations. It is also obvious that the coordinate system is not linked to the frame of reference, with the first being or not being attached to the second.

For $C_u = C_{pu} = 0$ it results $Q = 50 \, 1\,min^{-1}$, $d = 6$ mm, $R = 80$ mm,

$$\therefore \qquad |\mathbf{V}_r| = \frac{4Q}{3\pi d^2} = \frac{4 \times 50 \times 10^{-3}/60}{3 \times \pi \times \left(6.0 \times 10^{-3}\right)^2} = \mathbf{9.82\ m\,s^{-1}},$$

$$\therefore \qquad |\omega| = \frac{|\mathbf{V}_r| \cos\theta}{R} = \frac{9.82 \times \cos 30°}{80 \times 10^{-3}} = \mathbf{106.3\ s^{-1}}.$$

---

**Exercise 4.13** A circular water jet with a diameter of $d = 50$ mm flows out of the tank on a trolley in Fig. 4.26, with a relative speed of $V = 4.9 \, m\,s^{-1}$. The centroid of the outlet orifice has a depth of $z_G = (1500 + 20 \times C_u)$ mm.

- Calculate the force of the jet on the trolley, if the tank is at rest.
- Calculate the force of the jet on the trolley, if the trolley moves in the opposite direction to the jet with velocity $U = (1.2 + C_{pu}/10) \, m\,s^{-1}$ (assuming the relative velocity $V$ unchanged).
- Calculate, for the previous case, the work of the current on the trolley per unit time.

The contraction coefficient is unitary. Assume a stationary regime.

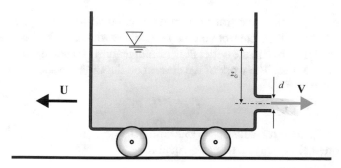

**Fig. 4.26** Tank on a trolley with outflow water jet

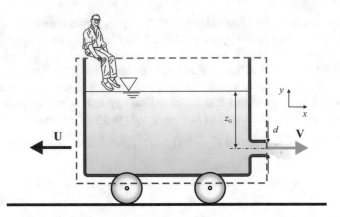

**Fig. 4.27** Control volume attached to the tank

**Solution** We choose the dashed control volume in Fig. 4.27.
The linear momentum balance projected along $x$ is

$$\cancel{G_x} + \Pi_x + \cancel{V_x} + \cancel{M_{1x}} - M_{2x} = 0.$$

Therefore, the surface forces on the control volume are equal to the outlet flow of momentum (which has only component according to the $x$-axis). For the principle of action and reaction, the force exerted by the fluid on the walls of the tank is equal to and opposite to the outlet flow of momentum,

$$F \equiv -\Pi_x = -M_{2x}.$$

The velocity of the jet is given and is always lower than the theoretical one $V = \sqrt{2gz_G}$, as it should be for the presence of dissipations. The force is equal to

$$F = -\rho V^2 \Omega \equiv -\rho V^2 \frac{\pi d^2}{4},$$

and it is pointing left. If the trolley is moving at constant speed, the force does not change. In fact, if the control volume is mobile with the tank, the inlet flow of momentum is still zero and the outlet flow of momentum is unchanged compared to the case of tank at rest. The work of the current on the trolley per unit time (power) is equal to

$$P = FU.$$

For $C_u = C_{pu} = 0$ it results $z_G = 1500$ mm, $U = 1.2$ m s$^{-1}$.
The theoretical velocity of the jet is equal to

$$V = \sqrt{2gz_G} = \sqrt{2 \times 9.806 \times 1.5} = \textbf{5.42 m s}^{-1},$$

the actual velocity is given and is equal to $V = 4.90 \text{ m s}^{-1}$.

The force has a magnitude equal to

$$\therefore \qquad F = -\rho V^2 \frac{\pi d^2}{4} = -1000 \times 4.90^2 \times \frac{\pi \times 0.05^2}{4} = \textbf{-47.1 N,}$$

and the power of the jet is equal to

$$\therefore \qquad P = FU = 47.1 \times 1.2 = \textbf{56.5 W.}$$

**Exercise 4.14**  The trolley in Fig. 4.28 moves without friction pushed by the water jet. The water jet can be controlled with a valve. Starting from the rest, the trolley, of mass $M = 3$ kg, must accelerate with constant acceleration and equal to $a = (2 + 0.1 \times C_{pu}) \text{ m s}^{-2}$. The exit angle of the jet is $\alpha = (90 + 10 \times C_u)°$ and the outflow velocity is constant and equal to $V = (10 + C_{pu}) \text{ m s}^{-1}$.

- Calculate the nozzle area function $A(t)$.
- Calculate the value of $A$ at the time $t = 3$ s.
- Calculate the power of the force acting on the trolley.

Assume that the outflow velocity is independent of the area of the jet, a unit coefficient of contraction, and neglect local inertia.

**Solution**  We choose the mobile frame of reference in Fig. 4.29 which, in the general case, is non-inertial. Neglecting the local inertia of the fluid, the force exerted by the jet is equal to the inflow momentum minus the outflow momentum:

$$F = \rho A(t) [V - U(t)]^2 (1 - \cos \alpha),$$

and it is null when $U(t) = V$.

**Fig. 4.28**  Water jet-propelled trolley

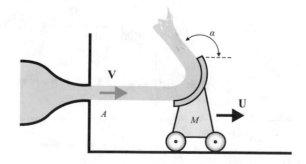

**Fig. 4.29** Mobile control volume attached to the trolley

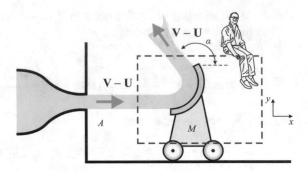

By applying Newton's second law, yields

$$\rho A(t) \left[V - U(t)\right]^2 (1 - \cos\alpha) = F \equiv Ma.$$

If the acceleration $a$ is constant, the velocity variation of the trolley is linear,

$$U(t) = U(0) + at = at.$$

By replacing the expression of $U(t)$ and inverting, the time function of the nozzle area is obtained:

$$A(t) = \frac{Ma}{\rho(V - at)^2 (1 - \cos\alpha)}, \quad 0 \le t < V/a.$$

The power of the force exerted by the jet on the trolley is equal to

$$P \equiv FU = Ma^2 t.$$

For $C_u = C_{pu} = 0$ it results $a = 2 \text{ m s}^{-2}$, $\alpha = 90°$, $V = 10 \text{ m s}^{-1}$,

$$\therefore \quad A(t) = \frac{Ma}{\rho(V - at)^2 (1 - \cos\alpha)} = \frac{6}{1000(10 - 2t)^2} \; \mathbf{m}^2 \quad (t \text{ in seconds}),$$

$$\therefore \quad A(3.0) = \frac{6}{1000 \times (10 - 2 \times 3)^2} = 3.75 \times 10^{-4} \; \mathbf{m}^2,$$

$$\therefore \quad P = Ma^2 t = 3 \times 2^2 \times t = (\mathbf{12 \times t}) \; \mathbf{W} \quad (t \text{ in seconds}).$$

**Fig. 4.30** Trolley pushed by a water jet

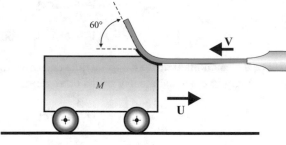

**Fig. 4.31** Mobile control volume attached to the trolley and coordinate system

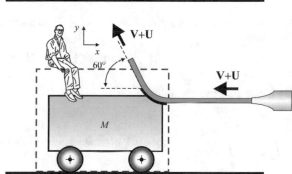

**Exercise 4.15** The trolley in Fig. 4.30, with mass $M = (10 + C_{pu})$ kg, moves without friction on a horizontal plane with an initial speed $U_0 = 12.5$ m s$^{-1}$. At a time $t = 0$ the trolley is hit by a jet of water coming out of a nozzle with a cross-sectional area $A = (900 + 10 \times C_u)$ mm$^2$, with an average speed $V = 8.3$ m s$^{-1}$. The jet is diverted upwards at an angle of 60° to the horizontal.

– Calculate the force of the jet on the trolley at the time $t = 0$.
– Calculate the stopping time of the trolley.
– Calculate the distance covered by the trolley from $t = 0$ to the stop.

Neglect inertia effects and assume a unit coefficient of contraction.

**Solution** We choose the control volume attached to the trolley and the coordinate system shown in Fig. 4.31. By applying the linear momentum balance and neglecting local inertia, the horizontal component of the force is

$$F_x = -\rho[V + U(t)]^2 A (1 - \cos 60°),$$

pointing to the left. The vertical component of the force is

$$F_y = -\rho[V + U(t)]^2 A \sin 60°,$$

pointing downwards. We have neglected the weight of the jet.
The initial force has components equal to

$$\begin{cases} F_x(t_0) = -\rho(V + U_0)^2 A \,(1 - \cos 60^\circ)\,, \\ \\ F_y(t_0) = -\rho(V + U_0)^2 A \sin 60^\circ. \end{cases}$$

For the calculation of the stopping time and of the space covered by the trolley, it is necessary to calculate the law of the time. Applying the equation of dynamics, it results

$$M\frac{dU}{dt} + \rho[V + U(t)]^2 A\,(1 - \cos 60^\circ) = 0.$$

Introducing the auxiliary variable $z = U + V$, results $dz = dU$ ($V$ is constant), and therefore

$$M\frac{dz}{dt} + \rho z^2 A\,(1 - \cos 60^\circ) = 0.$$

By separating the variables:

$$\int_{z_0}^{z} \frac{dz}{z^2} + \int_{t_0}^{t} \frac{\rho A}{M}\,(1 - \cos 60^\circ)\,dt = 0 \;\rightarrow\; \frac{1}{z} - \frac{1}{z_0} = \frac{\rho A}{M}\,(1 - \cos 60^\circ)\,(t - t_0)\,.$$

By imposing $t_0 = 0$ and replacing the original variables, yields

$$\frac{1}{V + U} - \frac{1}{V + U_0} = \frac{\rho A}{M}\,(1 - \cos 60^\circ)\,t,$$

or, in explicit form

$$U = \cfrac{1}{\cfrac{1}{V + U_0} + \cfrac{\rho A}{M}\,(1 - \cos 60^\circ)\,t} - V.$$

The stopping time of the trolley is calculated by setting $U = 0$, and it is equal to

$$t_{stop} = \cfrac{\cfrac{1}{V} - \cfrac{1}{V + U_0}}{\cfrac{\rho A}{M}\,(1 - \cos 60^\circ)}.$$

The distance covered before stopping is obtained by integrating the law of the time:

$$\frac{dx}{dt} = \frac{1}{\dfrac{1}{V + U_0} + \dfrac{\rho A}{M}(1 - \cos 60°)\, t} - V \rightarrow$$

$$\int_{x_0}^{x} dx = \int_{t_0}^{t} \left( \frac{1}{\dfrac{1}{V + U_0} + \dfrac{\rho A}{M}(1 - \cos 60°)\, t} - V \right) dt,$$

from which we obtain

$$x - x_0 = \frac{M}{\rho A\,(1 - \cos 60°)} \ln \left( \frac{\dfrac{1}{V + U_0} + \dfrac{\rho A}{M}(1 - \cos 60°)\, t}{\dfrac{1}{V + U_0} + \dfrac{\rho A}{M}(1 - \cos 60°)\, t_0} \right) - V\,(t - t_0).$$

For $t_0 = 0$ and for $t = t_{stop}$ it results

$$x_{stop} = \frac{M}{\rho A\,(1 - \cos 60°)} \left[ \ln \left( 1 + \frac{U_0}{V} \right) - \frac{U_0}{(U_0 + V)} \right].$$

For $C_u = C_{pu} = 0$ it results

$$\therefore \quad F_x(t_0) = -\rho(V + U_0)^2 A\,(1 - \cos 60°) =$$
$$- 1000 \times (8.3 + 12.5)^2 \times 900 \times 10^{-6} \times (1 - \cos 60°) = -\,\mathbf{194\ N},$$

$$\therefore \quad F_y(t_0) = -\rho(V + U_0)^2 A \sin 60° =$$
$$- 1000 \times (8.3 + 12.5)^2 \times 900 \times 10^{-6} \times \sin 60° = -\mathbf{337\ N},$$

$$\therefore \qquad\qquad t_{stop} = \mathbf{1.609\ s},$$

$$\therefore \qquad\qquad x_{stop} = \mathbf{7.07\ m}.$$

---

**Exercise 4.16** In the vertical reduction in Fig. 4.32 flows a fluid of specific gravity equal to 0.8. The flow rate is $Q = 0.6$ m$^3$ s$^{-1}$ and the gage pressure in the largest section is $p_1 = 20$ kPa.

– Calculate the force on the reduction taking into account the weight of the fluid.

  Assume $\gamma_w = 9806$ N m$^{-3}$.

**Fig. 4.32** Vertical reduction

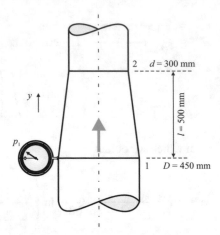

**Solution** The fluid velocity in section 1 is equal to

$$V_1 = \frac{4Q}{\pi D^2} = \frac{4 \times 0.6}{\pi \times 0.45^2} = \textbf{3.77 m s}^{-1},$$

and in section 2 it is equal to

$$V_2 = \frac{4Q}{\pi d^2} = \frac{4 \times 0.6}{\pi \times 0.3^2} = \textbf{8.49 m s}^{-1}.$$

The pressure in section 2 is calculated by applying Bernoulli's theorem (neglecting dissipations),

$$p_2 = p_1 + \gamma (z_1 - z_2) + \gamma \frac{V_1^2 - V_2^2}{2g} =$$

$$20\,000 - 0.8 \times 9806 \times 0.5 + 9806 \times 0.8 \times \frac{3.77^2 - 8.49^2}{2 \times 9.806} = \textbf{-7.1 kPa}.$$

In section 2, the pressure is therefore lower than the atmospheric pressure. Applying the linear momentum balance along the vertical, yields

$$G_y + \Pi_{0y} + \Pi_{1y} + \Pi_{2y} + \cancel{V_y} + M_{1y} - M_{2y} = 0,$$

and, by substitution,

$$-\gamma W_T - F_y + p_1 \frac{\pi D^2}{4} - p_2 \frac{\pi d^2}{4} + \rho Q V_1 - \rho Q V_2 = 0,$$

where $F_y = -\Pi_{0y}$ is the force exerted by the fluid in the control volume on the walls of the reduction. The volume of the truncated cone is equal to

$$W_T = \frac{1}{3}\left(\frac{\pi D^2}{4}3l - \frac{\pi d^2}{4}2l\right) =$$

$$\frac{\pi \times 0.45^2}{12} \times 3 \times 0.5 - \frac{\pi \times 0.3^2}{12} \times 2 \times 0.5 = \textbf{0.056 m}^3.$$

Substituting the numerical values, yields

$$\therefore \quad F_y = -0.8 \times 9806 \times 0.056 + 20\,000 \times \frac{\pi \times 0.45^2}{4}$$

$$+ 7100 \times \frac{\pi \times 0.3^2}{4} + 800 \times 0.6 \times 3.77 - 800 \times 0.6 \times 8.49 = \textbf{978 N},$$

pointing upwards.

---

**Exercise 4.17**  The trolley in Fig. 4.33 advances with an initial speed $U_0 = 10 \text{ m s}^{-1}$. The jet exerts a force in the opposite direction to the initial motion of the trolley.

– Calculate the initial force.
– Calculate the time required for stopping the trolley.

The local inertia of the fluid can be neglected.

**Solution** We choose a frame reference attached to the trolley. Applying the linear momentum balance and neglecting the local inertia, the horizontal component of the initial force of the jet, pointing to the left, is calculated as follows:

$$\therefore \quad F = -\rho\frac{\pi D^2}{4}(U_0 + V)|U_0 + V|(1 + \sin 45°) =$$

$$- 1000 \times \frac{\pi \times 0.05^2}{4} \times (10 + 20) \times |10 + 20| \times \left(1 + \frac{\sqrt{2}}{2}\right) = \textbf{-3016 N}.$$

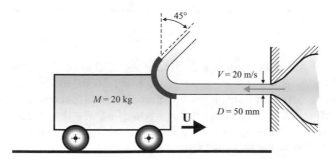

**Fig. 4.33** Trolley hit by a water jet

The vertical component is pointing downwards, it has the same magnitude of the horizontal component, and it is balanced by the reaction of the floor.

To calculate the stopping time, we need to integrate the following differential equation:

$$M\frac{dU}{dt} = -\rho\frac{\pi D^2}{4}(U+V)\,|U+V|\,(1+\sin 45°).$$

Introducing the auxiliary variable $z = U + V$, results $dz = dU$, since $V$ is constant. For $z > 0$ results

$$\frac{dz}{dt} + \underbrace{\frac{\rho\dfrac{\pi D^2}{4}}{M}(1+\sin 45°)}_{k} z^2 = 0 \;\rightarrow\; \frac{dz}{dt} + kz^2 = 0. \tag{4.5}$$

Equation (4.5) can be integrated between the initial null instant and the instant $t$, giving

$$\int_{z_0}^{z}\frac{dz}{z^2} = -\int_0^t k\,dt \;\rightarrow\; \frac{1}{z} - \frac{1}{z_0} = kt,$$

or

$$\frac{1}{U+V} - \frac{1}{U_0+V} = kt. \tag{4.6}$$

The stopping time is obtained by imposing $U = 0$ in Eq. (4.6):

$$\therefore\quad t = \frac{U_0}{kV(V+U_0)} \;\rightarrow\; t = \frac{MU_0}{\rho\dfrac{\pi D^2}{4}(1+\sin 45°)\,V\,(V+U_0)} =$$

$$\frac{20\times 10}{1000\times\dfrac{\pi\times 0.05^2}{4}\times(1+\sin 45°)\times 20\times(20+10)} = \mathbf{0.099\ s.}$$

---

**Exercise 4.18** A circular orifice with a diameter of $d = 0.10$ m is at the bottom of a tank filled with water, with water depth $h = 1.20$ m, see Fig. 4.34. The vena contracta is at a distance $d$ from the plane of the orifice, and the contraction coefficient is equal to $C_c = 0.62$. The jet impacts the free surface of a lower tank at a distance $H = 3.40$ m from the vena contracta. The lower tank is a cylinder with diameter $D = 1.20$ m, containing water with depth equal to $h/2$.

**Fig. 4.34** Water jet
outflowing from the upper
tank into the lower tank

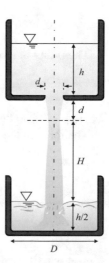

- Calculate the flow-rate through the orifice.
- Calculate the force at the bottom of the lower tank.

Assume $\gamma_w = 9806 \, \mathrm{N\,m^{-3}}$.

**Solution** Applying Bernoulli's theorem for a trajectory between the free surface
of the upper tank and the vena contracta, yields

$$V_c = \sqrt{2g\,(h+d)} = \sqrt{2 \times 9.806 \times (1.20 + 0.10)} = \mathbf{5.05 \ m\,s^{-1}}.$$

The flow rate through the orifice is equal to

$$\therefore \qquad Q = V_c C_c \frac{\pi d^2}{4} = 5.05 \times 0.62 \times \frac{\pi \times 0.10^2}{4} = \mathbf{24.6\,1\,s^{-1}}.$$

To calculate the force at the bottom of the lower tank, we apply the linear momentum balance in the vertical direction:

$$G + \cancel{\Pi_{1y}} - S_y + M_{1y} - \cancel{M_{2y}} = 0.$$

The weight is due to the fluid in the lower tank, and is equal to

$$G = \gamma_w \frac{\pi D^2}{4} \frac{h}{2} = 9806 \times \frac{\pi \times 1.20^2}{4} \times \frac{1.20}{2} = \mathbf{6654 \ N}.$$

In the hypothesis that the dissipations are negligible also for the water jet in air,
the velocity of arrival of the jet in the lower tank is still calculated using Bernoulli's
theorem, and it is equal to

$$V_F = \sqrt{2g\,(h+d+H)} = \sqrt{2 \times 9.806 \times (1.20 + 0.10 + 3.40)} = \mathbf{9.6\ m\,s^{-1}}.$$

The influx of momentum is equal to

$$M_{1y} = \rho Q V_F = 1000 \times 24.6 \times 10^{-3} \times 9.6 = \mathbf{236\ N},$$

where $\rho = 1000\ \mathrm{kg\,m^{-3}}$ is mass density of water.

The force of the fluid at the bottom of the lower tank is equal to:

$$\therefore \qquad\qquad S_y = G + M_{1y} = 6654 + 236 = \mathbf{6890\ N}.$$

---

**Exercise 4.19** In the reduction in Fig. 4.35 flows a fluid of specific gravity equal to 0.8. The gage pressure in the largest section is 88 kPa; the absolute pressure in the narrow section is 109 kPa. The volume of the reduction is $W_T = 0.6\ \mathrm{m^3}$.

- Calculate the force on the reduction taking into account the weight of the fluid.
- Draw the hydraulic grade line and the energy grade line.

Assume $\gamma_w = 9806\ \mathrm{N\,m^{-3}}$.

**Solution** We perform the calculations by considering the gage pressure first. The gage pressure in the narrow section is equal to

$$p_{2rel} = p_{2abs} - p_{atm} = 109\,000 - 101\,000 = 8\ \mathrm{kPa}.$$

We choose the coordinate system in Fig. 4.36.

The linear momentum balance projected in the $x$-direction for the hatched control volume in Fig. 4.36 is

**Fig. 4.35** Horizontal reduction

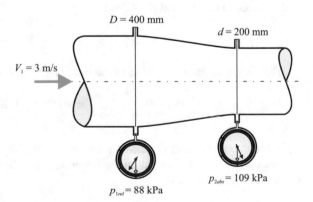

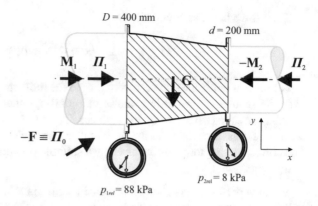

**Fig. 4.36** Control volume and coordinate system

$$\cancel{G_x} + \cancel{V_x} + \Pi_{1x} + \Pi_{2x} + \Pi_{0x} + M_{1x} - M_{2x} = 0,$$

i.e.,

$$F_x = -\Pi_{0x} = \left(p_{1rel} + \rho V_1^2\right) \frac{\pi D^2}{4} - \left(p_{2rel} + \rho V_2^2\right) \frac{\pi d^2}{4},$$

where $\rho = 0.8\rho_w = 0.8 \times 1000 = 800$ kg m$^{-3}$ is the mass density of the fluid, $\Pi_{0x}$ is the $x$-component of the action $\Pi_0$ exerted by the reduction on the control volume, equal and opposite to the unknown force **F** exerted on the reduction. The flow rate is equal to

$$Q = V_1 \frac{\pi D^2}{4} = 3 \times \frac{\pi \times 0.4^2}{4} = 0.377 \text{ m}^3 \text{ s}^{-1}.$$

The velocity in the narrow section is equal to

$$V_2 = \frac{4Q}{\pi d^2} = \frac{4 \times 0.377}{\pi \times 0.2^2} = 12.0 \text{ m s}^{-1}.$$

Hence,

$$\therefore \quad F_x = \left(p_{1rel} + \rho V_1^2\right) \frac{\pi D^2}{4} - \left(p_{2rel} + \rho V_2^2\right) \frac{\pi d^2}{4} =$$

$$\left(88\,000 + 800 \times 3^2\right) \times \frac{\pi \times 0.4^2}{4}$$

$$- \left(8000 + 800 \times 12^2\right) \times \frac{\pi \times 0.2^2}{4} = 8093 \text{ N}.$$

The linear momentum balance in the $y$-direction reduces to the weight component alone (ignoring the buoyancy due to the air):

$$\therefore \quad F_y = -\gamma W_T + \cancel{\gamma_a W_T} = -0.8 \times 9806 \times 0.6 = -4707 \text{ N},$$

where $\gamma = 0.8\gamma_w$.

Total force has magnitude

$$\therefore \qquad |\mathbf{F}| = \sqrt{F_x^2 + F_y^2} = \sqrt{8093^2 + 4707^2} = \mathbf{9362\ N}.$$

Notice that while the two vectors of flux of the momentum $\mathbf{M}_1$ and $\mathbf{M}_2$ are coaxial with the reduction (in the hypothesis of radial symmetry of the current velocity distribution), the two forces on the surfaces $\boldsymbol{\Pi}_1$ and $\boldsymbol{\Pi}_2$ are applied in the respective centres of pressure which, in the present case, are below the axis of the reduction. The corresponding torque is balanced by a suitable pressure distribution on the walls of the reduction.

We now perform the calculations considering the absolute value of the pressure. In this case it is necessary to include the force due to the atmospheric pressure acting on the external lateral surface of the reduction. The resultant of this force, neglecting the vertical gradient of air pressure, has only the component according to the $x$-axis, is pointing to the left and is equal to

$$-p_{atm}\left(\frac{\pi D^2}{4} - \frac{\pi d^2}{4}\right),$$

regardless of the shape of the reduction. The linear momentum balance projected in the $x$-direction is

$$F_x = \left(p_{1abs} + \rho V_1^2\right)\frac{\pi D^2}{4} - \left(p_{2abs} + \rho V_2^2\right)\frac{\pi d^2}{4} - p_{atm}\left(\frac{\pi D^2}{4} - \frac{\pi d^2}{4}\right) =$$
$$\left(p_{1rel} + \rho V_1^2\right)\frac{\pi D^2}{4} - \left(p_{2rel} + \rho V_2^2\right)\frac{\pi d^2}{4},$$

equivalent to the expression calculated with the gage pressure. As expected, the force on reduction is independent of the chosen pressure reference.

The head in the large cross-section, with respect to a horizontal barycentric axis, is equal to

$$H_1 = z_1 + \frac{p_1}{\gamma} + \frac{V_1^2}{2g} = \frac{88\,000}{0.8 \times 9806} + \frac{3^2}{2 \times 9.806} = \mathbf{11.68\ m}.$$

The head, with respect to the same axis, in the small cross-section is equal to

$$H_2 = z_2 + \frac{p_2}{\gamma} + \frac{V_2^2}{2g} = \frac{8000}{0.8 \times 9806} + \frac{12^2}{2 \times 9.806} = \mathbf{8.36\ m}.$$

The hydraulic head (gage) in the large cross-section is equal to

$$h_1 = z_1 + \frac{p_1}{\gamma} = \frac{88\,000}{0.8 \times 9806} = \mathbf{11.22\ m}.$$

**Fig. 4.37** Energy and
hydraulic grade lines

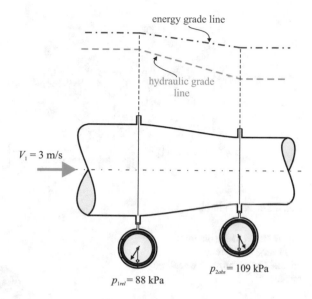

The hydraulic head (gage) in the small cross-section is equal to

$$h_2 = z_2 + \frac{p_2}{\gamma} = \frac{8000}{0.8 \times 9806} = \textbf{1.02 m}.$$

The drawing in Fig. 4.37 shows the qualitative trend of the heads in the reduction. The actual trend of the energy and hydraulic grade lines depends on the geometry of the reduction.

---

**Exercise 4.20** The Pelton turbine shown in Fig. 4.38, with a radius of $R = 1.0$ m, is driven by a cylindrical water jet with a diameter of $d = 5$ cm and an average velocity of the current $V = 60 \text{ m s}^{-1}$.

– Calculate the maximum force acting on the turbine bucket, if the exit angle of the jet is $\beta = 150°$.
– Calculate the power output at 200 rpm.
– Calculate the efficiency at 200 rpm.

**Solution** We consider a non-inertial frame of reference attached to the wheel. For the single bucket-shaped vane fitted to the wheel, the transformation of the velocities is given in the diagram shown in Fig. 4.39. $U$ is the peripheral velocity of the wheel.

The force exerted on the single bucket vane is equal to

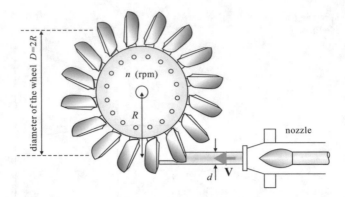

**Fig. 4.38** Pelton turbine

**Fig. 4.39** Velocity diagram
in a frame of reference
attached to the moving wheel
bucket

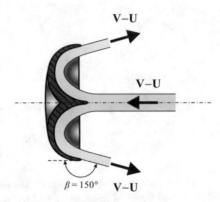

$$F_{vane} = \rho \frac{\pi d^2}{4} (V - U)^2 (1 - \cos \beta),$$

is maximum if the peripheral velocity of the wheel is zero, assuming the value

$$\therefore \quad F_{max} = \rho \frac{\pi d^2}{4} V^2 (1 - \cos \beta) =$$

$$1000 \times \frac{\pi \times 0.05^2}{4} \times 60^2 \times (1 - \cos 150°) = \mathbf{13.2 \ kN}.$$

Considering that there is always at least one vane hit by the water jet, it can be demonstrated that the average force on the system of vanes is equal to

$$F_{av} = \rho Q (V - U) (1 - \cos \beta),$$

i.e. all the flowrate $Q = V \pi d^2 / 4$, not only the reduced flowrate $(V - U) \pi d^2 / 4$ for an observer attached to the vanes, contributes to the force. The average output shaft power is equal to

$$P_{shaft} = \rho Q (V - U) (1 - \cos \beta) U.$$

At rotation rate $n = 200$ rpm, the peripheral velocity is equal to

$$U = \frac{2\pi n}{60} R = \frac{2 \times \pi \times 200}{60} \times 1.0 = \textbf{20.95 m s}^{-1},$$

and, therefore:

$$\therefore \quad P_{shaft} = \rho Q (V - U) (1 - \cos \beta) U = \rho \frac{\pi d^2}{4} V U (V - U) (1 - \cos \beta) =$$

$$1000 \times \frac{\pi \times 0.05^2}{4} \times 60 \times 20.95 \times (60 - 20.95) \times (1 - \cos 150°) = \textbf{180.0 kW}.$$

The efficiency is equal to the ratio between the shaft power and the power of the water jet (the latter is only kinetic power because the jet is in air, with an internal pressure equal to atmospheric pressure and a negligible contribution of gravity):

$$\therefore \quad \eta = \frac{P_{shaft}}{P_{jet}} = \frac{\rho Q (V - U) (1 - \cos \beta) U}{\gamma_w Q \frac{V^2}{2g}} = \frac{2 (V - U) (1 - \cos \beta) U}{V^2} =$$

$$\frac{2 \times (60 - 20.95) \times (1 - \cos 150°) \times 20.95}{60^2} = \textbf{0.85}.$$

---

**Exercise 4.21** The diffuser shown in Fig. 4.40 generates a radial water sheet with a thickness $t = 1.5$ mm. The output velocity is $V_2 = 10$ m s$^{-1}$, the radius is $R = 50$ mm and the sheet covers an angle of 180°. The supply pipe has a diameter of $D = 50$ mm and the manometer indicates a gage pressure $p_1 = 150$ kPa.

– Calculate the flow rate $Q$ that supplies the diffuser.
– Calculate the force and torque that load the flange.

Neglect the weight of the fluid. Assume a mass density of water $\rho = 1000$ kg m$^{-3}$.

**Solution** The flow rate exiting the diffuser is calculated by integration, and is equal to

$$\therefore \quad Q = \pi R t V_2 = \pi \times 50 \times 10^{-3} \times 1.5 \times 10^{-3} \times 10 = \textbf{2.36 l s}^{-1}.$$

Applying mass conservation, the velocity in the supply line is equal to

**Fig. 4.40**  Diffuser with
radial water sheet

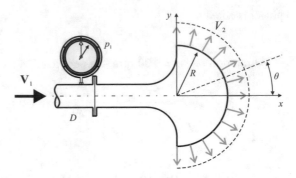

$$V_1 = \frac{4Q}{\pi D^2} = \frac{4 \times 2.36 \times 10^{-3}}{\pi \times \left(50 \times 10^{-3}\right)^2} = \mathbf{1.2\,ms^{-1}}.$$

In the calculation of the forces and torque we neglect the weight of the fluid.

In the coordinate system shown in Fig. 4.40, the equilibrium in the $y$-direction is satisfied by symmetry. The equilibrium in the $x$-direction requires that

$$\Pi_{0x} + \Pi_{1x} + \cancel{\Pi_{2x}} + M_{1x} - M_{2x} = 0,$$

or

$$-F_x + p_1 \frac{\pi D^2}{4} + \rho V_1^2 \frac{\pi D^2}{4} - \rho \int\limits_{-\pi/2}^{\pi/2} V_2^2 \cos\theta\, t\, R\, d\theta = 0 \rightarrow$$

$$F_x = p_1 \frac{\pi D^2}{4} + \rho V_1^2 \frac{\pi D^2}{4} - \rho V_2^2 t R \sin\theta\big|_{-\pi/2}^{\pi/2} =$$

$$p_1 \frac{\pi D^2}{4} + \rho V_1^2 \frac{\pi D^2}{4} - 2\rho V_2^2 t R,$$

where $F_x = -\Pi_{0x}$ is the force exerted by the fluid in the control volume on the diffuser device and, ultimately, on the flange. The torque is null by symmetry.

Substituting the numerical values, yields

$$\therefore \quad F_x = p_1 \frac{\pi D^2}{4} + \rho V_1^2 \frac{\pi D^2}{4} - 2\rho V_2^2 t R =$$

$$150 \times 10^3 \times \frac{\pi \times \left(50 \times 10^{-3}\right)^2}{4} + 1000 \times 1.2^2 \times \frac{\pi \times \left(50 \times 10^{-3}\right)^2}{4}$$

$$- 2 \times 1000 \times 10^2 \times 1.5 \times 10^{-3} \times 50 \times 10^{-3} = \mathbf{282\ N}.$$

**Fig. 4.41** Diffuser with
water sheet along a
generatrix

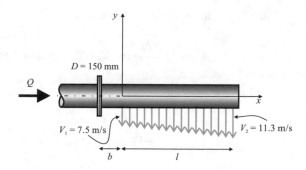

**Exercise 4.22** A pipe of diameter $D = 150$ mm ends with a diffuser connected with
a flange. The diffuser has a rectangular gap along the generatrix, $h = 15$ mm high
and $l = 1.0$ m long, through which the water flows out at a right angle to the axis,
with the velocity distribution shown in Fig. 4.41 which is due to an internal set of
vanes. The gap starts at a distance of $b = 0.2$ m from the flange. The gage pressure
in the flange section is $p_{in} = 30$ kPa.

- Calculate the flow rate $Q$ that supplies the diffuser.
- Calculate the forces and torque on the flange.

**Solution** In the coordinate system shown in Fig. 4.41, the velocity varies according
to the following equation:

$$V_y(x) = -V_1 - \frac{V_2 - V_1}{l}x.$$

The flowrate is equal to

$$\therefore \quad Q = \int_0^l |V_y(x)| h \, dx = \frac{V_1 + V_2}{2} lh = \frac{7.5 + 11.3}{2} \times 1 \times 0.015 = \mathbf{0.141 \ m^3 \, s^{-1}}.$$

The inlet velocity is equal to

$$V_{in} = \frac{4Q}{\pi D^2} = \frac{4 \times 0.141}{\pi \times 0.15^2} = \mathbf{7.98 \ m \, s^{-1}}.$$

The linear momentum balance projected in the $x$-direction gives

$$-F_x + p_{in} \frac{\pi D^2}{4} + \rho V_{in}^2 \frac{\pi D^2}{4} = 0.$$

Substituting the numerical values, yields

$$\therefore \quad F_x = \left(p_{in} + \rho V_{in}^2\right) \frac{\pi D^2}{4} = (30\,000 + 1000 \times 7.98^2) \times \frac{\pi \times 0.15^2}{4} = \mathbf{1655\ N}.$$

The linear momentum balance projected in the $y$-direction is

$$-F_y + \int_0^l \rho[V_y(x)]^2 h \, dx = 0,$$

hence

$$F_y = \int_0^l \rho\left(-V_1 - \frac{V_2 - V_1}{l}x\right)^2 h \, dx = \rho h \int_0^l \left(-V_1 - \frac{V_2 - V_1}{l}x\right)^2 dx =$$

$$\rho h \left[V_1^2 x + \left(\frac{V_2 - V_1}{l}\right)^2 \frac{x^3}{3} + V_1 \frac{V_2 - V_1}{l} x^2\right]\Bigg|_0^l =$$

$$\rho h \left[V_1^2 l + (V_2 - V_1)^2 \frac{l}{3} + V_1 (V_2 - V_1) l\right],$$

$$\therefore \quad F_y = \rho h \left(V_1^2 + V_2^2 + V_1 V_2\right) \frac{l}{3} =$$

$$1000 \times 0.015 \times \left[7.5^2 + 11.3^2 + 7.5 \times 11.3\right] \times \frac{1}{3} = \mathbf{1343\ N}.$$

The angular momentum balance projected in the $z$-direction, results in

$$-M_z + \int_0^l \rho\left[V_y(x)\right]^2 (x + b)h \, dx = 0.$$

Hence,

$$M_z = \int_0^l \rho\left(-V_1 - \frac{V_2 - V_1}{l}x\right)^2 (x + b)h \, dx =$$

$$\rho h b \int_0^l \left(-V_1 - \frac{V_2 - V_1}{l}x\right)^2 dx + \rho h \int_0^l \left(-V_1 - \frac{V_2 - V_1}{l}x\right)^2 x \, dx =$$

$$\rho h b \left[V_1^2 x + \left(\frac{V_2 - V_1}{l}\right)^2 \frac{x^3}{3} + V_1 \frac{V_2 - V_1}{l} x^2\right]\Bigg|_0^l$$

$$+ \rho h \left[ V_1^2 \frac{x^2}{2} + \left( \frac{V_2 - V_1}{l} \right)^2 \frac{x^4}{4} + 2V_1 \frac{V_2 - V_1}{l} \frac{x^3}{3} \right]\Bigg|_0^l =$$

$$\rho h b \left[ V_1^2 + V_2^2 + V_1 V_2 \right] \frac{l}{3} + \rho h \left[ \frac{V_1^2}{12} + \frac{V_2^2}{4} + \frac{1}{6} V_1 V_2 \right] l^2.$$

Substituting the numerical values, yields

$$\therefore \quad M_z = \rho h b \left[ V_1^2 + V_2^2 + V_1 V_2 \right] \frac{l}{3} + \rho h \left[ \frac{V_1^2}{12} + \frac{V_2^2}{4} + \frac{1}{6} V_1 V_2 \right] l^2 =$$

$$1000 \times 0.015 \times 0.2 \times \left[ 7.5^2 + 11.3^2 + 7.5 \times 11.3 \right] \times \frac{1}{3}$$

$$+ 1000 \times 0.015 \times \left[ \frac{7.5^2}{12} + \frac{11.3^2}{4} + \frac{1}{6} \times 7.5 \times 11.3 \right] \times 1^2 = \mathbf{1030 \ N \ m}.$$

---

**Exercise 4.23**   The air-cushion vehicle (ACV, hovercraft) in Fig. 4.42 is supported by pressurized air from a fan. The air fills a rectangular chamber $15 \times 20$ m$^2$ and escapes through a continuous slit of height $\delta = 7.5$ cm. The volume of the chamber is sufficiently large to neglect the average air velocity. The vehicle weight is $P = 45$ kN.

- Calculate the flow rate $Q$ required to support the vehicle.
- Calculate the fan power, if its efficiency is $\eta = 0.5$.
- Calculate the slit height, if the flow rate is reduced to 60% of the previously calculated value.

Neglect incoming momentum flux. The air mass density is equal to $\rho_{air} = 1.3$ kg m$^{-3}$.

**Solution**   The vehicle is supported by the air pressure in the chamber, which is in excess of the atmospheric pressure. The calculation is immediate because the outgoing momentum flux has an exclusively horizontal component and, by neglecting

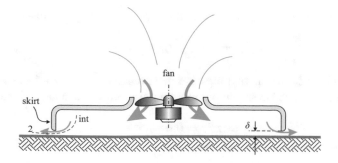

**Fig. 4.42**   Schematic of an air-cushion vehicle (ACV, hovercraft)

the incoming momentum flux due to the fan (because of the low speed and the low air mass density), it results

$$p_{int}ab = |\mathbf{P}|,$$

where $p_{int}$ is the relative pressure of the air inside the unit, and $a, b$ are the lengths of the sides of the skirt. Hence,

$$p_{int} = \frac{|\mathbf{P}|}{ab} = \frac{45\,000}{15 \times 20} = \mathbf{150\,Pa}.$$

Applying Bernoulli's theorem between the section inside the skirt and section 2 along the trajectory in Fig. 4.42, it results

$$p_{int} + \rho_{air}\frac{V_{int}^2}{2} = p_2 + \rho_{air}\frac{V_2^2}{2},$$

and therefore, since $V_{int} \approx 0$ and $p_2 = 0$, it results

$$V_2 = \sqrt{\frac{2p_{int}}{\rho_{air}}} = \sqrt{\frac{2 \times 150}{1.3}} = \mathbf{15.2\,m\,s^{-1}}.$$

Consequently, the volumetric flow rate must be equal to

$$\therefore \quad Q = V_2\delta(2a + 2b) = 15.2 \times 0.075 \times (2 \times 15 + 2 \times 20) = \mathbf{79.8\,m^3\,s^{-1}}.$$

The fan power is equal to:

$$\therefore \qquad P = \frac{\gamma_{air}QH}{\eta} \approx \frac{Qp_{int}}{\eta} = \frac{79.8 \times 150}{0.5} = \mathbf{24\,kW},$$

where $H$ is the head that, for the low speeds and for the low values of the air mass density, reduces to the pressure head component only:

$$H \approx \frac{p_{int}}{\gamma_{air}}.$$

If the flow rate is reduced to 60% of the value previously calculated, it results

$$V_2'\delta'\,(2a + 2b) = 0.6V_2\delta\,(2a + 2b).$$

Moreover, since the internal pressure must remain unchanged, from Bernoulli's theorem also the speed in the output section must remain unchanged, hence

$$\therefore \qquad\qquad \delta' = 0.6\delta = 0.6 \times 7.5 = \mathbf{4.5\,cm}.$$

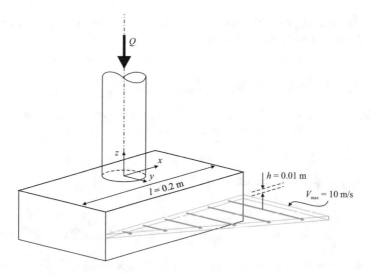

**Fig. 4.43**  Diffuser with horizontal water sheet

**Exercise 4.24**  In the device in Fig. 4.43, the water enters vertically and flows out laterally through a slot of length $l = 0.2$ m, with a linear velocity distribution between 0 and 10 m s$^{-1}$ and with constant sheet thickness $h = 0.01$ m. The diameter of the supply pipe is $D = 25$ mm. Calculate:

– the three components of the force exerted by the fluid on the device.
– Their centres of pressure.

Assume a uniform and permanent velocity distribution in the supply pipe. Neglect the effect of gravity. Assume a pressure at the inlet section equal to the atmospheric pressure. Assume water density $\rho = 1000$ kg m$^{-3}$.

**Solution**  The linear momentum balance in vector form reads

$$\mathbf{G} + \mathbf{\Pi} + \mathbf{I} + \mathbf{M_1} - \mathbf{M_2} = \mathbf{0}. \tag{4.7}$$

We neglect the weight. Local inertia is null in stationary regime. Projecting Eq. (4.7) in the $x$-direction results in
$$\Pi_x = 0,$$

hence no force is exerted along $x$.

Projecting Eq. (4.7) in the $y$-direction results in

$$F_y \equiv -\Pi_y = -M_{2y},$$

because the incoming momentum flux has null $y$-component. The velocity profile in
the slot varies according to the relation

$$V_y = V_{max} \frac{(x + l/2)}{l},$$

where $V_{max} = 10 \text{ m s}^{-1}$ is the maximum value of the output velocity, and $l = 0.2$ m
is the width of the slot. The outgoing momentum flux is equal to

$$M_{2y} = \int_A \rho V_y |\mathbf{V} \cdot \mathbf{n}| \, dS = \int_{-l/2}^{l/2} \rho h V_{max}^2 \left(\frac{x + l/2}{l}\right)^2 dx =$$

$$\frac{\rho h V_{max}^2}{3l^2} (x + l/2)^3 \Big|_{-l/2}^{l/2} = \frac{\rho h V_{max}^2 l}{3}.$$

Therefore, the force component in the $y$-direction is equal to

$$\therefore \qquad F_y = -\frac{\rho h V_{max}^2 l}{3} = -\frac{1000 \times 0.01 \times 10^2 \times 0.2}{3} = -66.7 \text{ N}.$$

Its centre of pressure is calculated by writing the balance equation of the angular
momentum about the $z$-axis:

$$\cancel{G_{mz}} + \Pi_{mz} + \cancel{I_{mz}} + \cancel{M_{1mz}} - M_{2mz} = 0.$$

The only terms contributing to the balance shall satisfy the following equation

$$\Pi_y x_c \equiv \Pi_{mz} = M_{2mz}, \qquad\qquad (4.8)$$

where $x_c$ is the abscissa of the centre of pressure of the $\Pi_y \equiv -F_y$ force component.
The $z$-component of the outgoing flux of angular momentum is equal to

$$M_{2mz} = \int_A \rho x V_y |\mathbf{V} \cdot \mathbf{n}| \, dS = \int_{-l/2}^{l/2} \rho h x V_{max}^2 \left(\frac{x + l/2}{l}\right)^2 dx = \frac{\rho h V_{max}^2 l^2}{12}.$$

Equation (4.8) reduces to

$$-F_y x_c = \frac{\rho h V_{max}^2 l^2}{12} \rightarrow$$

$$\therefore \qquad x_c = -\frac{\rho h V_{max}^2 l^2}{12 F_y} = \frac{l}{4} = \frac{0.2}{4} = \textbf{0.05 m}.$$

Projecting Eq. (4.7) in the $z$-direction, it results

$$F_z \equiv -\Pi_z = M_{1z},$$

because the outgoing momentum flux has null $z$-component. The incoming momentum flux is

$$M_{1z} = \int_A \rho V_z |\mathbf{V} \cdot \mathbf{n}| \, dS = -\rho \frac{4Q^2}{\pi D^2}.$$

The flow rate $Q$ is computed by integrating the velocity profile on the slot,

$$Q = \int_{-l/2}^{l/2} h V_{max} \left( \frac{x + l/2}{l} \right) dx = \frac{h V_{max} l}{2} = \frac{0.01 \times 10 \times 0.2}{2} = \textbf{0.01 m}^3 \, \textbf{s}^{-1}.$$

Substituting the numerical values, yields

$$\therefore \qquad F_z = -\rho \frac{4Q^2}{\pi D^2} = -1000 \times \frac{4 \times 0.01^2}{\pi \times 0.025^2} = \textbf{−203.7 N}.$$

$F_z$ is coaxial to the supply pipe.

---

**Exercise 4.25**  A boxcar moves at a speed of $V = 60 \text{ km h}^{-1}$ and is equipped with a vane as shown in Fig. 4.44, which is partially immersed in a water channel between the rails. The width of the vane between the rails is $l = 1.0$ m.

– Calculate the resistance to motion.

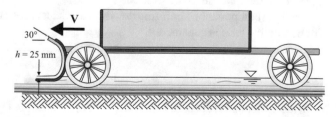

**Fig. 4.44**  Boxcar with a vane immersed in a channel between the rails

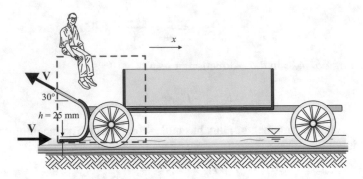

**Fig. 4.45** Mobile control volume attached to the boxcar

**Solution** We choose a control volume attached to the boxcar, as shown in Fig. 4.45. The water inlet speed is equal to $V = 60$ km h$^{-1} = 16.7$ m s$^{-1}$. Resistance to motion is calculated by writing the linear momentum balance for the control volume chosen in the mobile reference (which is inertial):

$$\mathbf{G} + \boldsymbol{\Pi} + \mathbf{I} + \mathbf{M}_1 - \mathbf{M}_2 = \mathbf{0}. \tag{4.9}$$

Projecting Eq. (4.9) in the $x$-direction yields

$$\cancel{G_x} + \Pi_x + \cancel{I_x} + M_{1x} - M_{2x} = 0 \rightarrow F_x \equiv -\Pi_x = M_{1x} - M_{2x},$$

where $F_x$ is the horizontal force exerted by the control volume to the boxcar.

Neglecting the dissipations and the variation of level head, applying Bernoulli's theorem results that the velocity of the outgoing water current is still equal to $V$. Hence,

$$\therefore \quad F_x = \rho V^2 lh + \rho V^2 lh \cos 30° =$$
$$1000 \times (16.7)^2 \times 1 \times 25 \times 10^{-3} \times (1 + \cos 30°) = \mathbf{13.0 \ kN},$$

pointing in the direction opposite to the motion.

There is also a vertical component $F_y$ pointing downwards, but it is not a resistance force, unless friction on the rails is considered.

---

**Exercise 4.26** In the horizontal reducer in Fig. 4.46 the inlet diameter is $D_1 = 8$ cm, the outlet diameter is $D_2 = 5$ cm. The outlet pressure $p_2$ coincides with the atmospheric pressure. The velocity of the inflowing water is $V_1 = 5$ m s$^{-1}$, and the reading of the mercury differential pressure gauge is $h = 58$ cm.

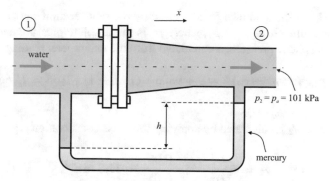

**Fig. 4.46** Reducer with a mercury differential pressure manometer

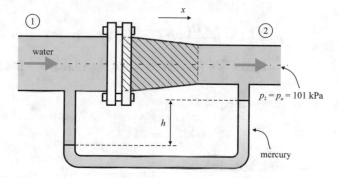

**Fig. 4.47** Control volume for calculating the force on the flanges

– Calculate the horizontal force on the flanges assuming that pipe after section 2 cannot apply forces on the reducer.

Assume $\gamma_w = 9806\,\text{N}\,\text{m}^{-3}$, $\rho = 1000\,\text{kg}\,\text{m}^{-3}$ and $\gamma_{Hg}/\gamma_w = 13.6$.

**Solution** The horizontal force is computed applying the linear momentum balance equation to the dashed control volume in Fig. 4.47.

The linear momentum balance in the $x$-direction is

$$G_x + \Pi_{0x} + \Pi_{1x} + \Pi_{2x} + I_x + M_{1x} - M_{2x} = 0,$$

where

$$G_x = 0, \quad \Pi_{1x} = p_{1g}\frac{\pi D_1^2}{4}, \quad \Pi_{2x} = -p_{2g}\frac{\pi D_2^2}{4}, \quad I_x = 0,$$

$$M_{1x} = \rho\frac{\pi D_1^2}{4}V_1^2, \quad M_{2x} = \rho\frac{\pi D_2^2}{4}V_2^2,$$

and where $\Pi_{0x}$ is the unknown force exerted by the reducer on the control volume. The gage pressure $p_{1g} \equiv p_1 - p_a$, where $p_1$ is the absolute pressure and $p_a$ is the absolute atmospheric pressure, is calculated from the manometer reading as follows:

$$p_{1g} = p_{2g} + \left(\gamma_{Hg} - \gamma_w\right)h = 0 + 9806 \times (13.6 - 1) \times 0.58 = \mathbf{71\ 660\ Pa},$$

where $p_{2g} \equiv p_2 - p_a = 0$.

The velocity $V_2$ is calculated by applying the mass conservation equation,

$$V_2 = V_1 \frac{D_1^2}{D_2^2} = 5 \times \frac{0.08^2}{0.05^2} = \mathbf{12.8\ m\,s^{-1}}.$$

Hence

$$\Pi_{0x} = p_{2g}\frac{\pi D_2^2}{4} - p_{1g}\frac{\pi D_1^2}{4} - \rho\frac{\pi D_1^2}{4}V_1^2 + \rho\frac{\pi D_2^2}{4}V_2^2 = 0 - 71\ 660 \times \frac{\pi \times 0.08^2}{4}$$
$$- 1000 \times \frac{\pi}{4} \times \left(0.08^2 \times 5^2 - 0.05^2 \times 12.8^2\right) = \mathbf{-164.2\ N}.$$

The force on the flanges is equal to

$$\therefore \qquad\qquad\qquad F_x = -\Pi_{0x} = \mathbf{164.2\ N},$$

and induces tensile stress in the bolts.

---

**Exercise 4.27** The nozzle in Fig. 4.48 diffuses a sheet of water of thickness $t = 0.03$ m for the whole semicircle, with a radial velocity $V = 15\ \mathrm{m\,s^{-1}}$ at a distance $R = 0.3$ m from the axis.

– Calculate the flow rate $Q$ and the force that must be applied in the $y$-direction to keep the nozzle in place.

Assume $\rho = 1000\ \mathrm{kg\,m^{-3}}$.

**Solution** We choose the control volume shown in Fig. 4.49. The mass conservation equation allows the calculation of the flow rate,

$$Q = \pi R t V = \pi \times 0.3 \times 0.03 \times 15 = \mathbf{0.424\ m^3\,s^{-1}}.$$

The linear momentum balance equation in the $y$-direction reads

$$\cancel{G_y} + \cancel{\Pi_{1y}} + \cancel{\Pi_{2y}} + \Pi_{0y} + \cancel{V_y} + \cancel{M_{1y}} - M_{2y} = 0.$$

**Fig. 4.48** Radial diffuser
nozzle

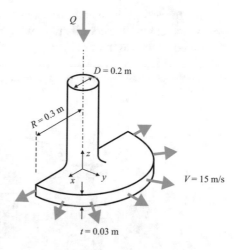

**Fig. 4.49** Control volume

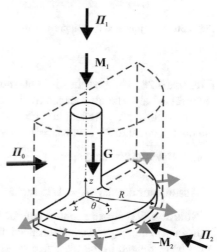

The unknown force $\Pi_{0y}$ is the only term balancing the outgoing momentum flux
($y$-component). In the polar coordinate system $\theta - R$ shown in Fig. 4.49, results

$$\therefore \quad \Pi_{0y} \equiv M_{2y} = \int_{A_e} \rho V_y \mathbf{V} \cdot \mathbf{n}\, dS =$$

$$\int_{-\frac{\pi}{2}}^{\frac{\pi}{2}} \rho |\mathbf{V}| \cos\theta\, |\mathbf{V}|\, Rt\, d\theta = \rho |\mathbf{V}|^2 Rt\, \sin\theta\big|_{-\frac{\pi}{2}}^{\frac{\pi}{2}} =$$

$$1000 \times 15^2 \times 0.3 \times 0.03 \times 2 = \mathbf{4.05\ kN}.$$

This force is positive and, for a symmetrical distribution of the elementary out-going momentum flux contributions with respect to a plane orthogonal to the axis $z$, it must be contained in this plane.

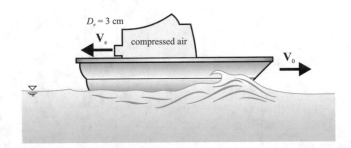

**Fig. 4.50**  Boat powered by compressed air

**Exercise 4.28** The boat in Fig. 4.50 moves at a constant speed driven by a jet of compressed air. The air escapes from a circular nozzle with a diameter $D_e = 3$ cm at atmospheric pressure and at a critical speed $V_e = 343$ m s$^{-1}$. The drag force on the boat is equal to $kV_0^2$ with $k = 19$ N s$^2$ m$^{-2}$.

– Calculate the constant speed $V_0$.

Assume air density $\rho_a = 1.25$ kg m$^{-3}$.

**Solution** We choose a control volume in the mobile reference frame, see Fig. 4.51. If the boat is moving at a constant speed, the reference frame is inertial. The velocities must be evaluated in the reference system of the control volume, i.e. in the mobile inertial reference to which we decide to attach the system of coordinates $X$.

The linear momentum balance projected on the $X$-axis is

$$\cancel{G_X} + \cancel{V_X} + \Pi_{0X} + \cancel{\Pi_{1X}} + \cancel{\Pi_{2X}} + \cancel{M_{1X}} - M_{2X} = 0,$$

where

$$\Pi_{0X} = -kV_0^2, \quad M_{2X} \equiv \int_{S_o} \rho V_X \mathbf{V} \cdot \mathbf{n} \, dS = -\rho_a V_e^2 \frac{\pi D_e^2}{4}.$$

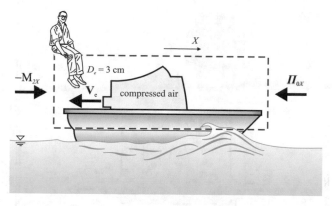

**Fig. 4.51** Mobile control volume attached to the boat

In equilibrium conditions, it results

$$\therefore \qquad V_0 = V_e D_e \sqrt{\frac{\rho_a \pi}{4k}} = 343 \times 0.03 \times \sqrt{\frac{1.25 \times \pi}{4 \times 19}} = 2.34 \text{ m s}^{-1}.$$

**Exercise 4.29** A cylindrical tank with a cross-section area $A = 0.09$ m$^2$ and height $h$, is on the floor of an elevator, see Fig. 4.52. The elevator is initially at rest. The empty tank has a mass $M = 2.5$ kg, is placed on a scale and is filled with water from above through a hole with cross-section area $A_1 = 0.009$ m$^2$. The water comes out laterally, near the bottom of the tank, trough two perfectly symmetrical holes of cross-section area $A_2 = A_3 = 2.22 \times 10^{-3}$ m$^2$. The water depth in the tank, in steady state, is $h_1 = 0.57$ m. Calculate:

– the reading on the scale if $V_1 = 1.5$ m s$^{-1}$.
– The minimum value of the coefficient of friction between the bottom of the tank and the scale plate required to prevent sliding.

At a certain moment the elevator moves upwards with an acceleration $a = 1$ m s$^{-2}$. Calculate:

– the new water level in the tank.
– The new minimum value of the coefficient of friction between the bottom of the tank and the scale plate required to prevent sliding.

At a later time, the lift moves with uniform motion ($a = 0$) and the flow through section 2 is interrupted. Calculate:

– the minimum value of $h$ necessary for the water not to reach the top of the container.

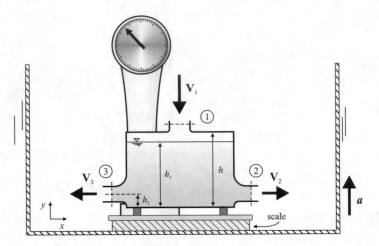

**Fig. 4.52**  Elevator with tank on a scale

– The minimum value of the coefficient of friction between the bottom of the tank
  and the scale plate required to prevent sliding.

Assume $\rho = 1000 \, \text{kg m}^{-3}$, a unit contraction coefficient for all sections and ignore
dissipation. Assume that the velocity of the fluid at free surface in the tank is negligible.

**Solution** After selecting the control volume in Fig. 4.53, the linear momentum
balance projected in the $y$-direction reads

$$G_y + \Pi_y + I_y + M_{1y} - M_{2y} = 0,$$

where $G_y = -\rho g A h_1$, $I_y = 0$, $M_{1y} = -\rho V_1^2 A_1$, $M_{2y} = 0$, and $\Pi_y$ is the unknown.
Hence, the force acting on the control volume is equal to

$$\Pi_y = \rho g A h_1 + \rho V_1^2 A_1.$$

The reading on the scale, including the weight of the container, is equal to

$$\therefore \quad \rho g A h_1 + \rho V_1^2 A_1 + Mg =$$
$$9806 \times 0.09 \times 0.57 + 1000 \times (1.5)^2 \times 0.009 + 2.5 \times 9.806 = \textbf{549 N}.$$

For the mass conservation and by symmetry, it also results

$$V_2 = V_3 = \frac{V_1 A_1}{2 A_2} = \frac{1.5 \times 0.009}{2 \times 2.22 \times 10^{-3}} = \textbf{3.04 m s}^{-1},$$

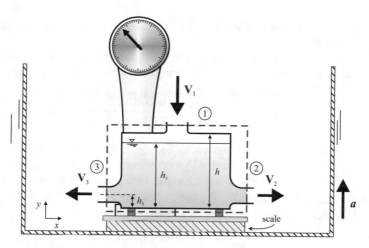

**Fig. 4.53** Control volume

when section 2 is still open.

The height of the centroid of section 3 (coinciding with the height of the centroid of section 2), measured from the bottom of the container, is calculated by applying Bernoulli's theorem between the free surface of the water in the tank (where the fluid velocity is null by hypothesis) and the outflow section, when the two orifices with cross-section area $A_2$ and $A_3$ are open and in the hypothesis that the vertical size of the two orifices is negligible with respect to the water depth. Hence,

$$V_2 = V_3 = \sqrt{2g\,(h_1 - h_3)},$$

from which yields

$$h_3 = h_1 - \frac{V_2^2}{2g} = 0.57 - \frac{(3.04)^2}{2 \times 9.806} = \mathbf{0.10\ m}.$$

Due to the symmetry, the linear momentum balance in the $x$-direction is satisfied even *without friction*.

In the presence of an upward acceleration, selecting a non-inertial control volume (identical to the previous one), the outflow velocity through sections 2 and 3 becomes

$$V_2 = V_3 = \sqrt{2\,(g + a)\,(h_1' - h_3)},$$

and the new level of the free surface (measured with respect to the bottom of the tank) becomes

$$\therefore \qquad h_1' = h_3 + \frac{V_2^2}{2\,(g + a)} = 0.10 + \frac{(3.04)^2}{2 \times (9.806 + 1)} = \mathbf{0.53\ m}.$$

The minimum coefficient of *friction is null* again, since linear momentum in the $x$-direction is self balanced.

If section 2 is occluded, in steady state regime ($a = 0$) the entire inlet flow rate must be evacuated through section 3. The conservation of the mass requires that

$$V_1 A_1 = V_3 A_3 \rightarrow V_3 = \frac{A_1 V_1}{A_3} = \frac{0.009 \times 1.5}{2.22 \times 10^{-3}} = \mathbf{6.08 \ m \, s^{-1}}.$$

The head measured with reference to the centroid of the outflow section 3, necessary to have an outflow velocity $V_3 = 6.08 \ \mathrm{m\,s^{-1}}$, is calculated by applying Bernoulli's theorem to a path starting from the free surface in the tank (where the velocity is negligible) and crossing the vena contracta of the outflow jet (null dissipations and unitary contraction coefficient by hypothesis):

$$V_3 = \sqrt{2g\,(h - h_3)} \rightarrow$$

$$\therefore \qquad h = h_3 + \frac{V_3^2}{2g} = 0.10 + \frac{(6.08)^2}{2 \times 9.806} = \mathbf{1.98 \ m}.$$

The linear momentum balance along the $x$-direction is

$$\cancel{G_x} + \Pi_x + \cancel{V_x} + \cancel{M_{1x}} - M_{2x} = 0,$$

where $M_{2x} = -\rho V_3^2 A_3$ and $\Pi_x$ is the unknown. Hence

$$\Pi_x = -\rho V_3^2 A_3 = -1000 \times (6.08)^2 \times 2.22 \times 10^{-3} = \mathbf{-82 \ N}.$$

The minimum value of the coefficient of friction required to prevent the container from sliding in the positive direction of the $x$-axis is obtained by imposing that

$$|\Pi_x| \leq f \left| \rho g h A + Mg + \rho V_1^2 A_1 \right| \rightarrow$$

$$\therefore \quad f \geq \frac{|\Pi_x|}{\left| \rho g h A + Mg + \rho V_1^2 A_1 \right|} =$$

$$\frac{|-82|}{\left| 1000 \times 9.806 \times 1.98 \times 0.09 + 2.5 \times 9.806 + 1000 \times (1.5)^2 \times 0.009 \right|} = \mathbf{0.046}.$$

**Exercise 4.30** The water jet in Fig. 4.54, of diameter $D = 20$ mm, impinges on the circular cap with speed $V = 30 \ \mathrm{m\,s^{-1}}$. The circular cap has an axial hole and moves

**Fig. 4.54** Water jet
impinging on a circular cap
with an axial hole

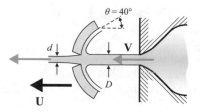

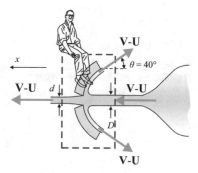

**Fig. 4.55** Control volume
attached to the circular cap

to the left with velocity $U = 10 \text{ m s}^{-1}$. The jet is partly diverted, and partly continues
with a diameter $d = 10$ mm.

– Calculate the force required to keep the cap in uniform motion.

Assume $\rho = 1000 \text{ kg m}^{-3}$.

**Solution** We choose a control volume in the mobile frame of reference attached to
the cap, see Fig. 4.55. If the cap moves at a constant velocity, the reference is inertial.
The velocities must be evaluated in the reference system of the control volume, that is
in the mobile inertial reference to which we decide to attach the system of coordinates
$x$.

The linear momentum balance projected along the $x$-axis is

$$\cancel{G_x} + \cancel{V_x} + \Pi_{0x} + \cancel{\Pi_{1x}} + \cancel{\Pi_{2x}} + M_{1x} - M_{2x} = 0,$$

where

$$\Pi_{0x} = -F_x,$$

being $F_x$ the action exerted by the fluid in the control volume on the circular cap.

If we neglect the dissipations, applying Bernoulli's theorem we can show that
the velocity of the current deflected by the cap and passing axially is equal to the
velocity of the incident current. In the relative mobile reference, this velocity is equal
to $V - U$.

The incoming momentum flux is equal to

$$M_{1x} = \rho(V - U)^2 \frac{\pi D^2}{4},$$

and the outgoing momentum flux is equal to

$$M_{2x} = \rho(V - U)^2 \frac{\pi d^2}{4} - \rho(V - U) Q_{div} \cos \theta,$$

where the first contribution is the momentum flux at the exit of the axial circular hole, the second contribution is the momentum flux of the diverted current. $Q_{div}$ represents the flow rate diverted for an observer attached to the cap, equal to the difference between the input flow rate and the flow rate through the circular axial hole:

$$Q_{div} = (V - U) \frac{\pi \left(D^2 - d^2\right)}{4}.$$

Hence,

$$M_{2x} = \rho(V - U)^2 \frac{\pi d^2}{4} - \rho(V - U)^2 \frac{\pi \left(D^2 - d^2\right)}{4} \cos \theta.$$

The horizontal component of force exerted by the jet on the circular cap is equal to:

$$\therefore \quad F_x = M_{1x} - M_{2x} \rightarrow$$

$$F_x = \rho(V - U)^2 \frac{\pi \left(D^2 - d^2\right)}{4} (1 + \cos \theta) =$$

$$1000 \times (30 - 10)^2 \times \frac{\pi \times \left(0.02^2 - 0.01^2\right)}{4} \times (1 + \cos 40°) = \mathbf{166 \ N},$$

and it is pointing to the left. To guarantee the uniform motion of the cap, it is necessary to apply a force equal to $-F_x$.

---

**Exercise 4.31** The jet in Fig. 4.56 impinges on the body of mass $M$. The dynamic friction coefficient between the body and the horizontal sliding plane is equal to $\mu_k = 0.30$. Calculate:

– the acceleration of the body, when its velocity is equal to $U = 10 \ \mathrm{m \, s^{-1}}$.
– The asymptotic velocity of the body.

Assume a mass density of the water $\rho = 999 \ \mathrm{kg \, m^{-3}}$.

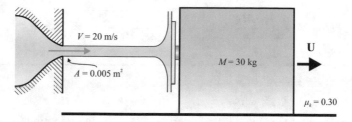

**Fig. 4.56**  Water jet impinging on a sliding body

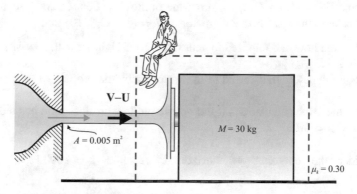

**Fig. 4.57**  Mobile control volume attached to the body

**Solution** We choose the control volume in Fig. 4.57, attached to the moving body. The force exerted by the jet is equal to the incoming momentum flux, neglecting the local inertia:

$$F = \rho A \, (V - U) \, |V - U|.$$

Considering all the forces acting on the body, it results

$$M \frac{dU}{dt} = \rho A \, (V - U) \, |V - U| - \mu_k M g,$$

Body acceleration for $U = 10 \text{ m s}^{-1}$ is equal to

$$\therefore \quad \frac{dU}{dt} = \frac{\rho A}{M} \, (V - U) \, |V - U| - \mu_k g =$$

$$\frac{999 \times 0.005}{30} \times (20 - 10) \times |20 - 10| - 0.3 \times 9.806 = \textbf{13.7 m s}^{-2}.$$

The asymptotic velocity of the body is reached when the acceleration is zero, that is

$$\therefore \qquad U = V - \sqrt{\frac{\mu_k M g}{\rho A}} = 20 - \sqrt{\frac{0.3 \times 30 \times 9.806}{999 \times 0.005}} = \mathbf{15.8 \, m \, s^{-1}}.$$

**Exercise 4.32**  A jet-ski moves in seawater, see Fig. 4.58. In navigation trip the water intake is inclined at an angle $\alpha = (30 + C_{pu})°$ to the horizontal, and has a cross-section area $\Omega_1 = (1.5 + C_{pu}/10) \, dm^2$; the outflow section, of cross-section area $\Omega_2$, is circular cylindrical with a diameter of $d = (8 + C_{pu}/10) \, cm$.

- Calculate the flow rate required to generate a 1300 N thrust with a jet-ski velocity of $V_1 = 7.0 \, m \, s^{-1}$.
- Calculate the power transferred to the water jet by the propeller to generate this thrust.
- Perform the same calculations if the nozzle outflow cross-section has a diameter of $2d$ and $\alpha = 90°$.

Consider the outgoing jet horizontal. Assume $\rho = 1000 \, kg \, m^{-3}$, $\gamma_w = 9806 \, N \, m^{-3}$.

**Solution**  We choose the control volume in Fig. 4.59, attached to the jet-ski. The linear momentum balance in the horizontal direction reads

$$\cancel{G_x} + \Pi_{0x} + \cancel{\Pi_{1x}} + \cancel{\Pi_{2x}} + \cancel{V_x} + M_{1x} - M_{2x} = 0.$$

The force exerted on the control volume is due to the distribution of stresses at the contact surface between the jet-ski and the fluid, equal and opposite to the action exerted on the jet-ski:

$$F_x \equiv -\Pi_{0x} = M_{1x} - M_{2x}.$$

**Fig. 4.58**  Jet-ski

**Fig. 4.59** Control volume

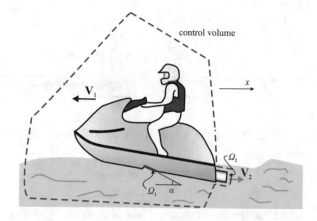

If the flow rate is $Q$, then

$$F_x = \rho Q \left( V_1 - V_2 \right),$$

where $V_2$ is the velocity of the jet relative to the jet-ski.

Since $Q = V_2 \Omega_2$, yields

$$F_x = \rho Q V_1 - \rho \frac{Q^2}{\Omega_2},$$

pointing left. Solving with respect to the flow rate, yields

$$Q = \frac{V_1 \Omega_2 + \sqrt{(V_1 \Omega_2)^2 - 4 F_x \dfrac{\Omega_2}{\rho}}}{2}.$$

The water enters the inlet with a relative speed equal to $V_1 \sin \alpha$. The power yielded by the pump to the jet is equal to the difference between the kinetic power of the output current and the kinetic power of the input current:

$$P_j = \gamma_w Q \left( \frac{V_2^2}{2g} - \frac{V_1^2 \sin^2 \alpha}{2g} \right).$$

The power transferred to the jet-ski is equal to:

$$P_{js} = F_x V_1 = \left( \rho \frac{Q^2}{\Omega_2} - \rho Q V_1 \right) V_1.$$

The efficiency is equal to

$$\eta = \frac{P_{js}}{P_j} = \frac{\left(\rho\frac{Q^2}{\Omega_2} - \rho Q V_1\right) V_1}{\gamma_w Q\left(\frac{V_2^2}{2g} - \frac{V_1^2\sin^2\alpha}{2g}\right)} = \frac{2\left(V_2 - V_1\right) V_1}{\left(V_2 - V_1\sin\alpha\right)\left(V_2 + V_1\sin\alpha\right)}.$$

If the axis of the intake is orthogonal to the direction of motion ($\alpha = 0$), the efficiency is

$$\eta = \frac{2\left(V_2 - V_1\right) V_1}{V_2^2},$$

which assumes a maximum value of 50% for $V_2 = 2V_1$.

If the axis of the intake is parallel to the direction of motion ($\alpha = 90°$), the efficiency is

$$\eta = \frac{2V_1}{V_2 + V_1},$$

which assumes a maximum theoretical value of 100% for $V_2 = V_1$. However, this value is meaningless since it would correspond to a null thrust.

If $V_2 = 2V_1$, for $\alpha = 90°$ the efficiency is 66%, higher than the maximum value calculated for $\alpha = 0$.

Doubling the diameter of the outflow nozzle, the area of its cross-section quadruples, and the flow rate becomes

$$Q' = \frac{V_1\Omega_2' + \sqrt{\left(V_1\Omega_2'\right)^2 - 4F_x\frac{\Omega_2'}{\rho}}}{2} =$$

$$\frac{4V_1\Omega_2 + \sqrt{(4V_1\Omega_2)^2 - 16F_x\frac{\Omega_2}{\rho}}}{2} > Q,$$

since $\Omega_2' = 4\Omega_2$.

The efficiency is equal to

$$\eta' = \frac{P_{js}}{P'_j} = \frac{\left(\rho\frac{Q'^2}{\Omega_2'} - \rho Q' V_1\right) V_1}{\gamma_w Q'\left(\frac{V_2'^2}{2g} - \frac{V_1^2}{2g}\right)} = \frac{2V_1}{\left(V_2' + V_1\right)}.$$

The efficiency increases for reduction of the speed $V_2'$, obtained by increasing the cross-section area of the outflow nozzle. At most, if the pump impeller is replaced by a propeller in water (traditional solution), a maximum efficiency is achieved. The ducted jet solution is frequently used for navigation in shallow water, to protect the propeller from impact against possible obstacles, or in jet-skis to ensure the safety of the driver in the event of falling into the water.

For $C_u = C_{pu} = 0$ it results $d = 8$ cm, $\alpha = 30°$ and

$$\Omega_2 = \frac{\pi d^2}{4} = \frac{\pi \times (8 \times 10^{-2})^2}{4} = 5.0265 \times 10^{-3} \text{ m}^2.$$

$$\therefore \quad Q = \frac{V_1 \Omega_2 + \sqrt{(V_1 \Omega_2)^2 - 4F_x \dfrac{\Omega_2}{\rho}}}{2} = \frac{7.0 \times 5.0265 \times 10^{-3}}{2}$$

$$+ \frac{1}{2}\sqrt{(7.0 \times 5.0265 \times 10^{-3})^2 - 4 \times (-1300) \times \frac{5.0265 \times 10^{-3}}{1000}} = \mathbf{0.100 \ m^3 \, s^{-1}},$$

$$V_2 = \frac{Q}{\Omega_2} = \frac{0.100}{5.0265 \times 10^{-3}} = \mathbf{19.89 \ m \, s^{-1}},$$

$$\therefore \quad P_j = \gamma_w Q \left( \frac{V_2^2}{2g} - \frac{V_1^2 \sin^2 \alpha}{2g} \right) =$$

$$9806 \times 0.100 \times \left( \frac{19.89^2}{2 \times 9.806} - \frac{7.0^2 \times \sin^2 30°}{2 \times 9.806} \right) = \mathbf{19.2 \ kW},$$

$$\therefore \quad \eta = \frac{2(V_2 - V_1)V_1}{(V_2 - V_1 \sin \alpha)(V_2 + V_1 \sin \alpha)} =$$

$$\frac{2 \times (19.89 - 7.0) \times 7.0}{(19.89 - 7.0 \times \sin 30°) \times (19.89 + 7.0 \times \sin 30°)} = \mathbf{47.1\%}.$$

For $d' = 2d = 16$ cm and $\alpha = 90°$ results

$$\Omega_2' = \frac{\pi d'^2}{4} = \frac{\pi \times (16 \times 10^{-2})^2}{4} = \mathbf{20.106 \times 10^{-3} \ m^2},$$

$$\therefore \quad Q' = \frac{V_1 \Omega_2' + \sqrt{(V_1 \Omega_2')^2 - 4F_x \dfrac{\Omega_2'}{\rho}}}{2} = \frac{7.0 \times 20.106 \times 10^{-3}}{2}$$

$$+ \frac{1}{2}\sqrt{(7.0 \times 20.106 \times 10^{-3})^2 - 4 \times (-1300) \times \frac{20.106 \times 10^{-3}}{1000}} = \mathbf{0.247 \ m^3 \, s^{-1}},$$

$$V_2' = \frac{Q'}{\Omega_2'} = \frac{0.247}{20.106 \times 10^{-3}} = 12.27 \text{ m s}^{-1},$$

$$\therefore \quad P_j' = \gamma_w Q' \left( \frac{V_2'^2}{2g} - \frac{V_1^2}{2g} \right) =$$

$$9806 \times 0.247 \times \left( \frac{12.27^2}{2 \times 9.806} - \frac{7.0^2}{2 \times 9.806} \right) = 12.5 \text{ kW},$$

$$\therefore \qquad \eta' = \frac{2V_1}{(V_2' + V_1)} = \frac{2 \times 7.0}{(12.27 + 7.0)} = 72.7\%.$$

# Chapter 5
# Pipeline Systems

Pipeline circuits are the most common application of Hydraulics and Fluid Mechanics in civil and industrial engineering. The liquid is conveyed into pipes of various diameters and roughness, usually in the presence of special components such as valves, curves, elbows, fittings. In most cases the problems can be solved by applying the energy balance in terms of Bernoulli's extended theorem with additional energy losses. Energy losses belong to the category of concentrated and distributed losses, both proportional to the velocity head (at least in turbulent conditions) with a coefficient that depends on the type of the special component, on the roughness of the duct and on the Reynolds number. In particular, the distributed energy losses are calculated with the Darcy formula $J = (\lambda/D)V^2/(2g)$, where the friction factor $\lambda$ is calculated with the Moody chart or, in case of turbulent flow, with the Colebrook–White equation, an equation in implicit form $f(\lambda, \text{Re}, \varepsilon/D) = 0$ of the friction factor, the relative roughness $\varepsilon/D$ and the number of Reynolds $\text{Re} = VD/v$. The Colebrook–White equation in the general case has no analytical solution, and a numerical procedure is required, if necessary with iterations.

For pipes in series the energy losses are additive, for pipes in parallel or with more branches, the energy losses must be calculated separately for each route, as they are per unit of weight (they are intensive) and are not affected by the division of the flow.

In most cases the liquid occupies the entire cross-section (at full-depth), in some cases it occupies part of it (at partial-depth), with air or vapour at vapour pressure on top.

An immediate graphic visualization of the energy balance in the pipes is represented by the line of the energy grade, a line whose vertical distance from a datum is the sum of the elevation head, the pressure (gage or absolute) head, and the velocity head. The energy grade line is always decreasing in the direction of flow, unless a pump or equivalent device capable of transferring energy to the current is inserted

---

$C_u$ and $C_{pu}$, that are two integer numbers between 0 and 9, for example, the last and second-last digit of the registration number.

© Springer Nature Switzerland AG 2021

S. Longo et al., *Problems in Hydraulics and Fluid Mechanics*, Springer Tracts in Civil Engineering, https://doi.org/10.1007/978-3-030-51387-0_5

in the circuit. The pressure regime is displayed by the hydraulic grade line, always below the energy grade line at a distance equal to the velocity head. The vertical distance between the pipe axis and the hydraulic grade line is the pressure head (gage or absolute) in the cross-section. If the hydraulic grade line (gage pressure) is above the pipe axis, the pressure in the cross-section is higher than the atmospheric pressure. Otherwise, the relative pressure in the cross-section is negative, with a minimum corresponding to the vapour pressure. The hydraulic grade line may increase in the direction of flow in the presence of an expansion, which induces a reduction of the velocity head.

For practical calculations, concentrated losses are negligible compared to distributed losses if they are less than a few percentage points of the total losses (1–4%). In this case, also the velocity head is negligible and energy grade line coincides with the hydraulic grade line.

**Exercise 5.1**  The siphon in Fig. 5.1 has a diameter $d = (0.2 + C_{pu}/100)$ m and has a local restriction. The pipe is supported by a floater and discharges water with a flow rate $Q = (0.06 + C_u/100)$ m$^3$ s$^{-1}$. The cross-section area of the cylindrical tank is $A = 10$ m$^2$.

– Calculate the coefficient $\xi$ of the head loss due to the restriction, neglecting distributed losses.
– Calculate the pressure at vertex A.
– Calculate the time required for a level lowering in the tank equal to 0.20 m.

Assume $H = 2.0$ m, $h = 1.5$ m, a loss coefficient at the inlet/outlet $\xi_{in} = 1.16$ and $\xi_{out} = 1$, $\gamma_w = 9806$ N m$^{-3}$.

**Fig. 5.1**  Schematic of the siphon with the inlet section on a floater

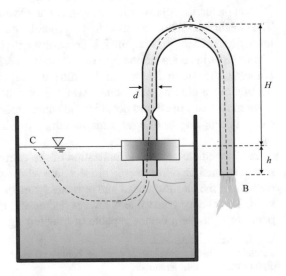

**Solution** The energy balance along the path from point C to the outflow section B, neglecting the distributed head losses and the losses in the curve, leads to the following relation:

$$z_C + \frac{p_C}{\gamma_w} + \frac{V_C^2}{2g} = z_B + \frac{p_B}{\gamma_w} + \xi_{out}\frac{V_B^2}{2g} + \xi_{in}\frac{V_B^2}{2g} + \xi\frac{V_B^2}{2g} \rightarrow$$

$$z_C - z_B \equiv h = \xi_{out}\frac{V_B^2}{2g} + \xi_{in}\frac{V_B^2}{2g} + \xi\frac{V_B^2}{2g},$$

which yields

$$\xi = \frac{2gh}{V_B^2} - \xi_{in} - \xi_{out}.$$

The average velocity of the current is equal to $V \equiv V_B = 4Q/(\pi d^2)$. Applying the energy balance between point C and vertex A, yields

$$z_C + \frac{p_C}{\gamma_w} + \frac{V_C^2}{2g} - (\xi + \xi_{in})\frac{V^2}{2g} = z_A + \frac{p_A}{\gamma_w} + \frac{V^2}{2g},$$

equivalent to

$$p_A = -\gamma_w\left[z_A - z_C + (\xi + \xi_{out} + \xi_{in})\frac{V^2}{2g}\right] = -\gamma_w\,(H + h).$$

A clarification is necessary on the meaning of $\xi_{out}$. In case of a cylindrical cross-section in B, the energy loss can be interpreted as the kinetic energy out-flowing with the exiting jet, and it results $\xi_{out} = 1$. The expression in terms of energy loss is more flexible since it also interprets cases where there is a divergent in the exiting section (to reduce energy loss), with $\xi_{out} < 1$; or a nozzle, with $\xi_{out} > 1$.

Since the available load $h$ is constant, the flow rate is constant and the time required to lower the water level in the tank by a value $\delta$ is equal to the ratio between the corresponding volume of water and the flow rate:

$$\Delta t = \frac{A\delta}{Q}.$$

For $C_u = C_{pu} = 0$ it results $d = 0.2$ m, $Q = 0.06$ m³ s⁻¹, $A = 10$ m², $H = 2.0$ m, $h = 1.5$ m, $\gamma_w = 9806$ N m⁻³,

$$V = \frac{4Q}{\pi d^2} = \frac{4 \times 0.06}{\pi \times 0.2^2} = \textbf{1.91 m s}^{-1},$$

$$\therefore \quad \xi = \frac{2gh}{V_B^2} - \xi_{in} - \xi_{out} = \frac{2 \times 9.806 \times 1.5}{1.91^2} - 2.16 = \textbf{5.90},$$

$$\therefore \qquad p_A = -\gamma_w \left( H + h \right) = -9806 \times \left( 2 + 1.5 \right) = -34.3 \text{ kPa},$$

$$\therefore \qquad \Delta t = \frac{A\delta}{Q} = \frac{10 \times 0.2}{0.06} = 33.3 \text{ s}.$$

---

**Exercise 5.2** The siphon in Fig. 5.2 is a pipe of diameter $D = 0.4$ m with roughness $\varepsilon = 0.4$ mm.

- Calculate the flow rate.
- Draw the energy grade line and the hydraulic grade line.

Assume an elbow loss coefficient $\xi_e = 0.3$ and inlet/outlet loss coefficients $\xi_{in} = \xi_{out} = 1$.

**Solution** The energy balance along a path connecting the two tanks reads

$$H = \xi_{in}\frac{V^2}{2g} + \xi_e\frac{V^2}{2g} + \xi_{out}\frac{V^2}{2g} + \lambda\frac{V^2}{2g}\frac{L_1 + L_2}{D}, \qquad (5.1)$$

where $\xi_{in}$, $\xi_e$ and $\xi_{out}$ are the coefficients of loss at the inlet, at the elbow, and at the outlet sections, respectively, and where $H = 8.5$ m is the available head, equal to the difference of level of the two tanks.

Solving Eq. (5.1) with respect to the velocity, yields

$$V = \sqrt{\frac{2gH}{\xi_{in} + \xi_e + \xi_{out} + \lambda\dfrac{L_1 + L_2}{D}}}. \qquad (5.2)$$

**Fig. 5.2** Siphon between two tanks

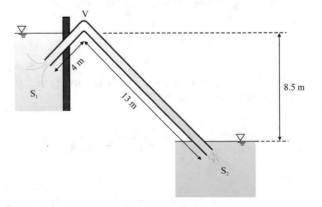

**Fig. 5.3** Energy and
hydraulic grade lines

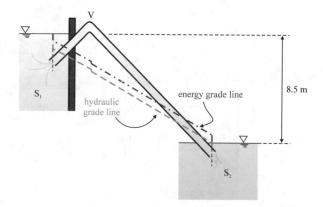

For the calculation of the distributed energy loss, we assume a value of the friction
factor $\lambda$ equal to the asymptotic value for fully developed turbulence. The asymptotic
value of the resistance index corresponding to a relative roughness $\varepsilon/D = 0.001$ is
equal to $\lambda_\infty = 0.0195$, see the Moody chart or the Prandtl-Nikuradse equation in the
Appendix.

Substituting the numerical values in Eq. (5.2), yields

$$V = \sqrt{\frac{2 \times 9.806 \times 8.5}{1 + 0.3 + 1 + 0.0195 \times \dfrac{4 + 13}{0.4}}} = \textbf{7.30 m s}^{-1}.$$

The corresponding Reynolds number is equal to:

$$\mathrm{Re} = \frac{VD}{\nu} = \frac{7.30 \times 0.4}{10^{-6}} = \textbf{2.92} \times \textbf{10}^{6},$$

where $\nu \approx 10^{-6}$ m$^2$ s$^{-1}$ is the kinematic viscosity of the water. The Moody chart
shows that the operating point of the pipeline is in conditions of fully developed
turbulence, with $\lambda \equiv \lambda_\infty$, confirming our hypothesis. The flow rate is equal to

$$\therefore \qquad Q = V\frac{\pi D^2}{4} = 7.30 \times \frac{\pi \times 0.4^2}{4} = \textbf{0.92 m}^3\,\textbf{s}^{-1}.$$

The energy and the hydraulic grade lines are shown in Fig. 5.3. The slope of the
straight lines upstream and downstream of the elbow is the same.

---

**Exercise 5.3** Water flows through the siphon in Fig. 5.4, with a diameter equal to
75 mm.

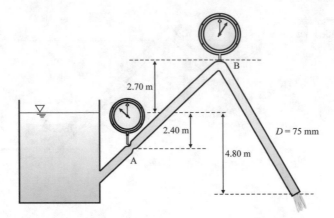

**Fig. 5.4** Siphon with a restriction

- Calculate the flow rate neglecting losses.
- Calculate the diameter of the restriction in section A if the two pressure gauges (in A and B) measure the same pressure, neglecting losses.

**Solution** The velocity of the current, neglecting losses, follows Torricelli's law:

$$V = \sqrt{2gh} = \sqrt{2 \times 9.806 \times 4.8} = \textbf{9.70 m s}^{-1}.$$

The flow rate is equal to

$$\therefore \qquad Q = V\frac{\pi D^2}{4} = 9.70 \times \frac{\pi \times 0.075^2}{4} = \textbf{42.85} \times \textbf{10}^{-3}\ \textbf{m}^3\,\textbf{s}^{-1}.$$

Applying Bernoulli's theorem between sections A and B, yields

$$z_A + \frac{p_A}{\gamma_w} + \alpha_A \frac{V_A^2}{2g} = z_B + \frac{p_B}{\gamma_w} + \alpha_B \frac{V_B^2}{2g},$$

where $V_B = V$.

Since $p_A = p_B$, if $\alpha_A = \alpha_B = 1$, it results

$$V_A^2 = V_B^2 + 2g\,(z_B - z_A),\tag{5.3}$$

and mass conservation equation gives

$$V_A \frac{\pi D_A^2}{4} = V_B \frac{\pi D_B^2}{4}.\tag{5.4}$$

Solving the system of Eqs.(5.3–5.4), gives

$$\frac{D_B}{D_A} = \left[1 + \frac{2g\,(z_B - z_A)}{V_B^2}\right]^{1/4} = \left[1 + \frac{2 \times 9.806 \times (2.4 + 2.7)}{9.7^2}\right]^{1/4} = \mathbf{1.20},$$

hence,

$$\therefore \quad D_A = \frac{D_B}{1.20} = \frac{0.075}{1.20} = \mathbf{62.5\,mm}.$$

---

**Exercise 5.4** The two tanks in Fig. 5.5 are connected by a siphon. The pipe is made of steel with a Gauckler-Strickler coefficient $k = 80$ m$^{1/3}$ s$^{-1}$. The ducts have length $L_{AB} = 8$ m, $L_{BM} = 32$ m, $L_{MC} = 45$ m and diameter $D = (200 + 10 \times C_{pu})$ mm. The free surface levels in tanks 1–2 are $z_1 = (11 + 0.5 \times C_u)$ m, and $z_2 = 8$ m, respectively, and the height of vertex M is $z_M = 19$ m.

- Calculate the flow rate.
- Draw the energy and the hydraulic grade lines.

Using a control system, it is possible to lower the level of the upstream tank $z_1$ at a rate equal to $\Delta W_{z_1} = 0.001$ m s$^{-1}$, whilst $z_2$ is constant and equal to 8 m.

- Calculate the time necessary for the current in the pipe to swap from a full- to a partial-depth flow.
- Calculate the flow rate at that time.

Assume a unitary loss coefficient at the inlet and at the outlet, and a loss coefficient $\xi_c = 0.1$ for each curve. The fluid is water at a temperature of 20 °C, with a vapour pressure of 2314 Pa and $\gamma_w = 9806$ N m$^{-3}$.

**Solution** The energy balance equation is

$$z_1 - z_2 = (L_{AB} + L_{BM} + L_{MC}) \frac{Q^2}{k^2 \left(\dfrac{D}{4}\right)^{4/3} \left(\dfrac{\pi D^2}{4}\right)^2}$$

$$+ (\xi_{in} + \xi_{out} + 2 \times \xi_c) \frac{Q^2}{2g \left(\dfrac{\pi D^2}{4}\right)^2}.$$

For $C_u = C_{pu} = 0$ it results $D = 200$ mm, $z_1 = 11$ m and the energy balance reads

**Fig. 5.5** Siphon between two tanks

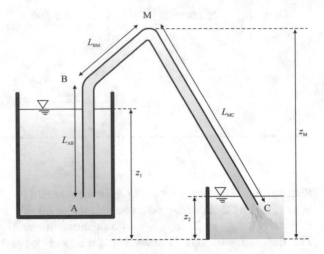

$$11 - 8 = (8 + 32 + 45) \times \frac{Q^2}{80^2 \times \left(\frac{0.2}{4}\right)^{4/3} \times \left(\frac{\pi \times 0.2^2}{4}\right)^2}$$

$$+ (1.0 + 1.0 + 2 \times 0.1) \times \frac{Q^2}{2 \times 9.806 \times \left(\frac{\pi \times 0.2^2}{4}\right)^2},$$

which admits the solution

$$\therefore \qquad Q = 0.060 \text{ m}^3 \text{ s}^{-1},$$

and $V = 1.91 \text{ m s}^{-1}$. The velocity head is equal to $V^2/2g = 0.18$ m.

In order to check whether the pipe is in full-depth flow condition, it is necessary to calculate the hydraulic head. By applying the energy balance equation, the data listed in Table 5.1 can be computed. The grade lines are shown in Fig. 5.6.

The hydraulic head line is always above the highest section of the pipe (plus the increment vapour pressure head): vertex M is at $+19.00$ m and the increase due to the vapour pressure is equal to 0.24 m. Hence, the pipe always operates in full-depth condition and the siphon must be primed.

If the level in the upstream tank is lowered, the flow rate is reduced because of the reduction in the available head. The flow becomes a partial-depth flow when the pressure at vertex M is equal to the vapour pressure. The two unknowns are the level $z_1$ and the flow rate, and the following equations hold:

**Table 5.1** Energy and hydraulic heads in some sections of interest. Pressure head contribution refers to absolute pressure

| Duct | Energy loss (m) | End of the duct | Energy head (m) | Hydraulic head (m) |
|------|-----------------|-----------------|-----------------|---------------------|
| Upstream duct | – | A | 21.33 | 21.33 |
| Inlet | $1.0 \times 0.18 = 0.18$ | A | 21.15 | 20.97 |
| AB | 0.24 | B | 20.91 | 20.73 |
| Curve No.1 | $0.1 \times 0.18 = 0.02$ | B | 20.89 | 20.71 |
| BM | 0.98 | M | 19.91 | 19.73 |
| Curve No.2 | 0.02 | M | 19.89 | 19.71 |
| MC | 1.38 | C | 18.51 | 18.33 |
| Outlet | $1.0 \times 0.18 = 0.18$ | C | 18.33 | 18.33 |

$$
\begin{cases}
z_1 - z_2 = (L_{AB} + L_{BM} + L_{MC}) \, \dfrac{Q'^2}{k^2 \left(\dfrac{D}{4}\right)^{4/3} \left(\dfrac{\pi D^2}{4}\right)^2} \\[4mm]
\qquad\qquad + (\xi_{in} + \xi_{out} + 2 \times \xi_c) \, \dfrac{Q'^2}{2g \left(\dfrac{\pi D^2}{4}\right)^2}, \\[8mm]
z_1 + \dfrac{p_{atm}}{\gamma_w} - \left[ z_M + \dfrac{p_{vap}}{\gamma_w} + \dfrac{Q'^2}{2g \left(\dfrac{\pi D^2}{4}\right)^2} \right] = \\[6mm]
(L_{AB} + L_{BM}) \, \dfrac{Q'^2}{k^2 \left(\dfrac{D}{4}\right)^{4/3} \left(\dfrac{\pi D^2}{4}\right)^2} + (\xi_{in} + \xi_c) \, \dfrac{Q'^2}{2g \left(\dfrac{\pi D^2}{4}\right)^2}.
\end{cases}
$$

By substitution, the following equation is obtained, which also represents the energy balance between vertex M and the downstream tank when the absolute pressure at vertex M is equal to the vapour pressure:

$$
z_M + \frac{p_{vap}}{\gamma_w} + \frac{Q'^2}{2g \left(\dfrac{\pi D^2}{4}\right)^2} - z_2 - \frac{p_{atm}}{\gamma_w} = L_{MC} \frac{Q'^2}{k^2 \left(\dfrac{D}{4}\right)^{4/3} \left(\dfrac{\pi D^2}{4}\right)^2}
$$
$$
+ (\xi_{out} + \xi_c) \frac{Q'^2}{2g \left(\dfrac{\pi D^2}{4}\right)^2}. \qquad (5.5)
$$

By substituting the numerical values, Eq. (5.5) becomes

$$19 + 0.24 + \frac{Q'^2}{2 \times 9.806 \times \left(\frac{\pi \times 0.2^2}{4}\right)^2} - 8 - 10.33 =$$

$$45 \times \frac{Q'^2}{80^2 \times \left(\frac{0.2}{4}\right)^{4/3} \times \left(\frac{\pi \times 0.2^2}{4}\right)^2}$$

$$+ (1.0 + 0.1) \times \frac{Q'^2}{2 \times 9.806 \times \left(\frac{\pi \times 0.2^2}{4}\right)^2},$$

with the solution

$$\therefore \qquad Q' = 0.048 \text{ m}^3 \text{ s}^{-1},$$

and $V' = 1.53 \text{ m s}^{-1}$, $z'_1 = 9.94$ m. The lowering of the level in the upstream tank requires a time equal to

$$\therefore \qquad \Delta t = \frac{z_1 - z'_1}{\Delta W_{z_1}} = \frac{11 - 9.94}{0.001} = 1060 \text{ s}.$$

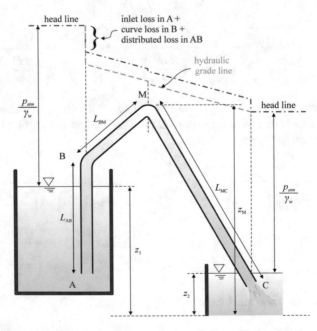

**Fig. 5.6** Energy and hydraulic (absolute pressure) grade lines. Notice that the grade lines are distorted

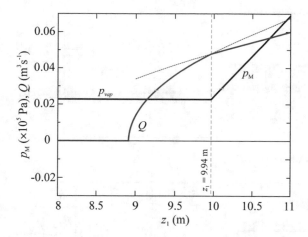

**Fig. 5.7** Rating curve (red bold line) and absolute pressure diagram (black bold line) at vertex M

Figure 5.7 shows the diagram of the flow rate and of the absolute pressure at vertex M as a function of the water level in the upstream tank.

---

**Exercise 5.5** The pipe in Fig. 5.8 is made of steel with diameter $D = (200 + C_{pu} \times 10)$ mm and roughness $\varepsilon = 0.2$ mm. The water flow rate is equal to $Q = (0.10 + C_u/100)$ m$^3$ s$^{-1}$ with $h = (100 + C_u \times 10)$ m. The ducts have length $L_1 = (500 + C_u \times 10)$ m, $L_2 = 150$ m, $L_3 = (200 + C_{pu} \times 10)$ m.

- Calculate the absolute pressure $p^*$ in the tank required for flow rate $Q$.
- Calculate the absolute pressure $p^*$ necessary for the incipient outflow, with the flow rate approaching zero.
- Draw the hydraulic grade line (with absolute pressure).

Neglect the velocity head and the local energy losses. The fluid is water with $\gamma_w = 9806$ N m$^{-3}$.

**Solution** The energy balance equation for the system shown in Fig. 5.8, reads

$$h + \frac{p^*}{\gamma_w} - L_2 - \frac{p^*_{atm}}{\gamma_w} - \frac{\cancel{V^2}}{\cancel{2g}} =$$

$$\lambda \frac{Q^2}{2g \left(\dfrac{\pi D^2}{4}\right)^2} \frac{1}{D} (L_1 + L_2 + L_3) + \sum \cancel{\xi_i \frac{V^2}{2g}}.$$

$$(5.6)$$

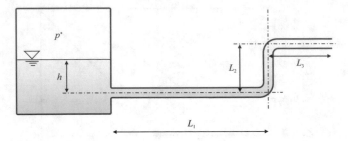

**Fig. 5.8**  Tank with pressurized air

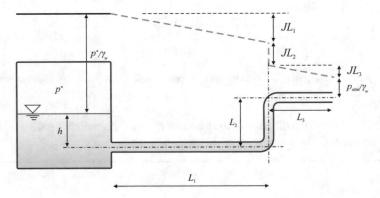

**Fig. 5.9**  Hydraulic grade line

The average velocity of water in the pipe is equal to $V = 4Q/(\pi D^2)$, the Reynolds number is equal to $\text{Re} = VD/\nu = 4Q/(\pi D \nu)$ and the relative roughness is equal to $\varepsilon/D$. The Moody chart allows to determine the friction factor $\lambda$ and, neglecting the concentrated energy losses, Eq. (5.6) becomes

$$p^* = p^*_{atm} + \gamma_w \left[ L_2 - h + \lambda \frac{V^2}{2g} \frac{1}{D} (L_1 + L_2 + L_3) \right].$$

For $Q \to 0$, Eq. (5.6) reduces to

$$p^* = p^*_{atm} + \gamma_w (L_2 - h),$$

where the finite size of the duct has been neglected.

Figure 5.9 shows the hydraulic grade line (absolute pressure head is considered).

For $C_u = C_{pu} = 0$ it results $D = 200$ mm, $\varepsilon = 0.2$ mm, $Q = 0.10$ m$^3$ s$^{-1}$, $h = 100$ m, $L_1 = 500$ m, $L_2 = 150$ m, $L_3 = 200$ m,

$$V = \frac{4Q}{\pi D^2} = \frac{4 \times 0.10}{\pi \times 0.2^2} = 3.18 \text{ m s}^{-1},$$

$$\text{Re} = \frac{VD}{\nu} = \frac{3.18 \times 0.2}{10^{-6}} = \mathbf{6.4 \times 10^5}.$$

For $\varepsilon/D = 0.2/200 = 10^{-3}$ it results $\lambda = \mathbf{0.02}$, and

$$\therefore \qquad p^* = p^*_{atm} + \gamma_w \left[ L_2 - h + \lambda \frac{V^2}{2g} \frac{1}{D} (L_1 + L_2 + L_3) \right]$$

$$= 10^5 + 9806 \times \left[ 150 - 100 + 0.02 \times \frac{3.18^2}{2 \times 9.806} \times \frac{1}{0.2} \right.$$

$$\left. \times (500 + 150 + 200) \right]$$

$$= \mathbf{1.02 \times 10^6 \ Pa}.$$

For incipient outflow ($Q \to 0$), it results

$$\therefore \qquad p^* = p^*_{atm} + \gamma_w (L_2 - h) = 10^5 + 9806 \times (150 - 100) = \mathbf{5.9 \times 10^5 \ Pa}.$$

---

**Exercise 5.6** The pipeline in Fig. 5.10 conveys water from tank A to tank B through a circular cross-section steel pipe of diameter $D = 200$ mm, length $L = (300 + 10 \times C_{pu})$ m, roughness $\varepsilon = 0.2$ mm. The water level difference between the two tanks is $Y = (30 + C_u/10)$ m.

- Calculate the flow rate $Q_1$.

In order to increase the flow rate to the value $Q_2 = 1.5 \times Q_1$, a replacement of the existing pipeline is planned with a new one made of material which is able to ensure the smooth-wall regime. The length of the new pipeline and the difference in level between the tanks are unchanged.

- Determine the diameter of the new pipeline from the commercial series listed below:

$D$ (mm) = 40, 50, 65, 80, 100, 125, 150, 200, 250, 300, 350, 400, 450, 500, 600, 700, 800, 900, 1000.

Assume $\xi_{in} = 0.5$, $\xi_{out} = 1$, $\nu = 10^{-6} \ m^2 \ s^{-1}$.

**Solution** The energy balance equation reads

$$Y = \frac{Q_1^2}{2g\Omega_1^2} \left[ \lambda_1 \frac{L}{D_1} + (\xi_{in} + \xi_{out}) \right],$$

where $\Omega_1 = \pi D_1^2/4$. The flow rate is equal to

**Fig. 5.10** Pipeline conveying water between two tanks

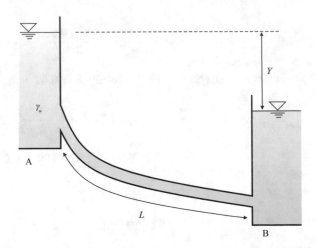

$$Q_1 = \sqrt{\frac{2g\Omega_1^2 Y}{\lambda_1 \dfrac{L}{D_1} + (\xi_{in} + \xi_{out})}}.$$

Using Moody chart, the friction coefficient $\lambda_1$ and the flow rate can be calculated by iteration, according to the following procedure.

We initially assume that the flow is fully turbulent. The initial value of $\lambda_1$ is the asymptotic value for $\mathrm{Re} \to \infty$, commonly defined $\lambda_\infty$, which only depends on the relative roughness $\varepsilon/D_1$:

$$\lambda_\infty = \left[ -2 \log_{10}\left( \frac{1}{3.71} \frac{\varepsilon}{D_1} \right) \right]^{-2}.$$

The flow rate of first approximation is equal to

$$Q_1' = \sqrt{\frac{2g\Omega_1^2 Y}{\lambda_\infty \dfrac{L}{D_1} + (\xi_{in} + \xi_{out})}}.$$

Once $Q_1'$ is known, the approximate Reynolds number is

$$\mathrm{Re}' = \frac{V_1' D_1}{\nu} = \frac{Q_1' D_1}{\dfrac{\pi D_1^2}{4} \nu}.$$

Entering in Moody chart with the values of the relative roughness $\varepsilon/D_1$ and of the Reynolds number $\mathrm{Re}'$, the value $\lambda_1'$ is estimated. If $\lambda_1' \approx \lambda_\infty$, the first approximation flow rate is correct and it can be assumed as the definitive value of $Q_1$. Otherwise, the new value of $\lambda_1'$ allows the calculation of the new flow rate. As an alternative to the Moody chart, the Colebrook–White equation can be solved numerically:

$$\lambda_1' = \left[ -2 \log_{10} \left( \frac{2.51}{Re' \sqrt{\lambda_1'}} + \frac{1}{3.71} \frac{\varepsilon}{D_1} \right) \right]^{-2}.$$

Once the updated value of $\lambda_1'$ has been computed, the flow rate of second approximation is calculated,

$$Q_1'' = \sqrt{\frac{2 g \Omega_1^2 Y}{\lambda_1' \dfrac{L}{D_1} + (\xi_{in} + \xi_{out})}}.$$

If $Q_1'$ and $Q_1''$ differ by a reasonable small amount (in the present case $10^{-4}$ m³ s⁻¹ may be sufficient), the iteration can be stopped, otherwise a new iteration is necessary repeating the procedure.

In order to obtain a flow rate $Q_2 = 1.5 \times Q_1$ a hydraulically smooth pipe is used, for which the resistance coefficient can be expressed by means of the Blasius' law:

$$\lambda = 0.3164 \, Re^{-0.25} \text{ for } Re < 10^5,$$

or by the Nikuradse's law:

$$\lambda = 0.0032 + 0.221 Re^{-0.237} \text{ for } Re \geq 10^5,$$

or by the Prandtl-Kármán's law:

$$\frac{1}{\sqrt{\lambda}} = -2 \log_{10} \left( \frac{2.51}{Re \sqrt{\lambda}} \right).$$

Assuming the Blasius' law and replacing the corresponding value of $\lambda$ in the balance equation, it results:

$$Y = 0.3164 \frac{(1.5 Q_1)^2}{2g \left( \dfrac{\pi D_2^2}{4} \right)^2} \frac{(1.5 Q_1 D_2)^{-0.25}}{\left( \dfrac{\pi D_2^2}{4} v \right)^{-0.25}} \frac{L}{D_2} + \frac{(1.5 Q_1)^2}{2g \left( \dfrac{\pi D_2^2}{4} \right)^2} (\xi_{in} + \xi_{out}). \quad (5.7)$$

As a first approximation, by neglecting the concentrated energy losses, Eq. (5.7) admits an analytical solution:

$$D_2 = \frac{0.860 \, 872 L^{4/19} Q_1^{7/19} v^{1/19}}{g^{4/19} Y^{4/19}}.$$

If the concentrated energy losses are not negligible, a different approach is suggested.

(i) A first attempt value of the diameter is calculated assuming an average velocity of $1.0 \text{ m s}^{-1}$, a value commonly accepted for a good operation of the plant.

On the basis of the definition of average velocity, $V = Q/\Omega$, a theoretical diameter equal to

$$D_{th} = \sqrt{\frac{4Q_2}{\pi \times 1.0}}$$

is computed, and the diameter $D_2'$ of the commercial series larger than $D_{th}$ is chosen.

(ii) The value of the head losses is calculated by assuming a flow rate equal to $Q_2$ in a pipe of diameter $D_2'$:

$$\Delta H_p = L \frac{\lambda V'^2}{2g D_2'} + \frac{V'^2}{2g} (\xi_{in} + \xi_{out}),$$

where $V' = 4Q_2/(\pi D_2'^2)$.

If $\Delta H_p > Y$, the diameter $D_2'$ is too small and it is necessary to select a larger commercial diameter. Again, it must be verified that the energy losses are not excessive by repeating the procedure indicated in point (ii).

If $\Delta H_p \ll Y$, it is necessary to choose a smaller commercial diameter. Also in this case it must be verified that the energy losses are not excessive.

The procedure stops when $\Delta H_p \leq Y$, having chosen a commercial diameter which is suitable from the hydraulic point of view.

For $C_u = C_{pu} = 0$ it results $L = 300$ m, $Y = 30$ m, $\varepsilon = 200$ mm, $D_1 = 200$ mm. Assuming a fully turbulent flow, we calculate, for $\varepsilon/D = 0.001$, $\lambda_1 = 0.020$, $Q_1 = $ **0.136 m$^3$ s$^{-1}$** and Re = **865 000**.

To increase the flow rate by 50%, in the case of hydraulically smooth pipe, we calculate a theoretical diameter equal to $D_2 = 0.211$ m, applying the Prandtl-Kármán's law, with $\lambda_2 = 0.011$. The commercial diameter of $D_2 = $ **250 mm** is chosen, but it is necessary to install a throttling valve to dissipate approximately 16.5 m of head in excess.

---

**Exercise 5.7** The pipeline in Fig. 5.11 has length $L = (4000 + 100 \times C_u)$ m, diameter $D = (200 + 10 \times C_{pu})$ mm, Gauckler-Strickler coefficient $k = 90$ m$^{1/3}$ s$^{-1}$. A Venturi meter is installed, with a differential pressure gauge indicating $\Delta p = (0.05 + 0.005 \times C_u) \times 10^5$ Pa. The discharge coefficient of the Venturi meter is $C_v = 0.981$. The diameter of the narrow section is $d = (150 + 5 \times C_{pu})$ mm, the diameter of the inlet section coincides with the actual diameter of the pipe.

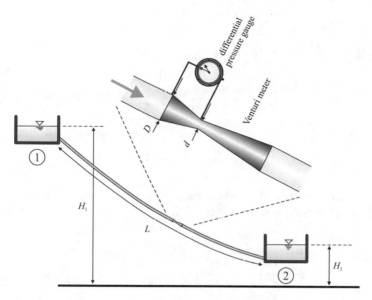

**Fig. 5.11** Pipeline connecting two tanks with a Venturi meter inserted in the circuit

- Calculate the flow rate when a permanent flow is established.
- Calculate the difference of water levels between the two tanks.

The fluid is water with $\gamma_w = 9806$ N m$^{-3}$. Neglect concentrated energy losses and the different elevation head between the intakes of the differential pressure gauge.

**Solution** The flow rate measured by the Venturi meter is equal to

$$Q = C_v \frac{\pi d^2}{4} \sqrt{\frac{2g\,\Delta h}{1 - \left(\dfrac{d}{D}\right)^4}} = C_v \frac{\pi d^2}{4} \sqrt{\frac{\dfrac{2g\,\Delta p}{\gamma_w}}{1 - \left(\dfrac{d}{D}\right)^4}}.$$

The difference in water levels of the upstream and downstream tanks must balance only the distributed losses (by hypothesis, we neglect the concentrated energy losses), hence

$$\Delta H \equiv H_1 - H_2 = JL = \frac{Q^2}{k^2 \left(\dfrac{D}{4}\right)^{4/3} \left(\dfrac{\pi D^2}{4}\right)^2} L,$$

where $J$ is the energy gradient.

For $C_u = C_{pu} = 0$ it results $L = 4000$ m, $D = 200$ mm, $\Delta p = 5 \times 10^3$ Pa, $d = 150$ mm,

$$\therefore\ Q = C_v \frac{\pi d^2}{4} \sqrt{\frac{\frac{2g\,\Delta p}{\gamma_w}}{1 - \left(\frac{d}{D}\right)^4}} =$$

$$0.981 \times \frac{\pi \times 0.150^2}{4} \times \sqrt{\frac{\frac{2 \times 9.806 \times 0.05 \times 10^5}{9806}}{1 - \left(\frac{0.150}{0.200}\right)^4}} = 66.3 \times 10^{-3}\ \text{m}^3\ \text{s}^{-1}.$$

The average velocity of the water in the pipeline is $2.11\ \text{m\,s}^{-1}$. The difference of water level between the two tanks is equal to

$$\therefore\ \Delta H = \frac{Q^2}{k^2 \left(\frac{D}{4}\right)^{4/3} \left(\frac{\pi D^2}{4}\right)^2} L =$$

$$\frac{(66.3 \times 10^{-3})^2}{90^2 \times \left(\frac{0.200}{4}\right)^{4/3} \times \left(\frac{\pi \times 0.200^2}{4}\right)^2} \times 4000 = 119.4\ \text{m}.$$

**Exercise 5.8** In the plant in Fig. 5.12, the free surface level in the three tanks is equal to $H_1 = (150 + 10 \times C_u)$ m, $H_2 = (100 + 5 \times C_u)$ m, $H_3 = (70 + 10 \times C_u)$ m. The connecting pipes have length $L_1 = L_2 = (4000 + 100 \times C_u)$ m and $L_3 = (1000 + 80 \times C_u)$ m, diameter $D_1 = D_2 = D_3 = D = (150 + 10 \times C_{pu})$ mm and Gauckler-Strickler coefficient equal to $k = 90\ \text{m}^{1/3}\ \text{s}^{-1}$. The elevation of node N is $z_N = 85$ m.

– Calculate the flow rate in the three pipes.
– Calculate the relative pressure in the node N.

The fluid is water with $\gamma_w = 9806\ \text{N\,m}^{-3}$. The plant meets the conditions for neglecting concentrated energy losses and velocity head.

**Solution** Assuming positive the flow direction indicated by the arrows in Fig. 5.12, we can write the following energy balance and mass conservation equations:

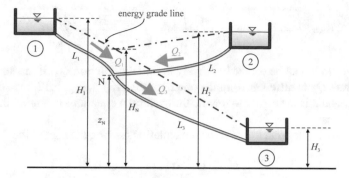

**Fig. 5.12** Plant with pipelines connecting three tanks

$$
\begin{cases}
H_1 - H_3 = \beta_1 \dfrac{Q_1^2}{D_1^5} L_1 + \beta_3 \dfrac{Q_3^2}{D_3^5} L_3, \\[2ex]
H_2 - H_3 = \beta_2 \dfrac{Q_2^2}{D_2^5} L_2 + \beta_3 \dfrac{Q_3^2}{D_3^5} L_3, \\[2ex]
Q_3 = Q_1 + Q_2,
\end{cases}
\tag{5.8}
$$

where

$$
\beta = \frac{4^3}{k^2 \pi^2 \left(\dfrac{D}{4}\right)^{1/3}}.
$$

The energy balance equation between tank 1 and tank 2,

$$
H_1 - H_2 = \beta_1 \frac{Q_1^2}{D_1^5} L_1 - \beta_2 \frac{Q_2^2}{D_2^5} L_2,
$$

is obtained by combining the first two equations in (5.8) and is automatically satisfied. From the first and second equations in (5.8), it results

$$
Q_1 = \sqrt{\frac{H_1 - H_3 - L_3 \beta_3 Q_3^2 D_3^5}{L_1 \beta_1 D_1^5}},
$$

$$
Q_2 = \sqrt{\frac{H_2 - H_3 - L_3 \beta_3 Q_3^2 D_3^5}{L_2 \beta_2 D_2^5}}.
$$

By substituting $Q_1$ and $Q_2$ in the third equation in (5.8), yields

$$Q_3 = \sqrt{\frac{H_1 - H_3 - L_3\beta_3 Q_3^2 D_3^5}{L_1\beta_1 D_1^5}} + \sqrt{\frac{H_2 - H_3 - L_3\beta_3 Q_3^2 D_3^5}{L_2\beta_2 D_2^5}}. \tag{5.9}$$

Equation (5.9) can be solved iteratively to find the solution $Q_3$. If the direction of the flow rate $Q_2$ is different from the one hypothesized in Fig. 5.12, the solution is imaginary and it is necessary to rewrite the system of equations changing the sign of the discharges.

An alternative approach is to write the equations saving the sign of the flow rates:

$$\begin{cases} H_1 - H_3 = \beta_1 \dfrac{Q_1|Q_1|}{D_1^5} L_1 + \beta_3 \dfrac{Q_3|Q_3|}{D_3^5} L_3, \\[2mm] H_2 - H_3 = \beta_2 \dfrac{Q_2|Q_2|}{D_2^5} L_2 + \beta_3 \dfrac{Q_3|Q_3|}{D_3^5} L_3, \\[2mm] Q_3 = Q_1 + Q_2, \end{cases}$$

which admits real positive or negative solutions. A negative value of the flow rate means that the water flows in the opposite direction.

Once the system of equations is solved, the head in node N is

$$H_N = H_1 - L_1\beta_1 Q_1|Q_1|/D_1^5,$$

and the pressure in the node N is

$$p_N = \gamma_w (H_N - z_N),$$

where we have neglected the velocity head.

For $C_u = C_{pu} = 0$ it results $\beta_1 = \beta_2 = \beta_3 = \beta = 0.00239$, and

$$\therefore \quad \begin{cases} H_N = \textbf{94.4 m} \\ Q_1 = \textbf{21.0 l s}^{-1} \\ Q_2 = \textbf{7.0 l s}^{-1} \\ Q_3 = \textbf{28.0 l s}^{-1} \\ p_N = \textbf{0.93} \times \textbf{10}^5 \textbf{ Pa} \end{cases}.$$

**Exercise 5.9** In the pipeline system shown in Fig. 5.13, the difference between the free surface level in tank A and tank B is equal to $\Delta H = (130 + 10 \times C_{pu})$ m. The pipe has diameter $D = 150$ mm and length $L = (8700 + 100 \times C_u)$ m, and the Gauckler-Strickler coefficient is $k = 90$ m$^{1/3}$ s$^{-1}$.

**Fig. 5.13** Pipeline between two tanks, with a bypass

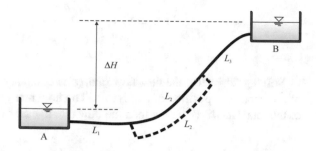

**Fig. 5.14** Schematic for the computation of the flow rate

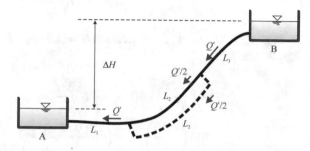

- Calculate the flow rate.
- Calculate the flow rate if a parallel pipeline (a bypass, the dashed pipeline in Fig. 5.13) with a length equal to 50% of the total length is installed. The bypass has the same diameter and roughness as the original pipeline.

The plant meets the conditions for neglecting concentrated energy losses and velocity head.

**Solution** The admitted energy loss per unit length is $J = \Delta H / L$, and the flow rate, in the case of the single pipeline and neglecting the concentrated energy losses, is equal to

$$Q = kR^{1/6}\Omega\sqrt{RJ} = k\left(\frac{D}{4}\right)^{2/3}\left(\frac{\pi D^2}{4}\right)\sqrt{\frac{\Delta H}{L}}.$$

If a bypass is installed, with pipeline characteristics identical to the ones of the original pipeline, by symmetry the flow rates in the two ducts assume the same value, equal to 50% of the total flow rate, see Fig. 5.14.

The energy balance equation becomes

$$\Delta H = \frac{Q'^2}{k^2\left(\frac{D}{4}\right)^{4/3}\left(\pi\frac{D^2}{4}\right)^2}(L_1 + L_3) + \frac{(Q'/2)^2}{k^2\left(\frac{D}{4}\right)^{4/3}\left(\pi\frac{D^2}{4}\right)^2}L_2,$$

with $L_1 + L_3 = L_2 = L/2$. Hence,

$$\Delta H = \frac{Q'^2}{k^2 \left(\dfrac{D}{4}\right)^{4/3} \left(\pi \dfrac{D^2}{4}\right)^2 \dfrac{5}{8}} L. \tag{5.10}$$

Solving Eq. (5.10), the new flow rate $Q'$ is computed, with a value independent of the geometric position of the bypass. The flow rate $Q'$ (with the bypass for 50% of the total length) is greater than the flow rate in case of a single pipe, and results

$$Q' = Q\sqrt{\frac{8}{5}}.$$

For $C_u = C_{pu} = 0$ it results $\Delta H = 130\,\text{m}$, $D = 150\,\text{mm}$, $L = 8700\,\text{m}$, $k = 90\,\text{m}^{1/3}\,\text{s}^{-1}$,

$$\therefore\ Q = k\left(\frac{D}{4}\right)^{2/3}\left(\frac{\pi D^2}{4}\right)\sqrt{\frac{\Delta H}{L}} =$$

$$90 \times \left(\frac{0.15}{4}\right)^{2/3} \times \left(\frac{\pi \times 0.15^2}{4}\right) \times \sqrt{\frac{130}{8700}} = \textbf{22.0}\,\textbf{l}\,\textbf{s}^{-1},$$

$$\therefore\qquad Q' = Q\sqrt{\frac{8}{5}} = 22.0 \times \sqrt{\frac{8}{5}} = \textbf{27.8}\,\textbf{l}\,\textbf{s}^{-1}.$$

---

**Exercise 5.10** The difference in level between the two tanks in Fig. 5.15 is $H = 12$ m.

– Calculate the flow rate in the pipelines.

Neglect concentrated energy losses and velocity head.

**Solution** The unknowns are the flow rates into the pipelines and the head in node C. We can write three energy balance equations and one mass conservation equation:

**Fig. 5.15** Pipeline network between two tanks

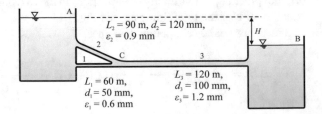

**Table 5.2** Summary of the operating characteristics of the pipes

| Pipe | L (m) | D (mm) | ε (mm) | Q ($1\,s^{-1}$) | V ($m\,s^{-1}$) | J ($m\,km^{-1}$) | λ (.) | Re (.) |
|------|-------|--------|--------|------|------|------|-------|--------|
| 1 | 60 | 50 | 0.6 | 1.76 | 0.89 | 34.04 | 0.042 | 44 500 |
| 2 | 90 | 120 | 0.9 | 13.94 | 1.23 | 22.70 | 0.035 | 147 600 |
| 3 | 120 | 100 | 1.2 | 15.70 | 2.00 | 82.97 | 0.041 | 200 000 |

$$\begin{cases} H_A - H_C = \lambda_1 \dfrac{Q_1^2}{2g\Omega_1^2} \dfrac{L_1}{D_1}, \\[2mm] H_A - H_C = \lambda_2 \dfrac{Q_2^2}{2g\Omega_2^2} \dfrac{L_2}{D_2}, \\[2mm] H_A - H_B = \lambda_2 \dfrac{Q_2^2}{2g\Omega_2^2} \dfrac{L_2}{D_2} + \lambda_3 \dfrac{Q_3^2}{2g\Omega_3^2} \dfrac{L_3}{D_3}, \\[2mm] Q_1 + Q_2 = Q_3, \end{cases} \qquad (5.11)$$

where $\Omega_1$, $\Omega_2$ and $\Omega_3$ are the areas of the cross-section of the pipelines. The flow direction is always from the tank with the highest head to the tank with the lowest head. The value of the head in node C must be in between the head in tank A and the head in tank B. By solving the system of Eq. (5.11), the results listed in Table 5.2 are obtained.

---

**Exercise 5.11** Tanks A and B in Fig. 5.16 are connected by two straight pipes of diameter $D = 0.10$ m, roughness $\varepsilon = 0.1$ mm, and are inclined at an angle $\pm\phi = 20°$ to the horizontal. The horizontal distance between the two tanks is $L = 5$ m and the difference of level height of the free surfaces is $H = 1$ m. The fluid has a specific weight $\gamma = 9$ kN m$^{-3}$ and dynamic viscosity $\mu = 0.89$ Pa s.

- Calculate the flow rate in the two pipes.
- Draw the energy and hydraulic grade lines.

The loss coefficient at the inlet is zero, since the pipe is well connected to the upstream tank. The loss coefficient at the outlet is equal to 1.0, if turbulent flow regime occurs, it is equal to 2.0 in laminar flow condition, since the Coriolis coefficient for kinetic head is equal to 2.0 in laminar flow.

**Solution** We first assume the laminar flow regime. The loss coefficient at the inlet is zero, the loss coefficient at the outlet is equal to 2.0. The length of the pipe is equal to $L/\cos\varphi$ and the energy balance equation reads

**Fig. 5.16** Pipelines
connecting two tanks

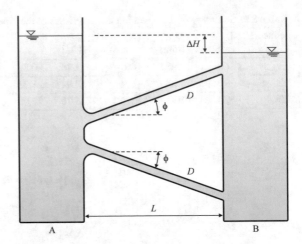

$$\Delta H = \frac{64}{\text{Re}} \frac{V^2}{2g} \frac{L}{\cos\varphi} \frac{1}{D} + 2\frac{V^2}{2g} = \frac{32\mu}{\rho D^2} \frac{V}{g} \frac{L}{\cos\varphi} + \frac{V^2}{g},$$

where $\rho$ is the mass density of the fluid, equal to $\gamma/g$. Defining $\chi = 32\mu L/(\rho D^2 \cos\varphi)$ the balance equation can be rewritten as follows:

$$V^2 + \chi V - g\Delta H = 0,$$

which admits the solution

$$V = -\frac{\chi}{2} + \sqrt{\frac{\chi^2}{4} + g\Delta H}.$$

Substituting the numerical values, yields

$$\chi = \frac{32\mu L}{\rho D^2 \cos\varphi} = \frac{32 \times 0.89 \times 5}{(9000/9.806) \times 0.10^2 \times \cos 20°} = \mathbf{16.51 \ m \ s^{-1}},$$

$$V = -\frac{\chi}{2} + \sqrt{\frac{\chi^2}{4} + g\Delta H} = -\frac{16.51}{2} + \sqrt{\frac{16.51^2}{4} + 9.806 \times 1.0} = \mathbf{0.57 \ m \ s^{-1}}.$$

The Reynolds number is equal to

$$\text{Re} = \frac{\rho V D}{\mu} = \frac{(9000/9.806) \times 0.57 \times 0.10}{0.89} = \mathbf{59} \ll 2000.$$

The flow regime is actually laminar, according to the first hypothesis. The flow rate is the same for the two pipes and sums up to

**Fig. 5.17** Energy and
hydraulic grade lines. Notice
that in laminar flow in
circular cross-section, the
velocity head is $2V^2/(2g)$

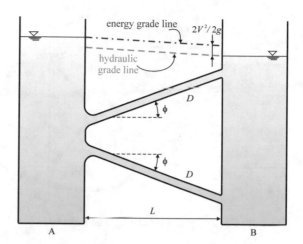

$$\therefore \qquad Q = 2V\frac{\pi D^2}{4} = 2 \times 0.57 \times \frac{\pi \times 0.10^2}{4} = 8.95\,\mathrm{l\,s^{-1}}.$$

The energy and the hydraulic grade lines shown in Fig. 5.17 are the same for the
two pipelines.

---

**Exercise 5.12** The steel pipes in Fig. 5.18 connect the three tanks, each of them with
constant free surface levels. The pipes have roughness $\varepsilon = 0.3$ mm.

– Calculate the flow rate in the three pipes and the head at node N.

Assume $L_{AN} = 1500$ m, $L_{CN} = 3000$ m, $L_{BN} = 700$ m, $D_{AN} = 1000$ mm,
$D_{CN} = 800$ mm, $D_{BN} = 500$ mm. Neglect concentrated energy losses.

**Solution** The unknowns are the flow rate in the three pipelines and the head in
node N. We can write a mass conservation equation for the node and three energy
balance equations:

$$\begin{cases}
Q_{AN} - Q_{BN} - Q_{CN} = 0, \\[2mm]
H_A - H_N = \lambda_{AN}\dfrac{Q_{AN}\,|Q_{AN}|}{2g\Omega_{AN}^2}\dfrac{L_{AN}}{D_{AN}}, \\[2mm]
H_B - H_N = -\lambda_{BN}\dfrac{Q_{BN}\,|Q_{BN}|}{2g\Omega_{BN}^2}\dfrac{L_{BN}}{D_{BN}}, \\[2mm]
H_C - H_N = -\lambda_{CN}\dfrac{Q_{CN}\,|Q_{CN}|}{2g\Omega_{CN}^2}\dfrac{L_{CN}}{D_{CN}}.
\end{cases} \qquad (5.12)$$

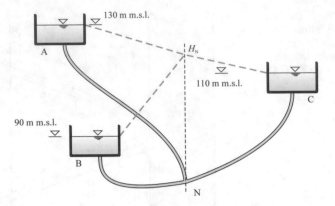

**Fig. 5.18**  Pipeline system connecting three tanks

**Table 5.3**  Summary of the operating characteristics of the pipes

| Pipe | L (m) | D (mm) | ε (mm) | Q (m³ s⁻¹) | V (m s⁻¹) | J (m km⁻¹) | λ (.) | Re (× 10⁶) |
|------|-------|--------|--------|------------|-----------|------------|-------|------------|
| AN   | 1500  | 1000   | 0.2    | **2.00**   | 2.55      | 5.07       | 0.015 | 2.50       |
| BN   | 700   | 500    | 0.2    | **0.99**   | 5.07      | 46.27      | 0.018 | 2.55       |
| CN   | 3000  | 800    | 0.2    | **1.01**   | 2.00      | 4.13       | 0.016 | 1.60       |

It is necessary to express the energy losses by introducing the absolute value of the flow rates, because the direction of the flow is not known a priori, although in the system in Fig. 5.18 the ambiguity holds only for the flow rate $Q_{CN}$. The system of Eq. (5.12) can be solved iteratively, obtaining the results listed in Table 5.3.

Positive flow rates indicate that the flow direction is consistent with the one assumed in Fig. 5.19. The head in node N is equal to

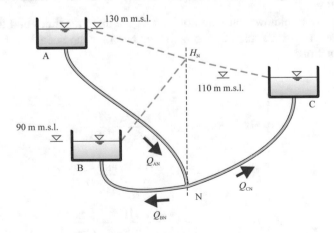

**Fig. 5.19**  Schematic of the flow rates

$$\therefore \quad H_N = H_A - \lambda_{AN} \frac{Q_{AN} |Q_{AN}|}{2g\Omega_{AN}^2} \frac{L_{AN}}{D_{AN}} =$$

$$130 - 0.015 \times \frac{2.00 \times |2.00|}{2 \times 9.806 \times \left(\frac{\pi \times 1.0^2}{4}\right)^2} \times \frac{1500}{1.0} = \textbf{122.5 m.}$$

**Exercise 5.13** The pipes in Fig. 5.20 are made of steel with roughness $\varepsilon = 0.36$ mm. A valve is installed on the duct NC.

- Calculate the flow rates to tanks B and C, if the valve is closed.
- Calculate the flow rates to tanks B and C, if the valve is open.

Neglect the concentrated energy losses. The fluid is water with $\gamma_w = 9806$ N m$^{-3}$.

**Solution** When the valve is closed, the flow rate to tank C is obviously zero. The flow rate to tank B is obtained by the following energy balance equation:

$$z_A - z_B = \lambda \frac{V^2}{2g} \frac{L}{D}.$$

We have neglected the concentrated energy losses. The first attempt friction factor value corresponds to Re $\to \infty$ for $\varepsilon/D = 3.6 \times 10^{-3}$, and is equal to $\lambda_\infty = 0.027$. The corresponding velocity is

$$V' = \sqrt{\frac{2g(z_A - z_B)D}{\lambda_\infty L}} = \sqrt{\frac{2 \times 9.806 \times (15 - 0) \times 0.1}{0.027 \times 120}} = \textbf{3.0 m s}^{-1}.$$

The Reynolds number is equal to Re $= V'D/\nu = 3.0 \times 0.1/10^{-6} = 300\,000$. Upon iteration, the approximate value of the friction factor is $\lambda = 0.028$, with a

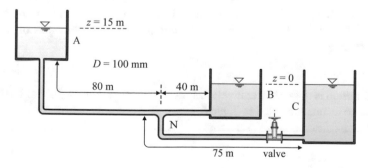

**Fig. 5.20** Pipeline network connecting three tanks, with a valve

**Table 5.4** Summary of the operating characteristics of the pipes

| Pipe | $V$ $(\mathrm{m\,s}^{-1})$ | $Q$ $(\mathrm{l\,s}^{-1})$ | $\lambda$ (.) |
|------|------|------|------|
| AN | 3.35 | **26.30** | 0.028 |
| NB | 1.94 | **15.23** | 0.028 |
| NC | 1.41 | **11.07** | 0.029 |

water velocity equal to

$$V = \sqrt{\frac{2g\,(z_\mathrm{A} - z_\mathrm{B})\,D}{\lambda L}} = \sqrt{\frac{2 \times 9.806 \times (15 - 0) \times 0.1}{0.028 \times 120}} = \mathbf{2.96\ m\,s^{-1}}.$$

The flow rate is equal to

$$\therefore \qquad Q = \frac{\pi D^2}{4} V = \frac{\pi \times 0.1^2}{4} \times 2.96 = \mathbf{23.2\,l\,s^{-1}}.$$

If the valve is open, the problem has four unknowns, namely the head in the common node N and the velocities in the three ducts. The following energy balance and mass conservation equations can be written:

$$\begin{cases} z_\mathrm{A} - z_\mathrm{N} = \lambda_\mathrm{AN} \dfrac{V_\mathrm{AN}^2}{2g} \dfrac{L_\mathrm{AN}}{D}, \\[2mm] z_\mathrm{N} - z_\mathrm{B} = \lambda_\mathrm{NB} \dfrac{V_\mathrm{NB}^2}{2g} \dfrac{L_\mathrm{NB}}{D}, \\[2mm] z_\mathrm{N} - z_\mathrm{C} = \lambda_\mathrm{NC} \dfrac{V_\mathrm{NC}^2}{2g} \dfrac{L_\mathrm{NC}}{D}, \\[2mm] Q_\mathrm{AN} = Q_\mathrm{NB} + Q_\mathrm{NC}. \end{cases}$$

The flows direction can only be from tank A to node N, and from node N to the two tanks B and C. Mass conservation equation in node N for pipes of the same diameter becomes

$$V_\mathrm{AN} = V_\mathrm{NB} + V_\mathrm{NC}.$$

The solution is obtained iteratively, searching for the correct values of the friction factors. The results are listed in Table 5.4.

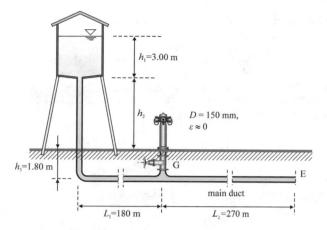

**Fig. 5.21**  Raised tank

**Exercise 5.14**  Calculate the minimum value of $h_2$ that is necessary to guarantee an outgoing flow rate of $28.3 \, \mathrm{l \, s^{-1}}$ from the tank of the plant in Fig. 5.21, with a relative pressure of $3.44 \times 10^5$ Pa in the terminal section. The flow rate is divided into two equal parts at the junction G. The duct is made of hydraulically smooth plastic and has a diameter of 150 mm.

Neglect the concentrated energy losses.

**Solution**  The energy balance equation between the tank and node G is

$$z_0 + \frac{p_0}{\gamma_w} + \frac{V_0^2}{2g} = z_G + \frac{p_G}{\gamma_w} + \frac{V_1^2}{2g} + \lambda_1 \frac{V_1^2}{2g} \frac{(L_1 + h_2 + h_3)}{D},$$

where $z_0$ and $z_G$ are the elevation head of the free surface in the tank and the elevation head of the node G, respectively. The energy balance equation between node G and the terminal section of the main duct is

$$z_G + \frac{p_G}{\gamma_w} + \frac{V_1^2}{2g} = z_E + \frac{p_E}{\gamma_w} + \frac{V_2^2}{2g} + \lambda_2 \frac{V_2^2}{2g} \frac{L_2}{D}.$$

Combining the two equations, yields

$$z_0 + \frac{p_0}{\gamma_w} + \frac{V_0^2}{2g} = z_E + \frac{p_E}{\gamma_w} + \frac{V_2^2}{2g} + \lambda_1 \frac{V_1^2}{2g} \frac{(L_1 + h_2 + h_3)}{D} + \lambda_2 \frac{V_2^2}{2g} \frac{L_2}{D}.$$

The average velocity of the water in the first duct is equal to

$$V_1 = \frac{4Q}{\pi D^2} = \frac{4 \times 28.3 \times 10^{-3}}{\pi \times (0.15)^2} = \mathbf{1.6 \, m \, s^{-1}}.$$

In the second duct, downstream of node G, the average velocity is equal to

$$V_2 = \frac{4\,(Q/2)}{\pi\,D^2} = \frac{4 \times (28.3/2) \times 10^{-3}}{\pi \times (0.15)^2} = \mathbf{0.8\ m\ s^{-1}}.$$

The friction factor in the first duct is obtained from the Moody chart or by solving the Coolebrook–White equation for smooth pipes:

$$\frac{1}{\sqrt{\lambda}} = -2\log_{10}\left(\frac{2.51}{\mathrm{Re}\sqrt{\lambda}}\right).$$

The Reynolds number in the two ducts is equal to

$$\mathrm{Re}_1 = \frac{V_1 D}{\nu} = \frac{1.6 \times 0.15}{10^{-6}} = \mathbf{240\ 000},$$

$$\mathrm{Re}_2 = \frac{V_2 D}{\nu} = \frac{0.8 \times 0.15}{10^{-6}} = \mathbf{120\ 000}.$$

The friction factors are

$$\lambda_1 = 0.015, \quad \lambda_2 = 0.0173.$$

By assuming $V_0 = 0$, $p_0 = 0$, and neglecting the velocity head in the second duct, it results

$$h_2 - \lambda_1 \frac{V_1^2}{2g}\frac{h_2}{D} = \frac{p_E}{\gamma_w} + \lambda_1 \frac{V_1^2}{2g}\frac{(L_1 + h_3)}{D} + \lambda_2 \frac{V_2^2}{2g}\frac{L_2}{D} - h_1 - h_3 \rightarrow$$

$$h_2 \times \left(1 - 0.015 \times \frac{1.6^2}{2 \times 9.806} \times \frac{1}{0.15}\right) = \frac{3.44 \times 10^5}{9806}$$

$$+0.015 \times \frac{1.6^2}{2 \times 9.806} \times \frac{180 + 1.8}{0.15} + 0.0173 \times \frac{0.8^2}{2 \times 9.806} \times \frac{270}{0.15} - 3.0 - 1.8.$$

$$(5.13)$$

Notice that $z_0 - z_E = h_1 + h_2 + h_3$. Equation (5.13) admits the solution

$$\therefore \qquad h_2 \approx \mathbf{34\ m}.$$

# Chapter 6
# Industrial Hydraulic Systems

Industrial hydraulic circuits are generally characterized by the presence of several curves, elbows, valves, junctions, manometers, flow meters, with dominance of concentrated losses on distributed losses. In most cases, pumps transfer power to the current, and one of the possible aim is transfer of power through the current. Calculations are similar to those for the cases of civil hydraulic circuits, paying more attention to the economic design of the components, in case of a new project. Some examples of hydraulic industrial circuits are also present in cars, as cooling circuits. In many cases, the circuits are closed, without variations of the elevation head.

**Exercise 6.1** The recirculation circuit shown in Fig. 6.1 works on a horizontal plane with a water flow rate $Q$. The pipes are made of steel with a Gauckler–Strickler coefficient $k$ and they have a uniform diameter $D$. The curves loss coefficient is equal to $\xi$. A pipe doubling, consisting of 2 tubes of diameter $d$, length $L_2$ with the same Gauckler–Strickler coefficient $k$, is realized by means of a T-collector. In the section immediately downstream of the pump, there is a pressure gauge indicating a pressure $p_n$. Determine:

- the power of the pump installed in the circuit, considering a total efficiency equal to $\eta$.
- The horizontal component of the force acting on the curve delimited by sections 1 and 2.

Assume $Q = (3 + 0.1 \times C_u) \ 1\,\mathrm{s}^{-1}$, $L = (10 + 0.1 \times C_{pu})$ m, $D = (40 + C_u)$ mm, $k = (80 + C_u)\ \mathrm{m}^{1/3}\,\mathrm{s}^{-1}$, $L_1 = (6 + 0.1 \times C_{pu})$ m, $\xi = 0.3$, $d = 15$ mm, $L_2 = (5 + 0.1 \times C_{pu})$ m, $p_n = 3 \times 10^5$ Pa, $\nu = 10^{-6}\,\mathrm{m}^2\,\mathrm{s}^{-1}$, $\gamma_w = 9806\,\mathrm{N\,m}^{-3}$.

**Solution** Since no variation of elevation head occurs, the pump prevalence balances only the concentrated and distributed energy losses:

---

$C_u$ and $C_{pu}$, that are two integer numbers between 0 and 9, for example, the last and second-last digit of the registration number.

© Springer Nature Switzerland AG 2021
S. Longo et al., *Problems in Hydraulics and Fluid Mechanics*, Springer Tracts in Civil Engineering, https://doi.org/10.1007/978-3-030-51387-0_6

**Fig. 6.1** Schematic of the
recirculation circuit

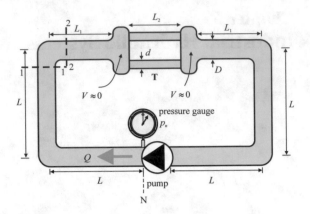

The power of the pump can be expressed as

$$ P = \frac{\gamma_w Q \, \Delta H}{\eta}. $$

$$ \Delta H = \underbrace{\frac{4^{10/3}}{\pi^2 k^2 D^{5.33}} (4L + 2L_1) \, Q^2}_{\substack{\text{distributed loss in the ducts} \\ \text{of diameter } D}} + \underbrace{\frac{4^{10/3}}{\pi^2 k^2 d^{5.33}} L_2 \left(\frac{Q}{2}\right)^2}_{\substack{\text{distributed loss in the ducts} \\ \text{of diameter } d}} $$

$$ + \underbrace{4\xi \frac{16 Q^2}{\left(\pi D^2\right)^2} \frac{1}{2g}}_{\substack{\text{concentrated loss} \\ \text{in the curves}}} + \underbrace{\frac{16 Q^2}{\left(\pi D^2\right)^2} \frac{1}{2g}}_{\substack{\text{outlet loss in the} \\ \text{left collector}}} $$

$$ + \underbrace{0.5 \frac{16 (Q/2)^2}{\left(\pi d^2\right)^2} \frac{1}{2g}}_{\substack{\text{inlet loss in the duct} \\ \text{of diameter } d}} + \underbrace{\frac{16 (Q/2)^2}{\left(\pi d^2\right)^2} \frac{1}{2g}}_{\substack{\text{outlet loss in the} \\ \text{right collector}}} + \underbrace{0.5 \frac{16 Q^2}{\left(\pi D^2\right)^2} \frac{1}{2g}}_{\substack{\text{inlet loss in the duct} \\ \text{of diameter } D}} . $$

The power of the pump can be expressed as

$$ P = \frac{\gamma_w Q \, \Delta H}{\eta}. $$

For the calculation of the force acting on the curve, we apply the linear momentum balance to the control volume delimited by sections 1 and 2 and by the walls of the curve, see Fig. 6.2:

$$ \cancel{G} + \Pi + \cancel{I} + M_1 - M_2 = 0 \rightarrow -F + \Pi_1 + \Pi_2 + M_1 - M_2 = 0, $$

where $-F$ is the force exerted by the curve on the control volume. The weight term **G** in the horizontal plane is null.

**Fig. 6.2**  Schematic for
calculating the forces acting
on the curve

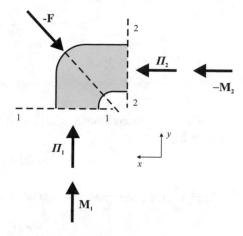

The calculation of the forces $\Pi_1$ and $\Pi_2$ requires the knowledge of the pressure in section 1 and section 2. We use the energy balance to evaluate the total head in section 1 starting from section N, where the pressure head is known:

$$H_1 = H_N - \underbrace{\frac{4^{10/3}}{\pi^2 k^2 D^{5.33}} (2l.) Q^2}_{\substack{\text{distributed loss in the ducts} \\ \text{of diameter } D}} - \underbrace{\xi \frac{16Q^2}{\left(\pi D^2\right)^2} \frac{1}{2g}}_{\substack{\text{concentrated loss} \\ \text{in the curve}}} \rightarrow$$

$$\cancel{z_1} + \frac{p_1}{\gamma_w} + \cancel{\frac{V_1^2}{2g}} = \cancel{z_n} + \frac{p_n}{\gamma_w} + \cancel{\frac{V_1^2}{2g}} - \underbrace{\frac{4^{10/3}}{\pi^2 k^2 D^{5.33}} (2L) Q^2}_{\substack{\text{distributed loss in the} \\ \text{ducts of diameter } D}} - \underbrace{\xi \frac{16Q^2}{\left(\pi D^2\right)^2} \frac{1}{2g}}_{\substack{\text{concentrated loss} \\ \text{in the curve}}} .$$

In a similar manner, the pressure acting in section 2 is calculated starting from the total head in section 1:

$$H_2 = H_1 - \underbrace{\xi \frac{16Q^2}{\left(\pi D^2\right)^2} \frac{1}{2g}}_{\substack{\text{concentrated loss} \\ \text{in the curve}}} \rightarrow$$

$$\cancel{z_2} + \frac{p_2}{\gamma_w} + \cancel{\frac{V_2^2}{2g}} = \cancel{z_1} + \frac{p_1}{\gamma_w} + \cancel{\frac{V_1^2}{2g}} - \underbrace{\xi \frac{16Q^2}{\left(\pi D^2\right)^2} \frac{1}{2g}}_{\substack{\text{concentrated loss} \\ \text{in the curve}}} \cdot$$

The incoming momentum flux $\mathbf{M}_1$ is equal, in magnitude, to the outgoing momentum flux $\mathbf{M}_2$:

$$|\mathbf{M}_1| = |\mathbf{M}_2| = \rho \frac{4Q^2}{\pi D^2}.$$

Applying the linear momentum balance in the $y$-direction, yields

$$F_y = p_1 \frac{\pi D^2}{4} + \rho \frac{4Q^2}{\pi D^2},$$

in the positive direction of the $y$ axis.

In the $x$ direction, yields

$$F_x = p_2 \frac{\pi D^2}{4} + \rho \frac{4Q^2}{\pi D^2},$$

in the positive direction of the $x$ axis.

For $C_u = C_{pu} = 0$ it results $Q = 31\,\text{s}^{-1}, L = 10\,\text{m}, D = 40\,\text{mm}, k = 80\,\text{m}^{1/3}\,\text{s}^{-1}$, $L_1 = 6\,\text{m},\ \xi = 0.3,\ d = 15\,\text{mm},\ L_2 = 5\,\text{m},\ p_n = 3 \times 10^5\,\text{Pa},\ \nu = 10^{-6}\,\text{m}^2\,\text{s}^{-1}$, $\gamma_w = 9806\,\text{N}\,\text{m}^{-3}, \rho = 1000\,\text{kg}\,\text{m}^{-3}$,

$$\Delta H = \underbrace{\frac{4^{10/3}}{\pi^2 \times 80^2 \times 0.04^{5.33}} \times (4 \times 10 + 2 \times 6) \times \left(3 \times 10^{-3}\right)^2}_{\text{distributed loss in the ducts of diameter } D}$$

$$+ \underbrace{\frac{4^{10/3}}{\pi^2 \times 80^2 \times 0.015^{5.33}} \times \left(\frac{3 \times 10^{-3}}{2}\right)^2 \times 5}_{\text{distributed loss in the ducts of diameter } d}$$

$$+ \underbrace{4 \times 0.3 \times \frac{16 \times \left(3 \times 10^{-3}\right)^2}{\left(\pi \times 0.04^2\right)^2} \times \frac{1}{2 \times 9.806}}_{\text{concentrated loss in the curves}} + \underbrace{\frac{16 \times \left(3 \times 10^{-3}\right)^2}{\left(\pi \times 0.04^2\right)^2} \times \frac{1}{2 \times 9.806}}_{\text{outlet loss in the left collector}}$$

$$+ \underbrace{0.5 \times \frac{16 \times \left(3 \times 10^{-3}/2\right)^2}{\left(\pi \times 0.015^2\right)^2} \times \frac{1}{2 \times 9.806}}_{\text{inlet loss in the duct of diameter } d} + \underbrace{\frac{16 \times \left(3 \times 10^{-3}/2\right)^2}{\left(\pi \times 0.015^2\right)^2} \times \frac{1}{2 \times 9.806}}_{\text{outlet loss in the right collector}}$$

$$+0.5 \times \underbrace{\frac{16 \times \left(3 \times 10^{-3}\right)^2}{\left(\pi \times 0.04^2\right)^2} \times \frac{1}{2 \times 9.806}}_{\text{inlet loss in the duct of diameter } D} =$$

$$21.26 + 95.24 + 0.35 + 0.29 + 1.84 + 3.67 + 0.14 = \mathbf{122.79\,m},$$

$$\therefore \qquad P = \frac{\gamma_w \, Q \, \Delta H}{\eta} = \frac{9806 \times 0.003 \times 122.79}{0.75} = \mathbf{4.82\,kW},$$

$$p_1 = p_n - \gamma_w \frac{4^{10/3}}{\pi^2 k^2 D^{5.33}} \, (2L) \, Q^2 - \gamma_w \xi \frac{16 Q^2}{\left(\pi D^2\right)^2} \frac{1}{2g} =$$

$$3 \times 10^5 - 9806 \times \frac{4^{10/3}}{80^2 \times 0.04^{5.33}} \times (2 \times 10) \times \left(3 \times 10^{-3}\right)^2$$

$$- 9806 \times 0.3 \times \frac{16 \times \left(3 \times 10^{-3}\right)^2}{\left(\pi \times 0.04^2\right)^2} \times \frac{1}{2 \times 9.806} = \mathbf{218\,973\,Pa},$$

$$p_2 = p_1 - \gamma_w \xi \frac{16 Q^2}{\left(\pi D^2\right)^2} \frac{1}{2g} =$$

$$218\,973 - 9806 \times 0.3 \times \frac{16 \times \left(3 \times 10^{-3}\right)^2}{\left(\pi \times 0.04^2\right)^2} \times \frac{1}{2 \times 9.806} = \mathbf{218\,118\,Pa},$$

$$\therefore \quad F_y = p_1 \frac{\pi D^2}{4} + \rho \frac{4 Q^2}{\pi D^2} =$$

$$218\,973 \times \frac{\pi \times 0.04^2}{4} + 1000 \times \frac{4 \times \left(3 \times 10^{-3}\right)^2}{\pi \times 0.04^2} = \mathbf{282\,N},$$

$$\therefore \quad F_x = p_2 \frac{\pi D^2}{4} + \rho \frac{4 Q^2}{\pi D^2} =$$

$$218\,118 \times \frac{\pi \times 0.04^2}{4} + 1000 \times \frac{4 \times \left(3 \times 10^{-3}\right)^2}{\pi \times 0.04^2} = \mathbf{281\,N}.$$

---

**Exercise 6.2**  In the plant in Fig. 6.3 the two tanks are pressurized and connected by a steel pipe with a diameter of $D = 50\,\text{mm}$. The reading of the manometers (gage pressure) is equal to $p_A = (15 + C_u/10) \times 10^5\,\text{Pa}$ and $p_B = (8 + C_{pu}/10) \times 10^5\,\text{Pa}$,

**Fig. 6.3**  Plant with two
pressurized tanks connected
by a pipeline with a control
valve

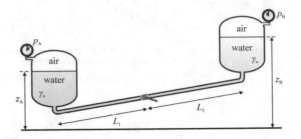

and the levels are equal to $z_A = (30 + C_u)$ m and $z_B = (40 + C_u)$ m, respectively.
A valve separates the pipeline into two ducts of length $L_1 = (35 + C_u)$ m and
$L_2 = (30 + C_u)$ m.

- Calculate the flow rate if the valve is fully open.
- Calculate the flow rate if the valve is partially open, with an opening degree
  $\eta = 0.5$.
- For the last configuration, draw the energy grade line.

For the pipeline, assume $k = 90 \, \text{m}^{1/3} \, \text{s}^{-1}$. The fluid is water with specific gravity $\gamma_w = 9800 \, \text{N} \, \text{m}^{-3}$. The concentrated energy losses coefficients are $\xi_{in} = 0.5$,
$\xi_{out} = 1$ and $\xi_c = 0.3$. The energy losses in the valve are related to the opening
degree of opening according to the following relationship: $\Delta H = \xi V^2/(2g)$, with
$\xi = (1/\eta - 1)^2$.

**Solution**  The energy balance equation for the system is

$$z_A + \frac{p_A}{\gamma_w} - z_B - \frac{p_B}{\gamma_w} = J_1 L_1 + J_2 L_2 + \left[ \xi_{in} + \xi_{out} + 2\xi_c + \left(\frac{1}{\eta} - 1\right)^2 \right] \frac{V^2}{2g},$$

where $\xi_{in}$, $\xi_{out}$ and $\xi_c$ are the energy loss coefficients at the inlet, outlet and in the
curve, respectively.

The energy loss in the valve is equal to

$$\Delta H_v = \left(\frac{1}{\eta} - 1\right)^2 \frac{V^2}{2g}.$$

The distributed energy loss per unit length is equal to

$$J_1 \equiv J_2 = \frac{4^{4/3} V^2}{k^2 D^{4/3}},$$

hence,

$$V = \sqrt{\frac{2g\left(z_A + \dfrac{p_A}{\gamma_w} - z_B - \dfrac{p_B}{\gamma_w}\right)}{\left[2g\left(L_1 + L_2\right)\dfrac{4^{4/3}}{k^2 D^{4/3}} + \xi_{in} + \xi_{out} + 2\xi_c + \left(\dfrac{1}{\eta} - 1\right)^2\right]}}.$$

The flow rate is equal to

$$Q = V\frac{\pi D^2}{4}.$$

For $C_u = C_{pu} = 0$ it results $D = 50\,\text{mm}$, $p_A = 1.5\times 10^6\,\text{Pa}$, $p_B = 8\times 10^5\,\text{Pa}$, $z_A = 30\,\text{m}$, $z_B = 40\,\text{m}$, $L_1 = 35\,\text{m}$, $L_2 = 30\,\text{m}$.

For $\eta = 1$ (completely open valve), it results

$$V = \sqrt{\frac{2g\left(z_A + \dfrac{p_A}{\gamma_w} - z_B - \dfrac{p_B}{\gamma_w}\right)}{\left[2g\left(L_1 + L_2\right)\dfrac{4^{4/3}}{k^2 D^{4/3}} + \xi_{in} + \xi_{out} + 2\xi_c + \left(\dfrac{1}{\eta} - 1\right)^2\right]}} =$$

$$\sqrt{\frac{2\times 9.806 \times \left(30 + \dfrac{15\times 10^5}{9800} - 40 - \dfrac{8\times 10^5}{9800}\right)}{\left[2\times 9.806 \times (35 + 30)\times \dfrac{4^{4/3}}{90^2 \times 0.05^{4/3}} + 0.5 + 1 + 2\times 0.3 + \left(\dfrac{1}{1} - 1\right)^2\right]}}$$

$$= 4.62\,\text{m s}^{-1}.$$

The flow rate is equal to

$$\therefore \qquad Q = V\frac{\pi D^2}{4} = 4.62\times \frac{\pi \times 0.05^2}{4} = 9.11\text{s}^{-1}.$$

For $\eta = 0.5$ (half-opened valve), results

$$V = \sqrt{\frac{2g\left(z_A + \dfrac{p_A}{\gamma_w} - z_B - \dfrac{p_B}{\gamma_w}\right)}{\left[2g\left(L_1 + L_2\right)\dfrac{4^{4/3}}{k^2 D^{4/3}} + \xi_{in} + \xi_{out} + 2\xi_c + \left(\dfrac{1}{\eta} - 1\right)^2\right]}} =$$

$$\sqrt{\frac{2\times 9.806 \times \left(30 + \dfrac{15\times 10^5}{9800} - 40 - \dfrac{8\times 10^5}{9800}\right)}{\left[2\times 9.806 \times (35 + 30)\times \dfrac{4^{4/3}}{90^2 \times 0.05^{4/3}} + 0.5 + 1 + 2\times 0.3 + \left(\dfrac{1}{0.5} - 1\right)^2\right]}}$$

$$= 4.58\,\text{m s}^{-1},$$

**Fig. 6.4** Energy grade line

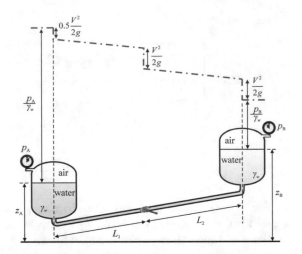

$$\therefore \qquad Q = V\frac{\pi D^2}{4} = 4.58 \times \frac{\pi \times 0.05^2}{4} = 9.01 \text{s}^{-1}.$$

The velocity head is equal to

$$\frac{V^2}{2g} = \frac{4.58^2}{2 \times 9.806} = 1.07 \text{ m}.$$

The energy loss in the valve is equal to

$$\Delta H_v = \left(\frac{1}{\eta} - 1\right)^2 \frac{V^2}{2g} = \left(\frac{1}{0.5} - 1\right)^2 \times \frac{4.58^2}{2 \times 9.806} = 1.07 \text{ m}.$$

The energy grade line is shown in Fig. 6.4.

---

**Exercise 6.3** In the closed circuit system in Fig. 6.5, which lays in the horizontal plane, a Venturi meter is installed for the measurement of the flow rate. The indication of the differential pressure gauge is equal to $\Delta p = (0.05 + 0.005 \times C_u) \times 10^5$ Pa and the Venturi meter velocity coefficient is equal to $C_v = 0.984$. The diameter of the throat is $d = (50 + 5 \times C_{pu})$ mm, the diameter of the entry section, coinciding with the actual diameter of the pipeline, is $D = (75 + 5 \times C_{pu})$ mm. The length of the steel duct, with a Gauckler–Strickler coefficient $k = 90 \text{ m}^{1/3} \text{ s}^{-1}$, is $L = (25 + 10 \times C_u)$ m. The concentrated loss coefficient for each curve is $\xi_c = 0.3$, the concentrated loss coefficient for the heat exchanger is $\xi_s = 1.65$, the energy loss in the Venturi meter is given by the following expression:

**Fig. 6.5** Closed circuit system with heat exchanger

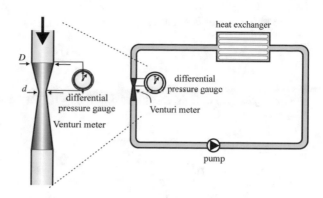

$$\Delta H_{Venturi} = \left[ 0.218 - 0.42\frac{d}{D} + 0.38\left(\frac{d}{D}\right)^2 \right] \frac{\Delta p}{\gamma},$$

where $\Delta p$ is the difference in pressure indicated by the differential pressure gauge.

- Calculate the flow rate in permanent flow condition.
- Calculate the power of the pump.

The fluid is water with $\gamma_w = 9806\ \text{N m}^{-3}$, the efficiency of the pump is $\eta = 0.8$.

**Solution** Applying Bernoulli's theorem between the entry section and the throat, we obtain the following expression of the flow rate through the Venturimeter:

$$Q = C_v \frac{\pi d^2}{4} \sqrt{\frac{\frac{2g\,\Delta p}{\gamma_w}}{1 - \left(\frac{d}{D}\right)^4}},$$

where $C_v$ is the discharge coefficient which embeds the dissipations, $\Delta p$ is the pressure difference at the pressure tappings. The pump prevalence must only balance the concentrated and distributed energy losses, and is equal to

$$\Delta H = JL + \sum \xi_c \frac{V^2}{2g} + \xi_s \frac{V^2}{2g} + \Delta H_{Venturi} =$$

$$\frac{4^{4/3} Q^2}{k^2 D^{4/3} \left(\frac{\pi D^2}{4}\right)^2} L + (4\xi_c + \xi_s) \frac{Q^2}{2g \left(\frac{\pi D^2}{4}\right)^2}$$

$$+ \left[ 0.218 - 0.42\frac{d}{D} + 0.38\left(\frac{d}{D}\right)^2 \right] \frac{\Delta p}{\gamma_w}.$$

The pump power is equal to

$$P = \frac{\gamma_w Q \, \Delta H}{\eta}.$$

For $C_u = C_{pu} = 0$ it results $\Delta p = 0.05 \times 10^5$ Pa, $d = 50$ mm, $D = 75$ mm, $L = 25$ m,

$$\therefore \quad Q = C_v \frac{\pi d^2}{4} \sqrt{\frac{\dfrac{2g \, \Delta p}{\gamma_w}}{1 - \left(\dfrac{d}{D}\right)^4}} =$$

$$0.984 \times \frac{\pi \times 0.05^2}{4} \times \sqrt{\frac{\dfrac{2 \times 9.806 \times 0.05 \times 10^5}{9806}}{1 - \left(\dfrac{0.05}{0.075}\right)^4}} = 6.82 \times 10^{-3} \, \text{m}^3 \, \text{s}^{-1}.$$

The average velocity of the water in the pipeline is $V = 4Q/(\pi D^2) = 4 \times 6.82 \times 10^{-3}/(\pi \times 0.075^2) = 1.54 \, \text{m s}^{-1}$. The energy loss in the Venturi meter is equal to

$$\Delta H_{Venturi} = \left[0.218 - 0.42 \frac{d}{D} + 0.38 \left(\frac{d}{D}\right)^2\right] \frac{\Delta p}{\gamma_w} =$$

$$\left[0.218 - 0.42 \times \frac{0.05}{0.075} + 0.38 \times \left(\frac{0.05}{0.075}\right)^2\right] \times \frac{0.05 \times 10^5}{9806} = 0.05 \, \text{m},$$

and the energy loss in the circuit is equal to

$$\Delta H = \frac{4^{4/3} Q^2}{k^2 D^{4/3} \left(\dfrac{\pi D^2}{4}\right)^2} L + (4\xi_c + \xi_s) \frac{Q^2}{2g \left(\dfrac{\pi D^2}{4}\right)^2} + \Delta H_{Venturi} =$$

$$\frac{Q^2}{\left(\dfrac{\pi D^2}{4}\right)^2} \left(\frac{4^{4/3} L}{k^2 D^{4/3}} + \frac{4\xi_c + \xi_s}{2g}\right) + \Delta H_{Venturi} =$$

$$\frac{\left(6.82 \times 10^{-3}\right)^2}{\left(\dfrac{\pi \times 0.075^2}{4}\right)^2} \times \left(\frac{4^{4/3} \times 25}{90^2 \times 0.075^{4/3}} + \frac{4 \times 0.3 + 1.65}{2 \times 9.806}\right) + 0.05 = 1.87 \, \text{m}.$$

The pump power is equal to

$$\therefore \qquad P = \frac{\gamma_w Q \, \Delta H}{\eta} = \frac{9806 \times 6.82 \times 10^{-3} \times 1.87}{0.8} = \mathbf{156 \, W}.$$

---

**Exercise 6.4** In the plant in Fig. 6.6 water is pumped into a conduit 60 m long, with a diameter $D = 30$ mm and a relative roughness equal to 0.01. The water passes through a filter with a loss coefficient of 12.0, and through five elbows with a loss coefficient of 1.5 for each elbow. The loss coefficients at the inlet and at the outlet are equal to 0.8 and 1.0, respectively. The valve has a loss coefficient of 6.0. Calculate:

– the flow rate in the circuit, if the power transferred to the fluid by the pump is 270 W.
– The ratio of concentrated versus distributed energy losses.

   **Solution** The pump prevalence must balance only the distributed and concentrated energy losses in the circuit, hence

$$\Delta H = \left( \xi_{in} + 5\xi_c + \xi_{filter} + \xi_v + \xi_{out} \right) \frac{V^2}{2g} + \lambda \frac{V^2}{2g} \frac{L}{D},$$

and, as a function of the flow rate,

$$\Delta H = \left( \xi_{in} + 5\xi_c + \xi_{filter} + \xi_v + \xi_{out} + \lambda \frac{L}{D} \right) \frac{Q^2}{2g\Omega^2},$$

where $\Omega = \pi D^2 / 4$.

   The power transferred to the current is equal to

$$P = \gamma_w Q \, \Delta H,$$

**Fig. 6.6** Filtering circuit

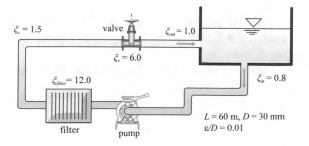

hence,

$$P = \gamma_w Q \, \Delta H \equiv \left( \xi_{in} + 5\xi_c + \xi_{filter} + \xi_v + \xi_{out} + \lambda \frac{L}{D} \right) \frac{\gamma_w Q^3}{2g\Omega^2}.$$

Solving with respect to $Q$, yields

$$Q = \left[ \frac{2g\Omega^2 P}{\gamma_w \left( \xi_{in} + 5\xi_c + \xi_{filter} + \xi_v + \xi_{out} + \lambda \frac{L}{D} \right)} \right]^{1/3},$$

where the friction factor $\lambda$ depends on $Q$. The equation can be solved iteratively by assuming a first attempt value of the friction factor equal to its asymptotic value $(\text{Re} \to \infty)$,

$$\lambda = \lambda_\infty = 0.038.$$

By substituting the numerical values, yields

$$\therefore \quad Q =$$

$$\left[ \frac{2 \times 9.806 \times \left( \pi \times 0.25 \times 0.03^2 \right)^2 \times 270}{9806 \times \left( 0.8 + 5 \times 1.5 + 12.0 + 6.0 + 1.0 + 0.038 \times \frac{60}{0.03} \right)} \right]^{1/3}$$

$$= 1.381 \, \text{s}^{-1}.$$

The average velocity in the pipe is equal to

$$V = \frac{4Q}{\pi D^2} = \frac{4 \times 0.00138}{\pi \times 0.03^2} = 1.95 \, \text{m s}^{-1},$$

with a Reynolds number

$$\text{Re} = \frac{VD}{\nu} = \frac{1.95 \times 0.03}{10^{-6}} = 58\,500.$$

The flow regime is transitional. The correct friction factor is calculated by iteration and it assumes a value equal to $\lambda = 0.039$, with a corresponding flow rate $Q = 1.371 \, \text{s}^{-1}$.

The ratio between the concentrated and distributed energy losses is equal to

$$\therefore \quad \frac{\xi_{in} + 5\xi_c + \xi_{filter} + \xi_v + \xi_{out}}{\lambda \frac{L}{D}} = \frac{0.8 + 5 \times 1.5 + 12.0 + 6.0 + 1.0}{0.039 \times \frac{60}{0.03}} = 0.35.$$

# Chapter 7
# Circuits with Hydraulic Machines: Pumps and Turbines

In some hydraulic circuits, power is exchanged between the current and the machines. Pumps supply energy to the current and are called *energy absorption devices*, and are inserted in order to increase the head; turbines subtract energy from the current, usually transferring it to an electric generator, and are called *energy production devices*. The energy grade line has a jump across the machine. The characteristics of the machine are suggested by the general characteristics of the plant and the design is optimal, with the aim of minimizing costs and maximizing efficiency.

In the energy balance, the pump increases the head according to a *performance curve* (or *characteristic curve*), a function that relates head, flow rate, speed of rotation, with an efficiency closely linked to the mechanical design of the components. The operating point is obtained as the intersection between the system curve and the performance curve: the system curve is representative of the head required to modify the static pressure, the dynamic pressure, the elevation of the fluid and to overcome the losses. However, for simplicity it is often assumed, in the exercises, that the pump is ideal, with a constant efficiency.

The turbines reduce the head and generate energy according to an efficiency curve, where the rotation rate is set according to the characteristics of the electric generator and the frequency of generated alternating current (50 Hz in Europe, 60 Hz in the USA). A distinction is made between the *gross head*, which is the difference in level between the free surface level of the upstream tank and the free surface level of the downstream channel (or lake, or tank); and the *net head*, which is the difference between the immediately upstream section energy grade line and the outlet section of the draught pipe (or, in some cases, the immediately downstream section energy grade line). The draft pipe is designed as a diffuser to reduce energy losses. The difference between the gross head and the neat head is due to energy losses in the external hydraulic circuit. This difference is a function of the characteristics of the

---

$C_u$ and $C_{pu}$, that are two integer numbers between 0 and 9, for example, the last and second-last digit of the registration number.

© Springer Nature Switzerland AG 2021
S. Longo et al., *Problems in Hydraulics and Fluid Mechanics*, Springer Tracts in Civil Engineering, https://doi.org/10.1007/978-3-030-51387-0_7

external circuits, which is designed in order to achieve the maximum economic advantage.

**Exercise 7.1** A cast-iron pipe ($\varepsilon = 0.25\,\text{mm}$) consisting of a series of two ducts, with length $L_1 = (20 + C_u)\,\text{m}$, $L_2 = (35 + C_{pu})\,\text{m}$, and diameter $D_1 = 80\,\text{mm}$, $D_2 = 60\,\text{mm}$, delivers water from the tank in Fig. 7.1. Notice that the two ducts represent the suction and the discharge lines of a booster pump.

The tank contains water at $15\,°\text{C}$, with $\rho = 999.1\,\text{kg m}^{-3}$, $\mu = 1.138 \times 10^{-3}\,\text{Pa s}$, and with depth $h = (20 + C_u)\,\text{m}$.

– Calculate the pump head that is required for a flow rate $Q = 16\,\text{l s}^{-1}$.
– Calculate the corresponding pump power, assuming an efficiency $\eta = 0.75$.
– Calculate the pressure in the pipeline immediately upstream of the pump.

Assume loss coefficients at the inlet and outlet sections equal to $\xi_{in} = 0.5$ and $\xi_{out} = 1.4$, respectively. For the interpretation of $\xi_{out}$, see the comment in Exercise 5.1.

**Solution** The energy balance from the tank to the outflow section is

$$h + \Delta H_{pump} = \xi_{in}\frac{V_1^2}{2g} + \lambda_1\frac{V_1^2}{2g}\frac{L_1}{D_1} + \lambda_2\frac{V_2^2}{2g}\frac{L_2}{D_2} + \xi_{out}\frac{V_2^2}{2g},$$

where $\Delta H_{pump}$ is the head of the pump. The velocities in the two pipe sections can be expressed as a function of the flow rate as

$$V_1 = \frac{4Q}{\pi D_1^2}, \quad V_2 = \frac{4Q}{\pi D_2^2}.$$

For the calculation of the friction factors, the Reynolds numbers to be used in the Moody chart, or in the Colebrook–White equation, are:

$$\text{Re}_1 = \frac{\rho V_1 D_1}{\mu}, \quad \text{Re}_2 = \frac{\rho V_2 D_2}{\mu}.$$

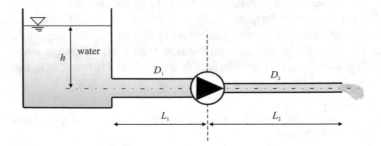

**Fig. 7.1** Tank with booster pump

Once the required pump head has been calculated, the power is equal to

$$P = \frac{\gamma_w Q \Delta H_{pump}}{\eta}.$$

For the calculation of the pressure in the pipeline in the section immediately upstream of the pump, we apply the energy balance equation between the tank and that section:

$$\underbrace{z_t + \frac{p_t}{\gamma_w} + \frac{V_t^2}{2g}}_{h} - \xi_{in}\frac{V_1^2}{2g} - \lambda_1\frac{V_1^2}{2g}\frac{L_1}{D_1} = z_u + \frac{p_u}{\gamma_w} + \frac{V_1^2}{2g} \rightarrow$$

$$p_u = \gamma_w\left[h - (1 + \xi_{in})\frac{V_1^2}{2g} - \lambda_1\frac{V_1^2}{2g}\frac{L_1}{D_1}\right],$$

where $z_t$ and $p_t/\gamma_w$ are the elevation head and pressure head for a particle in the tank moving on a pathline.

For $C_u = C_{pu} = 0$ it results $\varepsilon = 0.25\,\text{mm}$, $L_1 = 20\,\text{m}$, $L_2 = 35\,\text{m}$, $D_1 = 80\,\text{mm}$, $D_2 = 60\,\text{mm}$, $\rho = 999.1\,\text{kg m}^{-3}$, $\gamma_w \equiv g\rho = 9.806 \times 999.1\,\text{N m}^{-3}$, $\mu = 1.138 \times 10^{-3}\,\text{Pa s}$, $h = 20\,\text{m}$.

$$V_1 = \frac{4Q}{\pi D_1^2} = \frac{4 \times 16 \times 10^{-3}}{\pi \times 0.08^2} = 3.18\,\text{m s}^{-1},$$

$$V_2 = \frac{4Q}{\pi D_2^2} = \frac{4 \times 16 \times 10^{-3}}{\pi \times 0.06^2} = 5.66\,\text{m s}^{-1},$$

$$Re_1 = \frac{\rho V_1 D_1}{\mu} = \frac{999.1 \times 3.18 \times 0.08}{1.138 \times 10^{-3}} = 223\,000,$$

$$Re_2 = \frac{\rho V_2 D_2}{\mu} = \frac{999.1 \times 5.66 \times 0.06}{1.138 \times 10^{-3}} = 298\,000.$$

By solving the Colebrook–White equation:

$$\frac{1}{\sqrt{\lambda}} = -2\log_{10}\left(\frac{2.51}{Re\sqrt{\lambda}} + \frac{1}{3.71}\frac{\varepsilon}{D}\right),$$

yields

$$\lambda_1 = 0.027, \quad \lambda_2 = 0.029.$$

Hence,

$$\Delta H_{pump} = -h + \xi_{in}\frac{V_1^2}{2g} + \lambda_1\frac{V_1^2}{2g}\frac{L_1}{D_1} + \lambda_2\frac{V_2^2}{2g}\frac{L_2}{D_2} + \xi_{out}\frac{V_2^2}{2g} =$$

$$- 20 + 0.5 \times \frac{3.18^2}{2 \times 9.806} + 0.027 \times \frac{3.18^2}{2 \times 9.806} \times \frac{20}{0.08}$$

$$+ 0.029 \times \frac{5.66^2}{2 \times 9.806} \times \frac{35}{0.06} + \frac{5.66^2}{2 \times 9.806} = \mathbf{13.00\ m,}$$

$$\therefore \quad P = \frac{\gamma_w Q \Delta H_{pump}}{\eta} = \frac{9.806 \times 999.1 \times 16 \times 10^{-3} \times 13.00}{0.75} = \mathbf{2718\ W,}$$

$$\therefore \quad p_u = \gamma_w\left[h - (1 + \xi_{in})\frac{V_1^2}{2g} - \lambda_1\frac{V_1^2}{2g}\frac{L_1}{D_1}\right] = 9.806 \times 999.1$$

$$\times \left(20 - 1.5 \times \frac{3.18^2}{2 \times 9.806} - 0.027\frac{3.18^2}{2 \times 9.806} \times \frac{20}{0.08}\right) = \mathbf{1.54 \times 10^5\ Pa.}$$

---

**Exercise 7.2**  In the pumping plant shown in Fig. 7.2, the tank B is pressurized. All the characteristics of the pipes, the power and the efficiency of the pump are known.

– Calculate the flow rate.
– Draw the energy and the hydraulic grade lines.

The manometer indicates the gage pressure. Numerical data: $\Delta H_g = (25 + C_{pu})$ m, $p_m = (1 + C_u \times 0.1) \times 10^5$ Pa, $D_a = D_m = 300$ mm, $\varepsilon_a = \varepsilon_m = 0.2$ mm, $L_a = 20$ m, $L_m = 400$ m, $P = (50 + C_u)$ kW, $\eta = 0.75$, $\gamma_w = 9800$ N m$^{-3}$, $\xi_{in} = 0.5$, $\xi_{out} = 1$, $v = 10^{-6}$ m$^2$ s$^{-1}$.

**Solution**  The power required is equal to

$$P = \frac{\gamma_w Q\ \Delta H}{\eta},$$

where $\Delta H$ is the net head, given by the sum of the elevation head (due to the different levels of the two tanks), the pressure head (due to the difference of pressure between the two tanks), and the friction losses:

$$\Delta H = \Delta H_g + \lambda_a\frac{L_a}{D_a}\frac{V_a^2}{2g} + \lambda_m\frac{L_m}{D_m}\frac{V_m^2}{2g} + \xi_{in}\frac{V_a^2}{2g} + \xi_{out}\frac{V_m^2}{2g} + \frac{p_m}{\gamma_w},$$

or:

$$\Delta H = \Delta H_g + \left(\lambda \frac{L_a + L_m}{D} + \xi_{in} + \xi_{out}\right) \frac{Q^2}{2g\left(\pi D^2/4\right)^2} + \frac{p_m}{\gamma_w},$$

where $p_m$ is the reading of the manometer (gage). Hence,

$$P = \frac{\gamma_w Q \, \Delta H}{\eta} =$$

$$\frac{1}{\eta}\gamma_w Q\left[\Delta H_g + \left(\lambda \frac{L_a + L_m}{D} + \xi_{in} + \xi_{out}\right) \frac{Q^2}{2g\left(\pi D^2/4\right)^2} + \frac{p_m}{\gamma_w}\right].$$

$$(7.1)$$

Equation (7.1) can be solved numerically. As a first approximation, we assume a fully turbulent flow and we estimate the friction factor with the Moody chart, $\text{Re} \to \infty$, or solving the Prandtl–Nikuradse equation:

$$\frac{1}{\sqrt{\lambda_\infty}} = -2\log_{10}\left(\frac{1}{3.71}\frac{\varepsilon}{D}\right).$$

Introducing $\lambda_\infty$ into Eq. (7.1), we can evaluate $Q$, $V$, Re and than we check if the hypothesis of fully turbulent flow is verified. Otherwise, the friction factor is recalculated on the basis of the Reynolds number with the first iteration values, using the Moody chart or solving the Colebrook–White equation:

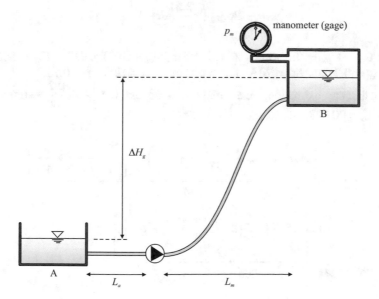

**Fig. 7.2** Pumping plant

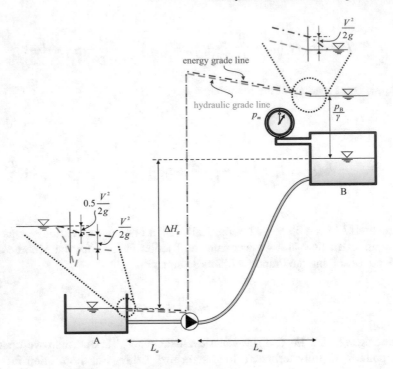

**Fig. 7.3** Energy and hydraulic grade lines

$$\frac{1}{\sqrt{\lambda}} = -2\log_{10}\left(\frac{2.51}{\mathrm{Re}\sqrt{\lambda}} + \frac{1}{3.71}\frac{\varepsilon}{D}\right).$$

For $C_u = C_{pu} = 0$ it results $\Delta H_g = 25$ m, $p_m = 10^5$ Pa, $D_a = D_m = 300$ mm, $\varepsilon_a = \varepsilon_m = 0.2$ mm, $L_a = 20$ m, $L_m = 400$ m, $P = 50$ kW, $\eta = 0.75$.

As a first approximation, we assume $\mathrm{Re} \to \infty$ and, for $\dfrac{\varepsilon}{D} = \dfrac{0.2}{300} = 6.6 \times 10^{-4}$, results $\lambda_\infty = 0.018$. Hence,

$$P = \frac{\gamma_w Q\left[\Delta H_g + \left(\lambda_\infty \dfrac{L_a + L_m}{D} + 1.5\right)\dfrac{Q^2}{2g\left(\pi D^2/4\right)^2} + \dfrac{p_m}{\gamma_w}\right]}{\eta} \rightarrow$$

$$\frac{9800 \times Q \times \left[\begin{array}{l} 25 + \left(0.018 \times \dfrac{20 + 400}{0.3} + 1.5\right) \\[2mm] \quad\times \dfrac{Q^2}{2 \times 9.806 \times \left(\pi \times 0.3^2/4\right)^2} \\[2mm] + \dfrac{10^5}{9800} \end{array}\right]}{0.75} = 50\,000.$$

By solving the cubic equation in the flow rate by iteration, yields

$$\therefore Q = 100 \, \mathrm{l\,s^{-1}},$$

and $V = 1.42 \, \mathrm{m\,s^{-1}}$, Re $= 427\,000$. Substituting Re and $\varepsilon/D = 6.6 \times 10^{-4}$ in the Colebrook–White equation, yields $\lambda = 0.0187$, which is less than 4% greater than $\lambda_\infty$, hence it is not necessary an iteration.

The energy and the hydraulic grade lines are shown in Fig. 7.3. It should be noted that, in the present case, ignoring the concentrated energy losses leads to a negligible overestimation of the flow rate of 0.3%.

---

**Exercise 7.3** The pumping plant shown in Fig. 7.4 should convey the flow rate $Q$ from tank A to tank B, with a difference of level $Y$. A pump with a power $P$ is installed.

– Determine the minimum diameter of the plastic pipeline that is necessary in order to meet the design requirements. Select the pipe diameter from the commercial series below listed.

Consider a constant efficiency of the pump $\eta = 0.80$. Numerical data: $Y = (10 + C_u)$ m, $P = (5 + C_{pu}/10)$ kW, $L_1 = (2 + C_{pu}/9)$ m, $L_2 = (25 + C_u)$ m, $\xi_{in} = 0.5$, $\xi_{out} = 1.0$, total coefficient $\xi$ for concentrated losses in the valves is $\xi = 5.0$, $Q = (15 + C_{pu})$ l s$^{-1}$, $\gamma_w = 9806 \, \mathrm{N\,m^{-3}}$. Commercial diameters in mm (the nominal diameter is the same as the internal diameter): 50, 65, 80, 100, 125, 150, 200, 250.

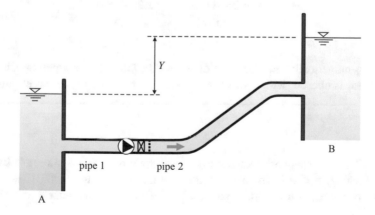

Fig. 7.4 Pumping plant

**Solution** The power of the pump is equal to

$$P = \frac{\gamma_w Q \, \Delta H}{\eta},$$

and the energy balance yields

$$\Delta H = Y + \frac{4^{10/3}}{\pi^2} \frac{Q^2}{k^2 D^{16/3}} (L_1 + L_2) + \frac{8Q^2}{\pi^2 D^4 g} (\xi_{in} + \xi_{out} + \xi).$$

Combining the two expressions, yields

$$Y + \frac{4^{10/3}}{\pi^2} \frac{Q^2}{k^2 D^{16/3}} (L_1 + L_2) + \frac{8Q^2}{\pi^2 D^4 g} (\xi_{in} + \xi_{out} + \xi) - \frac{\eta P}{\gamma_w Q} = 0.$$

From this relationship, by numerical iteration we evaluate the theoretical diameter $D$. The commercial pipe with a diameter immediately greater than the theoretical one will be the solution.

For $C_u = C_{pu} = 0$ it results $Y = 10\,\text{m}$, $P = 5\,\text{kW}$, $L_1 = 2\,\text{m}$, $L_2 = 25\,\text{m}$, $\xi = 5.0$, $Q = 15\,\text{1s}^{-1}$. We assume $k = 100\,\text{m}^{1/3}\,\text{s}^{-1}$ for the plastic pipe.

$$Y + \frac{4^{10/3}}{\pi^2} \frac{Q^2}{k^2 D^{16/3}} (L_1 + L_2) + \frac{8Q^2}{\pi^2 D^4 g} (\xi_{in} + \xi_{out} + \xi) - \frac{\eta P}{\gamma_w Q} = 0 \rightarrow$$

$$10 + \frac{4^{10/3}}{\pi^2} \times \frac{0.015^2}{100^2 \times D^{16/3}} \times (2 + 25) + \frac{8 \times 0.015^2}{\pi^2 \times D^4 \times 9.806} \times (0.5 + 1.0 + 5.0)$$

$$- \frac{0.80 \times 5000}{9806 \times 0.015} = 0 \rightarrow$$

$$\frac{6.2534 \times 10^{-6}}{D^{16/3}} + \frac{1.2089 \times 10^{-4}}{D^4} - 17.194 = 0$$

$$(7.2)$$

Solving numerically Eq. (7.2), yields $D = \mathbf{0.067\,m}$. The commercial diameter $D = \mathbf{80\,mm}$ is chosen, with a valve to reduce the flow rate to the target value $Q$.

---

**Exercise 7.4** In the network shown in Fig. 7.5, the pump absorbs a power of 8 kW and has an efficiency $\eta = 0.7$. The friction factor of the pipeline is $\lambda = 0.01$. The water flows out of the upper tank through a circular orifice with a diameter $d = 100\,\text{mm}$.
  Calculate:

– the free surface level in the upper tank.
– The flow rate.

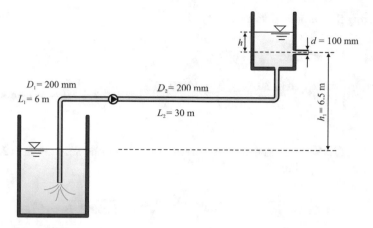

**Fig. 7.5** Pumping plant

Assume a unitary efflux coefficient for the circular orifice. Assume a unit loss coefficient at the inlet and outlet, and a loss coefficient equal to 0.1 for each curve.

**Solution** The pump power is expressed as

$$P = \frac{\gamma_w Q \, \Delta H}{\eta},$$

and the required net head is

$$\Delta H = \left( \lambda \frac{L_1 + L_2}{D} + \sum \xi_i \right) \frac{Q^2}{2g\Omega^2} + h_1 + h,$$

where $\Omega$ is the cross-section area of the pipe with diameter $D_1 = D_2 = D = 200$ mm. In steady condition, the inflow rate in the upper tank is equal to the outflow rate through the circular orifice, which can be expressed as a function of the level $h$ as follows:

$$Q = C_Q \omega \sqrt{2gh}.$$

$\omega$ is the cross-section area of circular orifice and $C_Q$ is the outflow coefficient. Substituting, yields

$$\gamma_w C_Q \omega \sqrt{2gh} \left[ \left( \lambda \frac{L_1 + L_2}{D} + \sum \xi_i \right) \frac{C_Q^2 \omega^2}{\Omega^2} h + h_1 + h \right] = \eta P.$$

Inserting the numerical values, results in:

$$9806 \times 1 \times \frac{\pi \times (0.1)^2}{4} \times \sqrt{2 \times 9.806 \times h}$$

$$\times \left[ \left( 0.01 \times \frac{6 + 30}{0.2} + 2.2 \right) \times \frac{1 \times (0.1)^4}{(0.2)^4} \times h + 6.5 + h \right] = 0.7 \times 8000.$$

$$(7.3)$$

Equation (7.3) can be numerically solved and admits the solution $h = \mathbf{2.74\ m}$. The flow rate is equal to

$$\therefore \qquad Q = C_Q \omega \sqrt{2gh} = 1 \times \frac{\pi \times (0.1)^2}{4} \times \sqrt{2 \times 9.806 \times 2.74} = \mathbf{57.6\ l s^{-1}}.$$

---

**Exercise 7.5** In the pumping plant in Fig. 7.6, tank A is pressurized. All the characteristics of the pipes, the flow rate $Q$ and the efficiency of pump $\eta$ are known.

– Calculate the power of the pump.
– Draw the energy and the hydraulic grade lines.

Assume $\Delta H_g = (25 + C_{pu})$ m, $p_m = (1 + C_u \times 0.1) \times 10^5$ Pa, $D_a = D_m = 300$ mm, $\varepsilon_a = \varepsilon_m = 0.2$ mm, $L_a = 20$ m, $L_m = 400$ m, $Q = (80 + C_u)$ l s$^{-1}$, $\eta = 0.75$.

**Solution** The power of the pump is equal to

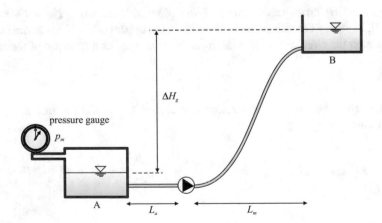

**Fig. 7.6** Pumping plant

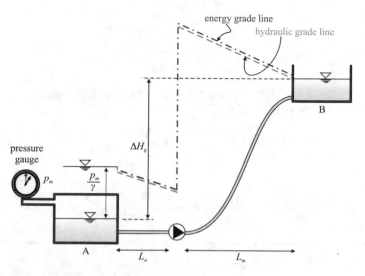

**Fig. 7.7** Energy and hydraulic grade lines

$$P = \frac{\gamma_w Q \, \Delta H}{\eta},$$

and the required head is

$$\Delta H = \Delta H_g + \lambda_a \frac{L_a}{D_a} \frac{V_a^2}{2g} + \lambda_m \frac{L_m}{D_m} \frac{V_m^2}{2g} + \xi_{in} \frac{V_a^2}{2g} + \xi_{out} \frac{V_m^2}{2g} - \frac{p_m}{\gamma_w}.$$

Friction factors are calculated from the Moody chart or solving the Colebrook–White equation. The Reynolds number is equal to

$$\mathrm{Re} = \frac{VD}{\nu} = \frac{4Q}{\nu \pi D},$$

and the relative roughness is equal to $\varepsilon/D$.

The energy and the hydraulic grade lines are drawn in Fig. 7.7.

For $C_u = C_{pu} = 0$ it results $\Delta H_g = 25\,\mathrm{m}$, $p_m = 10^5\,\mathrm{Pa}$, $D_a = D_m = 300\,\mathrm{mm}$, $\varepsilon_a = \varepsilon_m = 0.2\,\mathrm{mm}$, $L_a = 20\,\mathrm{m}$, $L_m = 400\,\mathrm{m}$, $Q = 80\,\mathrm{l\,s^{-1}}$, $\eta = 0.75$.

$$\mathrm{Re} = \frac{QD}{\nu \left(\dfrac{\pi D^2}{4}\right)} = \frac{0.08 \times 0.3}{10^{-6} \times \left(\dfrac{\pi \times 0.3^2}{4}\right)} = \mathbf{340\,000},$$

$$\frac{\varepsilon}{D} = \frac{0.2}{300} = 6.6 \times 10^{-4},$$

$$\lambda = 0.018,$$

$$V_a = V_m = \frac{4Q}{\pi D^2} = \frac{4 \times 0.08}{\pi \times 0.3^2} = \textbf{1.13 m s}^{-1},$$

$$\Delta H = \Delta H_g + \lambda_a \frac{L_a}{D_a} \frac{V_a^2}{2g} + \lambda_m \frac{L_m}{D_m} \frac{V_m^2}{2g} + \xi_{in} \frac{V_a^2}{2g} + \xi_{out} \frac{V_m^2}{2g} - \frac{p_m}{\gamma_w} =$$

$$25 + \left( 0.018 \times \frac{20 + 400}{0.3} + 1.5 \right) \times \frac{1.13^2}{2 \times 9.806} - \frac{10^5}{9806} = \textbf{16.5 m},$$

$$\therefore \qquad P = \frac{\gamma_w Q \, \Delta H}{\eta} = \frac{9806 \times 0.08 \times 16.5}{0.75} = \textbf{17.3 kW}.$$

---

**Exercise 7.6** In the pumping plant in Fig. 7.8, the tank A is pressurized. All the characteristics of the pipes, the power $P$ of the pump and its efficiency are known.

– Calculate the flow rate.

Assume $\Delta H_g = (25 + C_{pu})$ m, $p_m = (1 + C_u \times 0.1) \times 10^5$ Pa, $D_a = D_m = 300$ mm, $\lambda_a = 0.020$, $\lambda_m = 0.015$, $L_a = 20$ m, $L_m = 400$ m, $P = 17$ kW, $\eta = 0.75$. The fluid is water with $\gamma_w = 9806$ N m$^{-3}$.

**Solution** The required net head is

$$\Delta H = \Delta H_g + \lambda_a \frac{L_a}{D_a} \frac{V_a^2}{2g} + \lambda_m \frac{L_m}{D_m} \frac{V_m^2}{2g} + \xi_{in} \frac{V_a^2}{2g} + \xi_{out} \frac{V_m^2}{2g} - \frac{p_m}{\gamma_w}. \qquad (7.4)$$

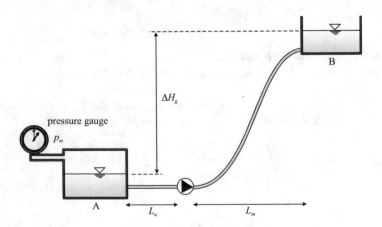

**Fig. 7.8** Pumping station

Equation (7.4) is rewritten as function of the flow rate as follows:

$$\Delta H = \Delta H_g + \left(\lambda_a \frac{L_a}{D} + \lambda_m \frac{L_m}{D} + \xi_{in} + \xi_{out}\right) \frac{Q^2}{2g} \frac{1}{\left(\pi D^2/4\right)^2} - \frac{p_m}{\gamma_w}.$$

The pump power is equal to

$$P = \frac{\gamma_w Q \, \Delta H}{\eta}.$$

Replacing the expression of the net head as a function of the flow rate, one can obtain the following cubic equation, which can be analytically or numerically solved:

$$\gamma_w \left(\lambda_a \frac{L_a}{D} + \lambda_m \frac{L_m}{D} + \xi_{in} + \xi_{out}\right) \frac{Q^3}{2g} \frac{1}{\left(\pi D^2/4\right)^2}$$

$$+ \gamma_w Q \left(\Delta H_g - \frac{p_m}{\gamma_w}\right) - \eta P = 0 \rightarrow$$

$$9806 \times \left(0.02 \times \frac{20}{0.3} + 0.015 \times \frac{400}{0.3} + 1.5\right) \times \frac{Q^3}{2 \times 9.806} \times \frac{1}{\left(\pi \times 0.3^2/4\right)^2}$$

$$+ 9806 \times Q \times \left(25 - \frac{1 \times 10^5}{9806}\right) - 0.75 \times 17\,000 = 0 \rightarrow$$

$$2\,284\,938 \times Q^3 + 145\,150 \times Q - 12\,750 = 0.$$

The only physically acceptable solution for the flow rate is $Q = \mathbf{79.8\,ls^{-1}}$.

---

**Exercise 7.7** In the hydraulic plant shown in Fig. 7.9, the two tanks are supplied by two pipes of length $L_{NB} = L_{NC} = 300$ m and diameter $D = 200$ mm. The pipes are made of steel with roughness $\varepsilon = 0.1$ mm. The concentrated energy losses and the distributed energy losses from tank A to node N can be ignored.

– Calculate the flow rate to each tank, if the pump net head at the operating point is $\Delta H = 26$ m.

The fluid is water with $\gamma_w = 9806$ N m$^{-3}$ and $\nu = 10^{-6}$ m$^2$ s$^{-1}$.

**Solution** The total head in node N is equal to the sum of the free surface level in the supply tank A and of the pump head:

$$H_N = \Delta H + z_A.$$

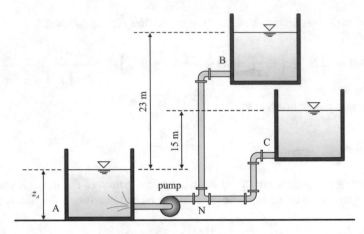

**Fig. 7.9** Pumping station with multiple tanks

The following energy balance equation applies for the pipeline NB (neglecting the concentrated energy losses):

$$\Delta H + z_A - z_B = \lambda_{NB} \frac{U_{NB}^2}{2g} \frac{L_{NB}}{D},$$

and for the pipeline NC:

$$\Delta H + z_A - z_C = \lambda_{NC} \frac{U_{NC}^2}{2g} \frac{L_{NC}}{D}.$$

Assuming an asymptotic friction factor $\lambda_{NB\infty} = 0.0170$ for $\varepsilon/D = 5 \times 10^{-4}$, yields

$$U'_{NB} = \sqrt{\frac{2gD}{L_{NB}\lambda_{NB\infty}} (\Delta H + z_A - z_B)} =$$

$$\sqrt{\frac{2 \times 9.806 \times 0.2}{300 \times 0.0170} \times (26 - 23)} = \mathbf{1.52\ m\,s^{-1}}.$$

The Reynolds number is equal to

$$Re = \frac{UD}{\nu} = \frac{1.52 \times 0.2}{10^{-6}} = \mathbf{304\,000}.$$

After the first iteration the friction factor is equal to $\lambda''_{NB} = 0.0181$, to which corresponds an average velocity in the pipe equal to

$$U''_{NB} = \sqrt{\frac{2gD}{L_{NB}\lambda''_{NB}}} (\Delta H + z_A - z_B) =$$

$$\sqrt{\frac{2 \times 9.806 \times 0.2}{300 \times 0.0181}} \times (26 - 23) = \mathbf{1.47\ m\,s^{-1}},$$

and a flow rate

$$\therefore \qquad Q_{NB} = \frac{U''_{NB}\pi D^2}{4} = \frac{1.47 \times \pi \times 0.2^2}{4} = \mathbf{46.2\ 1s^{-1}}.$$

The relative velocity difference in the two iterations is equal to $(1.52 - 1.47)/1.52 = 3.2\%$. If such an approximation is considered acceptable (usually it is), any further iteration is unnecessary.

For the pipeline NC, assuming an asymptotic friction factor $\lambda_{NC\infty} = 0.0170$ for $\varepsilon/D = 5 \times 10^{-4}$, yields

$$U'_{NC} = \sqrt{\frac{2gD}{L_{NC}\lambda_{NC\infty}}} (\Delta H + z_A - z_C) =$$

$$\sqrt{\frac{2 \times 9.806 \times 0.2}{300 \times 0.017}} \times (26 - 15) = \mathbf{2.91\ m\,s^{-1}}.$$

The Reynolds number is equal to

$$Re = \frac{UD}{\nu} = \frac{2.91 \times 0.2}{10^{-6}} = \mathbf{582\,000}.$$

After the first iteration the friction factor is equal to $\lambda''_{NC} = 0.0175$, with an average velocity in the pipe equal to

$$U''_{NC} = \sqrt{\frac{2gD}{L_{NC}\lambda''_{NC}}} (\Delta H + z_A - z_C) =$$

$$\sqrt{\frac{2 \times 9.806 \times 0.2}{300 \times 0.0175}} \times (26 - 15) = \mathbf{2.87\ m\,s^{-1}},$$

and a flow rate

$$\therefore \qquad Q_{NC} = \frac{U''_{NC}\pi D^2}{4} = \frac{2.87 \times \pi \times 0.2^2}{4} = \mathbf{90.11s^{-1}}.$$

The relative velocity difference in the two iterations is equal to $(2.91 - 2.87)/2.91 = 1.3\%$. Again, any further iteration is unnecessary.

The total outflow rate from tank A is equal to

$$\therefore \qquad Q_A = Q_{NC} + Q_{NB} = 90.1 + 46.2 = \mathbf{136.3 \ 1s^{-1}}.$$

---

**Exercise 7.8** Water from a lake is pumped into a pressurized tank, which feeds two pipes of length $L_1 = 210\,\mathrm{m}$, $D_1 = 150\,\mathrm{mm}$ and $L_2 = 360\,\mathrm{m}$, $D_2 = 125\,\mathrm{mm}$, according to the schematic in Fig. 7.10. The roughness of the pipes is $\varepsilon = 0.1\,\mathrm{mm}$ and the characteristic curve of the pump has the following expression:

$$H_d = 13.5 + 305 \times Q - 22\,200 \times Q^2, \quad H_d \text{ in metres and } Q \text{ in m}^3\,\mathrm{s}^{-1}.$$

-  Calculate the flow rate in the pipes.
-  Calculate the pump power, if the efficiency is constant and equal to $\eta = 0.6$.

Neglect the difference of level between the lake and the outlet section and neglect the concentrated energy losses, including the outlet losses. Distributed energy losses from the lake to the tank can also be neglected.

**Solution** The pump head is necessary to balance the energy losses, which have the same value in each of the two branches. The following system of equations holds:

$$\begin{cases} H_d = 13.5 + 305 \times Q - 22\,200 \times Q^2, \\[2mm] H_d = \dfrac{\lambda_1 Q_1 \, |Q_1|}{2g\Omega_1^2} \dfrac{L_1}{D_1}, \\[2mm] H_d = \dfrac{\lambda_2 Q_2 \, |Q_2|}{2g\Omega_2^2} \dfrac{L_2}{D_2}, \\[2mm] Q = Q_1 + Q_2. \end{cases} \qquad (7.5)$$

The first attempt value of the friction factors is their asymptotic value for $\mathrm{Re} \to \infty$: for $\varepsilon/D_1 = 6.6 \times 10^{-4}$ results $\lambda_{1\infty} = 0.0178$; for $\varepsilon/D_2 = 8 \times 10^{-4}$ results

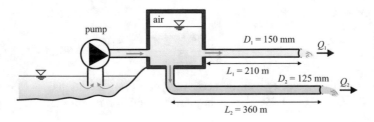

**Fig. 7.10** Pumping station

$\lambda_{2\infty} = 0.0185$. The solution of the system of equations is better achieved by reducing it to an equation in a single unknown. By choosing $Q_1$ as the main unknown and introducing the symbols

$$\beta_1 = \frac{\lambda_1}{2g\Omega_1^2}\frac{L_1}{D_1}, \quad \beta_2 = \frac{\lambda_2}{2g\Omega_2^2}\frac{L_2}{D_2},$$

removing also the absolute value since the direction of the flow is unambiguous, it is possible to write

$$H_d = \beta_1 Q_1^2, \quad \text{and} \quad H_d = \beta_2 Q_2^2 \rightarrow Q_2 = Q_1\sqrt{\frac{\beta_1}{\beta_2}}.$$

By eliminating $H_d$ from the first two equations of the system (7.5) and inserting the continuity equation, the system becomes

$$\begin{cases} \beta_1 Q_1^2 = 13.5 + 305 \times (Q_1 + Q_2) - 22\,200 \times (Q_1 + Q_2)^2, \\ Q_2 = Q_1\sqrt{\dfrac{\beta_1}{\beta_2}}. \end{cases}$$

Finally, by eliminating $Q_2$ from the two equations, the following quadratic equation in the unknown $Q_1$ is obtained:

$$\beta_1 Q_1^2 = 13.5 + 305 \times Q_1\left(1 + \sqrt{\frac{\beta_1}{\beta_2}}\right) - 22\,200 \times Q_1^2 \times \left(1 + \sqrt{\frac{\beta_1}{\beta_2}}\right)^2. \qquad (7.6)$$

By introducing the variable $\chi = 1 + \sqrt{\dfrac{\beta_1}{\beta_2}}$, Eq. (7.6) admits the solutions

$$Q_1 = \frac{305 \times \chi \pm \sqrt{305^2 \times \chi^2 + 4 \times 13.5 \times (22\,200 \times \chi^2 + \beta_1)}}{2 \times (22\,200 \times \chi^2 + \beta_1)}.$$

One of the two solutions is always positive and the other is always negative, and the first one is admissible. By inserting the numerical values, it results:

$$\beta_1 = \frac{\lambda_1}{2g\Omega_1^2}\frac{L_1}{D_1} = \frac{0.0178}{2 \times 9.806 \times \left(\pi \times 0.15^2/4\right)^2} \times \frac{210}{0.15} = 4069\ \text{s}^2\,\text{m}^{-5},$$

$$\beta_2 = \frac{\lambda_2}{2g\Omega_2^2}\frac{L_2}{D_2} = \frac{0.0185}{2 \times 9.806 \times \left(\pi \times 0.125^2/4\right)^2} \times \frac{360}{0.125} = 18\ 039\ \text{s}^2\ \text{m}^{-5},$$

$$\chi = 1 + \sqrt{\frac{\beta_1}{\beta_2}} = 1 + \sqrt{\frac{4069}{18\ 039}} = \mathbf{1.475}.$$

Finally, it results

$$\begin{cases} Q_1 = \mathbf{0.021\,m^3\,s^{-1}}, \\[4mm] Q_2 = Q_1\sqrt{\dfrac{\beta_1}{\beta_2}} = 0.021 \times \sqrt{\dfrac{4069}{18\ 039}} = \mathbf{0.010\ m^3\,s^{-1}}. \end{cases}$$

The corresponding velocities are:

$$V_1 = \frac{4Q_1}{\pi D_1^2} = \frac{4 \times 0.021}{\pi \times 0.15^2} = \mathbf{1.20\ m\,s^{-1}},$$

$$V_2 = \frac{4Q_2}{\pi D_2^2} = \frac{4 \times 0.010}{\pi \times 0.125^2} = \mathbf{0.80\ m\,s^{-1}},$$

and Reynolds numbers are equal to:

$$\text{Re}_1 = \frac{V_1 D_1}{\nu} = \frac{1.20 \times 0.15}{10^{-6}} = \mathbf{180\ 000},$$

$$\text{Re}_2 = \frac{V_2 D_2}{\nu} = \frac{0.8 \times 0.125}{10^{-6}} = \mathbf{100\ 000}.$$

The friction factors of a second iteration are equal to $\lambda_1 = \mathbf{0.0198}$ and $\lambda_2 = \mathbf{0.0215}$, and

$$\beta_1 = \frac{\lambda_1}{2g\Omega_1^2}\frac{L_1}{D_1} = \frac{0.0198}{2 \times 9.806 \times \left(\pi \times 0.15^2/4\right)^2} \times \frac{210}{0.15} = \mathbf{4526\ s^2\ m^{-5}},$$

$$\beta_2 = \frac{\lambda_2}{2g\Omega_2^2}\frac{L_2}{D_2} = \frac{0.0215}{2 \times 9.806 \times \left(\pi \times 0.125^2/4\right)^2} \times \frac{360}{0.125} = \mathbf{20\ 965\ s^2\ m^{-5}},$$

$$\chi = 1 + \sqrt{\frac{\beta_1}{\beta_2}} = 1 + \sqrt{\frac{4526}{20\ 965}} = \mathbf{1.465}.$$

Finally, it results:

$$\begin{cases} Q_1 = 0.0209 \text{ m}^3\text{ s}^{-1}, \\[2mm] Q_2 = 0.0097 \text{ m}^3\text{ s}^{-1}. \end{cases}$$

Subsequent iterations do not significantly change the result. The pump flow rate is

$$\therefore \qquad Q = Q_1 + Q_2 = 0.0307 \text{ m}^3\text{ s}^{-1},$$

with a pump head $H_d = 1.98 \text{ m}$. The required power is equal to:

$$\therefore \qquad P = \frac{\gamma_w Q H_d}{\eta} = \frac{9806 \times 0.0307 \times 1.98}{0.6} = 993 \text{ W}.$$

**Exercise 7.9** In the hydroelectric plant in Fig. 7.11, the gross head is $Y = (200 + 5 \times C_u)$ m, and the flow rate is $Q$. The pipeline has an initial section of length $L_1 = (400 + 10 \times C_{pu})$ m, diameter $D_1 = 800$ mm, roughness $\varepsilon_1 = 0.2$ mm, and a terminal section of length $L_2 = (25 + C_u)$ m, diameter $D_2 = 800$ mm and roughness $\varepsilon_2 = 0.2$ mm. The entrance and exit are rounded, with a loss coefficient of 0.06. The water is discharged in a tank from which it flows away through a rectangular slot, with height $a = 0.20$ m and width $b = 1.0$ m. The vena contracta has a contraction coefficient $C_c = 0.6$ and a velocity coefficient $C_v = 0.98$. The contraction is only in the upper edge and any lateral contraction is suppressed. In steady condition, the water level in the tank is equal to $h + a = 12.20$ m.

- Calculate the flow rate.
- Calculate the power of the turbine, if the efficiency of the turbine is $\eta_t = 0.80$.
- Calculate the annual revenue with an alternator efficiency $\eta_a = 0.92$, if the plant works 24 h a day, 365 days a year, and the selling price of the electricity is $0.08 \text{ €kWh}^{-1}$.

**Solution** The flow rate can be calculated taking into account the outflow process from the slot. In steady condition, it results:

$$Q = C_c C_v ab \sqrt{2g \, (h + a - C_c a)}.$$

The energy balance equation reads

$$Y - Q^2 \left( \lambda_1 \frac{1}{2g\Omega_1^2} \frac{L_1}{D_1} + \lambda_2 \frac{1}{2g\Omega_2^2} \frac{L_2}{D_2} + \xi_{in} \frac{1}{2g\Omega_1^2} + \xi_{out} \frac{1}{2g\Omega_2^2} \right) - \Delta H = 0,$$

where $Y$ is the *gross head*, $\Delta H$ is the *net head* (corresponding to the head that is actually available for the turbine). The power of the turbine is equal to

$$P = \eta_t \gamma_w Q \, \Delta H,$$

and the electrical power is equal to

$$P_e = \eta_a P,$$

with an annual revenue equal to

$$R = P_e T C_{en}/1000,$$

where $P_e$ is expressed in watts, the operating time per year $T$ is in hours, $C_{en}$ is the selling price expressed in € kWh$^{-1}$.

For $C_u = C_{pu} = 0$ it results $Y = 200$ m, $L_1 = 400$ m, $D_1 = 800$ mm, $\varepsilon_1 = 0.2$ mm, $L_2 = 25$ m, $D_2 = 800$ mm, $\varepsilon_2 = 0.2$ mm, $\xi_{in} = \xi_{out} = 0.06$, $a = 0.20$ m, $b = 1.0$ m, $C_c = 0.6$, $C_v = 0.98$, $h = 12$ m.

$$\Omega_1 = \Omega_2 = \frac{\pi D_1^2}{4} = \frac{\pi \times 0.8^2}{4} = \mathbf{0.503 \ m^2},$$

$$Q = C_c C_v ab \sqrt{2g \, (h + a - C_c a)} = 0.6 \times 0.98 \times 0.20 \times 1.0$$
$$\times \sqrt{2 \times 9.806 \times (12 + 0.2 - 0.6 \times 0.2)} = \mathbf{1.81 \ m^3 \, s^{-1}}.$$

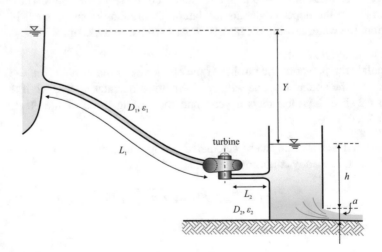

**Fig. 7.11**  Hydroelectric plant

The average water velocity in the pipe is equal to

$$V = \frac{4Q}{\pi D_1^2} = \frac{4 \times 1.81}{\pi \times 0.8^2} = 3.60 \text{ ms}^{-1},$$

the Reynolds number is equal to

$$\text{Re} = \frac{V D_1}{\nu} = \frac{3.60 \times 0.8}{10^{-6}} = 2.88 \times 10^6,$$

the relative roughness is equal to

$$\frac{\varepsilon_1}{D_1} = \frac{0.2}{800} = 2.5 \times 10^{-4}.$$

From the Colebrook–White equation (or from Moody chart) results a friction factor $\lambda = 0.0147$. The net head is equal to:

$$\Delta H = Y - Q^2 \left( \lambda_1 \frac{1}{2g\Omega_1^2} \frac{L_1}{D_1} + \lambda_2 \frac{1}{2g\Omega_2^2} \frac{L_2}{D_2} + \xi_{in} \frac{1}{2g\Omega_1^2} + \xi_{out} \frac{1}{2g\Omega_2^2} \right) =$$

$$200 - 1.81^2 \times \left( \begin{array}{l} 0.0147 \times \dfrac{1}{2 \times 9.806 \times 0.503^2} \times \dfrac{400}{0.8} \\[2mm] +0.0147 \times \dfrac{1}{2 \times 9.806 \times 0.503^2} \times \dfrac{25}{0.8} \\[2mm] +0.06 \times \dfrac{1}{2 \times 9.806 \times 0.503^2} \\[2mm] +0.06 \times \dfrac{1}{2 \times 9.806 \times 0.503^2} \end{array} \right) = 194.8 \text{ m},$$

the power is

$$P = \eta_t \gamma_w Q \, \Delta H = 0.80 \times 9806 \times 1.81 \times 194.8 = 2.76 \text{ MW},$$

the electric power is

$$P_e = \eta_a P = 0.92 \times 2.76 = 2.54 \text{ MW},$$

and the annual revenue (gross) is

$$\therefore \qquad R = P_e T C_{en} = 2540 \times 24 \times 365 \times 0.08 = 1.78 \text{ M€ a}^{-1}.$$

**Exercise 7.10** In the hydroelectric plant in Fig. 7.12, the pressure drop across the turbine is equal to $\Delta p = (5 + C_{pu}/10) \times 10^5$ Pa. The gross head is equal to $Y = (70 + C_u)$ m. The pipelines have lengths $L_1 = (2000 + 10 \times C_u)$ m, $L_2 = 200$ m, $L_3 = 100$ m, and diameter $D_1 = (500 + 10 \times C_u)$ mm, $D_2 = 300$ mm, $D_3 = 300$ mm. The friction factors for the pipelines are $\lambda_1 = 0.02, \lambda_2 = \lambda_3 = 0.015$. Calculate:

– the flow rate.
– The output power of the turbine, if the efficiency of the turbine is 98%.
– The ratio between the theoretical available power and the power of the plant for a
  given flow rate.

   **Solution** The head at the inlet of the turbine (ignoring concentrated losses) is equal to:

$$H_u = z_A - L_1 J_1 - L_2 J_2,$$

where $z_A$ is free surface level in tank A measured from a datum. The head at the outlet section of the turbine is

$$H_d = z_B + L_3 J_3,$$

where $z_B$ is the free surface level in tank B. The head drop across the turbine can be expressed as follows:

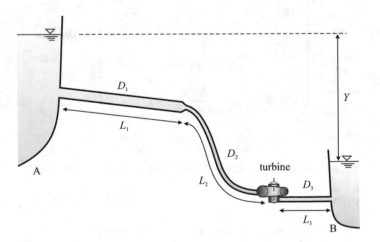

**Fig. 7.12** Hydroelectric plant

$$H_u - H_d = z_u + \frac{p_u}{\gamma_w} + \frac{V_2^2}{2g} - z_d - \frac{p_d}{\gamma_w} - \frac{V_3^2}{2g}.$$

Since $z_u \approx z_d$, and $\dfrac{V_2^2}{2g} = \dfrac{V_3^2}{2g}$, it results

$$H_u - H_d = \frac{p_u}{\gamma_w} - \frac{p_d}{\gamma_w} = \frac{\Delta p}{\gamma_w},$$

which represents the net head $\Delta H$. By neglecting the concentrated energy losses, the energy balance reads

$$Y = Q^2 \left( \lambda_1 \frac{1}{2g\Omega_1^2} \frac{L_1}{D_1} + \lambda_2 \frac{1}{2g\Omega_2^2} \frac{L_2}{D_2} + \lambda_3 \frac{1}{2g\Omega_3^2} \frac{L_3}{D_3} \right) + \frac{\Delta p}{\gamma_w} \rightarrow$$

$$Q = \sqrt{ \frac{2g \left( Y - \dfrac{\Delta p}{\gamma_w} \right)}{\lambda_1 \dfrac{1}{\Omega_1^2} \dfrac{L_1}{D_1} + \lambda_2 \dfrac{1}{\Omega_2^2} \dfrac{L_2}{D_2} + \lambda_3 \dfrac{1}{\Omega_3^2} \dfrac{L_3}{D_3}} },$$

where $Y = z_A - z_B$ is the gross head.

The available power for the turbine is equal to

$$P = \eta \gamma_w Q \frac{\Delta p}{\gamma_w},$$

and the theoretical available power is equal to

$$P_t = \gamma_w Q Y.$$

The ratio between the available theoretical power and the real power is equal to

$$\frac{\gamma_w Q Y}{\eta \gamma_w Q \left( \Delta p / \gamma_w \right)} \equiv \frac{\gamma_w Y}{\eta \Delta p}$$

and it is greater than the unit.

For $C_u = C_{pu} = 0$ it results $\Delta p = 5 \times 10^5$ Pa, $Y = 70$ m, $L_1 = 2000$ m, $L_2 = 200$ m, $L_3 = 100$ m, $D_1 = 500$ mm, $D_2 = 300$ mm, $D_3 = 300$ mm, $\lambda_1 = 0.02$, $\lambda_2 = \lambda_3 = 0.015$.

$$\therefore \quad Q = \sqrt{\frac{2g\left(Y - \dfrac{\Delta p}{\gamma_w}\right)}{\lambda_1 \dfrac{1}{\Omega_1^2}\dfrac{L_1}{D_1} + \lambda_2 \dfrac{1}{\Omega_2^2}\dfrac{L_2}{D_2} + \lambda_3 \dfrac{1}{\Omega_3^2}\dfrac{L_3}{D_3}}} =$$

$$= \sqrt{\frac{2 \times 9.806 \times \left(70 - \dfrac{5 \times 10^5}{9806}\right)}{\left[\begin{array}{c} 0.02 \times \dfrac{16}{\left(\pi \times 0.5^2\right)^2} \times \dfrac{2000}{0.5} + 0.015\dfrac{16}{\left(\pi \times 0.3^2\right)^2} \times \dfrac{200}{0.3} \\[4mm] +0.015 \times \dfrac{16}{\left(\pi \times 0.3^2\right)^2} \times \dfrac{100}{0.3} \end{array}\right]}} =$$

$$= \mathbf{0.271\ m^3\,s^{-1}},$$

$$\therefore \qquad P = \eta \gamma_w Q \frac{\Delta p}{\gamma_w} = 0.98 \times 9806 \times 0.271 \times \frac{5 \times 10^5}{9806} = \mathbf{132.8\ kW},$$

$$\therefore \qquad \frac{\gamma_w Y}{\eta\,\Delta p} = \frac{9806 \times 70}{0.98 \times 5 \times 10^5} = \mathbf{1.40}.$$

**Exercise 7.11** In the hydroelectric plant in Fig. 7.13, the gross head is $Y = (100 + 5 \times C_u)$ m, the pipe is made of steel and has diameter $D = 800\,\text{mm}$ and length $L = (400 + 10 \times C_u)$ m. The net head is $\Delta H = (30 + 5 \times C_{pu})$ m. The draft tube (diffuser) is horizontal, with inlet diameter $D_i = D$ and outlet diameter $D_o = 1.5 \times D$. The depth of the centroid of the outlet section is $z_G = 12$ m, with respect to the free surface of the downstream tank.

– Calculate the flow rate $Q$.
– Calculate the maximum output power with an efficiency $\eta = 0.85$.
– Calculate the resulting force acting on the diffuser neglecting the weight of the fluid.

The fluid is water with $\gamma_w = 9806\,\mathrm{N\,m^{-3}}$. Assume $\xi_{in} = 0.5$ and $\xi_{out} = 0.06$, while the distributed energy losses in the diffuser are negligible.

**Solution** The energy balance, ignoring the distributed losses in the diffuser, is

$$Y = \Delta H + \lambda \frac{V^2}{2g} \frac{L}{D} + \xi_{in} \frac{V^2}{2g} + \xi_{out} \frac{V_o^2}{2g},$$

where $\Delta H$ is the net head and $V_o$ is the velocity in the outlet section, calculated on the basis of the mass conservation equation:

$$V_o \frac{\pi D_o^2}{4} = V \frac{\pi D^2}{4} \rightarrow V_o = V \frac{D^2}{D_o^2}.$$

For used steel pipes with slight rust, we assume an equivalent homogeneous roughness $\varepsilon = 0.3\,\mathrm{mm}$. The friction factor is calculated, by iteration, using the Moody chart, or the Colebrook–White equation:

$$\frac{1}{\sqrt{\lambda}} = -2\log_{10}\left(\frac{2.51}{Re\sqrt{\lambda}} + \frac{1}{3.71} \frac{\varepsilon}{D}\right).$$

Since Re is unknown, it is necessary to initially formulate the hypothesis of fully developed turbulent flow, adopting the asymptotic value of the friction factor (Prandtl–Nikuradse's equation):

$$\lambda_\infty = \left[-2\log_{10}\left(\frac{1}{3.71} \frac{\varepsilon}{D}\right)\right]^{-2}.$$

The value of $\lambda_\infty$ allows a first estimate of the velocity:

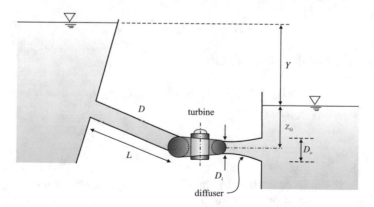

**Fig. 7.13** Hydroelectric plant

$$V = \sqrt{\frac{2g\,(Y - \Delta H)}{\left(\lambda_\infty \dfrac{L}{D} + \xi_{in} + \xi_{out} \dfrac{D^4}{D_o^4}\right)}},$$

which, in turn, allows the evaluation of the Reynolds number and then the value of the first correction of $\lambda$. The correct value of the velocity can be obtained by means of further iterations.

Assuming a fully developed turbulent flow, one can alternatively adopt the Chézy formula with Gauckler–Strickler roughness coefficient, rewriting the energy balance equation in the following form:

$$Y = \Delta H + \left[\frac{1}{k^2\left(\dfrac{D}{4}\right)^{4/3}}L + \frac{\xi_{in}}{2g} + \frac{\xi_{out}}{2g}\frac{D^4}{D_o^4}\right]V^2 \rightarrow$$

$$V = \sqrt{\frac{Y - \Delta H}{\dfrac{1}{k^2\left(\dfrac{D}{4}\right)^{4/3}}L + \dfrac{\xi_{in}}{2g} + \dfrac{\xi_{out}}{2g}\dfrac{D^4}{D_o^4}}}.$$

For used steel pipes with slight rust, the Gauckler–Strickler coefficient can be assumed to be equal to $k = 100\,\mathrm{m}^{1/3}\,\mathrm{s}^{-1}$ or less. The maximum output power is equal to

$$P = \gamma_w Q\,\Delta H\,\eta.$$

The force on the diffuser is calculated by applying the momentum balance to the dashed control volume in Fig. 7.14:

$$\mathbf{G} + \mathbf{\Pi} + \mathbf{I} + \mathbf{M_1} - \mathbf{M_2} = \mathbf{0}.$$

In the horizontal direction, it results

$$\Pi_{0x} + \Pi_{1x} - \Pi_{2x} + M_{1x} - M_{2x} = 0,$$

that is:

$$\Pi_{0x} = -p_{Gu}\Omega_u + p_{Gd}\Omega_d - \rho\frac{Q^2}{\Omega_u} + \rho\frac{Q^2}{\Omega_d}.$$

$\Pi_{0x} = -F_x$ is the force exerted by the diffuser on the control volume, opposite to the force $F_x$ exerted on the diffuser. We have assumed the coefficients of the momentum flux equal to 1. The pressure values refer to the centroids of the upstream and downstream sections, in the hypothesis of rectilinear and parallel trajectories and,

**Fig. 7.14** Schematic for calculating the force on the diffuser

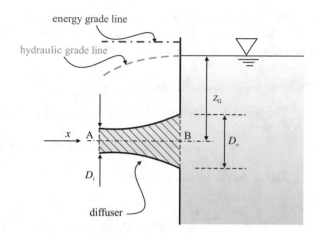

therefore, of hydrostatic distribution of the pressure. The pressure in the upstream section is calculated by applying Bernoulli's theorem to any trajectory between the upstream section (section A) and the downstream section (section B):

$$z_u + \frac{p_u}{\gamma_w} + \frac{V_u^2}{2g} = z_d + \frac{p_d}{\gamma_w} + \frac{V_d^2}{2g},$$

and, for a trajectory through both centroids:

$$p_{Gu} = p_{Gd} + \rho \frac{V_d^2}{2} - \rho \frac{V_u^2}{2} = p_{Gd} + \rho \frac{Q^2}{2} \left( \frac{1}{\Omega_d^2} - \frac{1}{\Omega_u^2} \right).$$

The pressure in the centroid of the downstream section is equal to

$$p_{Gd} = \gamma_w z_G.$$

Rigorously, the pressure forces are applied in the centres of pressure, which are deeper than the centroids, while the momentum flows are barycentric vectors. The residual torque is balanced by a suitable pressure distribution at the side walls.

For $C_u = C_{pu} = 0$ it results $Y = 100\,$m, $D = 800\,$mm, $L = 400\,$m, $\Delta H = 30\,$m.

$$\Omega_u = \frac{\pi D^2}{4} = \frac{\pi \times 0.8^2}{4} = 0.502\,\text{m}^2,$$

$$\Omega_d = \frac{\pi (1.5D)^2}{4} = \frac{\pi \times 1.2^2}{4} = 1.131\,\text{m}^2.$$

The first attempt value of $V$ is calculated assuming $\lambda_\infty = 0.0157$, and results $V = 12.54$ m s$^{-1}$. After some iterations, we obtain $\lambda = 0.016$ and $V = \mathbf{12.52}$ **m s$^{-1}$**, and $Q = \mathbf{6.29}$ **m$^3$ s$^{-1}$** is then calculated. The power is

$$\therefore \qquad P = \gamma_w Q\,\Delta H \eta = 9806 \times 6.29 \times 30 \times 0.85 = \mathbf{1.57\ MW}.$$

The pressure in the centroid of the downstream section is equal to

$$p_{Gd} = \gamma_w z_G = 9806 \times 12 = \mathbf{1.176 \times 10^5\ Pa}.$$

The upstream pressure is equal to

$$p_{Gu} = p_{Gd} + \rho\frac{Q^2}{2}\left(\frac{1}{\Omega_d^2} - \frac{1}{\Omega_u^2}\right) =$$

$$117\,600 + 1000 \times \frac{6.29^2}{2}\left(\frac{1}{1.131^2} - \frac{1}{0.502^2}\right) = \mathbf{0.553 \times 10^5\ Pa},$$

$$\tag{7.7}$$

which is less than the downstream pressure. The horizontal force on the diffuser is equal to

$$\therefore \quad F_x \equiv -\Pi_{0x} = p_{Gu}\Omega_u - p_{Gd}\Omega_d + \rho\frac{Q^2}{\Omega_u} - \rho\frac{Q^2}{\Omega_d} =$$

$$55\,280 \times 0.502 - 117\,600 \times 1.131 + 1000 \times \frac{6.29^2}{0.502} - 1000 \times \frac{6.29^2}{1.131} =$$

$$- \mathbf{61.87\,kN},$$

pointing to the left.

Similar results are obtained by calculating the flow rate with the Chézy formula.

---

**Exercise 7.12**  In the hydroelectric power plant in Fig. 7.15 there is a pipe of diameter $D = (200 + 10 \times C_{pu})$ mm and roughness $\varepsilon = 0.2$ mm. The flow rate is equal to $Q = (50 + 2 \times C_u)$ l s$^{-1}$. The pipe sections have lengths $L_1 = (400 + 10 \times C_u)$ m, $L_2 = (300 + 20 \times C_u)$ m, $L_3 = 30$ m, respectively. The elevations of the pump, of the pipeline vertex, of the turbine and of the free surface in the downstream reservoir are equal to $z_P = 25$ m, $z_v = 250$ m, $z_T = 16$ m, $z_s = 15$ m, respectively. The pump has an efficiency $\eta_P = 0.85$ and the turbine has an efficiency $\eta_T = 0.75$. The pressure upstream of the pump is $p_u = 1.5 \times 10^5$ Pa.

– Design the pump so that, at vertex V, the pressure is $p_v \geq 0.5 \times 10^5$ Pa.
– Calculate the maximum power of the turbine.

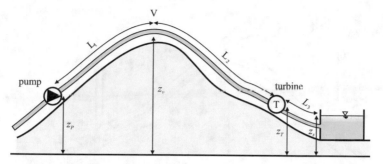

**Fig. 7.15** Hydraulic circuit with pump and turbine

– Draw the hydraulic and the energy grade lines.

The fluid is water with $\gamma_w = 9806\,\mathrm{N\,m^{-3}}$. Neglect the concentrated pressure losses, except for the outlet loss in the tank.

**Solution** To calculate the power of the pump, we need to evaluate the head $H_d$ necessary to meet the required condition in the vertex. The energy balance yields

$$z_P + \frac{p_u}{\gamma_w} + \frac{V^2}{2g} + H_d = z_v + \frac{p_v}{\gamma_w} + \frac{V^2}{2g} + \lambda \frac{V^2}{2g} \frac{L_1}{D}.$$

The average water velocity in the pipe is equal to

$$V = \frac{4Q}{\pi D^2},$$

and the friction factor is calculated from Moody chart or from the Colebrook–White equation. The power of the pump is equal to

$$P_P = \frac{\gamma_w Q H_d}{\eta_P}.$$

The net head, equal to the difference between the head upstream and downstream of the turbine, is equal to

$$\Delta H = \underbrace{z_v + \frac{p_v}{\gamma_w} + \frac{V^2}{2g} - \lambda \frac{V^2}{2g} \frac{L_2}{D}}_{\text{upstream head}} - \underbrace{\left( z_s + \lambda \frac{V^2}{2g} \frac{L_3}{D} + \frac{V^2}{2g} \right)}_{\text{downstream head}}.$$

The output power is equal to

$$P_T = \eta_T \gamma_w Q \,\Delta H.$$

For $C_u = C_{pu} = 0$ it results $D = 200\,\mathrm{mm}$, $Q = 50\,\mathrm{l\,s^{-1}}$, $L_1 = 400\,\mathrm{m}$, $L_2 = 300\,\mathrm{m}$, $L_3 = 30\,\mathrm{m}$.

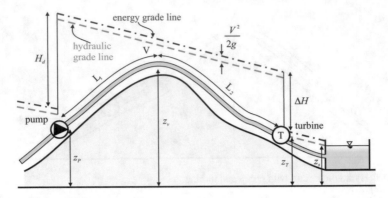

**Fig. 7.16** Hydraulic and energy grade lines

The average velocity in the pipe is equal to

$$V = \frac{4Q}{\pi D^2} = \frac{4 \times 50 \times 10^{-3}}{\pi \times 0.2^2} = 1.6 \text{ m s}^{-1},$$

the Reynolds number is

$$\text{Re} = \frac{VD}{\nu} = \frac{1.6 \times 0.2}{10^{-6}} = 320\,000.$$

From Moody chart, for $\varepsilon/D = 0.2/200 = 10^{-3}$ and $\text{Re} = 320\,000$ results $\lambda = 0.0205$. The required net head is equal to

$$H_d = z_v + \frac{p_v}{\gamma_w} + \frac{V^2}{2g} + \lambda \frac{V^2}{2g} \frac{L_1}{D} - \left( z_P + \frac{p_u}{\gamma_w} + \frac{V^2}{2g} \right) =$$

$$250 + \frac{0.5 \times 10^5}{9806} + 0.0205 \times \frac{1.6^2}{2 \times 9.806} \times \frac{400}{0.2}$$

$$- \left( 25 + \frac{1.5 \times 10^5}{9806} \right) = 220.15 \text{ m}.$$

The power of the pump must be equal to

$$\therefore \qquad P_P = \frac{\gamma_w Q H_d}{\eta_P} = \frac{9806 \times 0.05 \times 220.15}{0.85} = 127.0 \text{ kW}.$$

The net head is

$$\Delta H = \underbrace{z_v + \frac{p_v}{\gamma_w} + \frac{V_v^2}{2g} - \lambda \frac{V^2}{2g}\frac{L_2}{D}}_{\text{upstream head}} - \underbrace{\left(z_s + \lambda \frac{V^2}{2g}\frac{L_3}{D} + \frac{V_s^2}{2g}\right)}_{\text{downstream head}} =$$

$$250 + \frac{0.5 \times 10^5}{9806} - 0.0205 \times \frac{1.6^2}{2 \times 9.806} \times \frac{300}{0.2}$$

$$- \left(15 + 0.0205 \times \frac{1.6^2}{2 \times 9.806} \times \frac{30}{0.2}\right) = \textbf{235.72 m}.$$

The maximum power of the turbine is equal to

$$\therefore \qquad P_T = \eta_T \gamma_w Q \, \Delta H = 0.75 \times 9806 \times 0.05 \times 235.72 = \textbf{86.7 kW}.$$

The hydraulic and the energy grade lines are shown in Fig. 7.16.

This plant is often used for aqueducts that have to cross a hill with a strong difference in level: the energy produced by the turbine is used to partially power the pump, reducing operating costs.

---

**Exercise 7.13** In the plant in Fig. 7.17, the gross head is $Y = (100 + 5 \times C_u)$ m, the pipes are made of steel with diameter $D = 800$ mm and length $L_1 = (400 + 10 \times C_u)$ m and $L_2 = (100 + 8 \times C_u)$ m. The net head is $\Delta H = (30 + 5 \times C_{pu})$ m.

- Calculate the flow rate $Q$.
- Calculate the maximum output power with an efficiency $\eta = 0.85$.
- Draw the hydraulic and the energy grade lines.

The fluid is water. Assume $\xi_{in} = 0.5$, $\xi_{out} = 1.0$.

**Solution** The required gross head is

$$Y = \Delta H + \lambda_1 \frac{V_1^2}{2g}\frac{L_1}{D_1} + \lambda_2 \frac{V_2^2}{2g}\frac{L_2}{D_1} + \xi_{in}\frac{V_1^2}{2g} + \xi_{out}\frac{V_2^2}{2g},$$

where $\Delta H$ is the net head. If the diameter of the pipes is uniform, the energy balance equation reduces to

$$Y = \Delta H + \left(\lambda \frac{L_1 + L_2}{D} + \xi_{in} + \xi_{out}\right)\frac{V^2}{2g}.$$

For steel pipes in use with slight rust, we assume an equivalent homogeneous roughness $\varepsilon = 0.3$ mm. The friction factor is calculated by iteration using either the Moody chart or the Colebrook–White equation:

$$\frac{1}{\sqrt{\lambda}} = -2\log_{10}\left(\frac{2.51}{\text{Re}\sqrt{\lambda}} + \frac{1}{3.71}\frac{\varepsilon}{D}\right).$$

Being Re unknown, it is necessary to initially formulate the hypothesis of fully developed turbulence adopting the asymptotic value of the friction factor:

$$\lambda_\infty = \left[-2\log_{10}\left(\frac{1}{3.71}\frac{\varepsilon}{D}\right)\right]^{-2}.$$

This value gives a first estimate of the velocity,

$$V = \sqrt{\frac{2g\,(Y - \Delta H)}{\left(\lambda_\infty\dfrac{L_1 + L_2}{D} + \xi_{in} + \xi_{out}\right)}},$$

which will be used to evaluate the Reynolds number and to calculate a new value of $\lambda$. The process is iterated to reach the correct value of the velocity. Assuming a fully developed turbulent flow, we can alternatively adopt the Chézy formula and rewrite the energy balance equation in the following form:

$$Y = \Delta H + \left[\frac{1}{k^2(D/4)^{4/3}}(L_1 + L_2) + \frac{\xi_{in} + \xi_{out}}{2g}\right]V^2.$$

For steel pipes in use with slight rust, the Gauckler–Strickler coefficient can be assumed to be equal to $k = 100\,\text{m}^{1/3}\,\text{s}^{-1}$. The maximum output power is equal to

$$P = \gamma_w Q\,\Delta H \eta.$$

For $C_u = C_{pu} = 0$ it results:

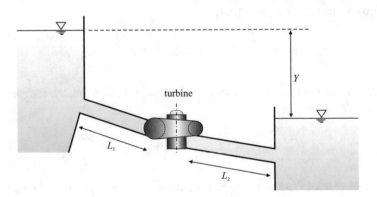

**Fig. 7.17** Hydroelectric plant

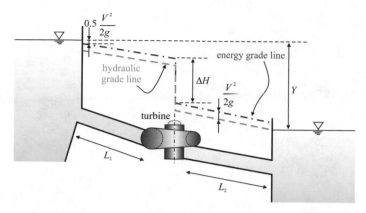

**Fig. 7.18** Hydraulic and energy grade lines

- $\lambda = 0.0157$, $V = \mathbf{11.0\ m\,s^{-1}}$, $Q = \mathbf{5.53\ m^3\,s^{-1}}$, $P = \mathbf{1.384\ MW}$, applying the Colebrook–White formula.
- $V = \mathbf{11.79\ m\,s^{-1}}$, $Q = \mathbf{5.92\ m^3\,s^{-1}}$, $P = \mathbf{1.481\ MW}$, applying Chézy formula with Gauckler–Strickler coefficient.

The hydraulic grade line is shown in Fig. 7.18.

# Chapter 8
# Hydraulic Transients

The flow regime in hydraulic plants is often unsteady with frequent variations of flow rate due to opening/closing valves, pumps, turbines. The inertia of the fluid must be included in computations, and the induced variations of pressure can be so high to excite the fluid compressibility. According to the dominant aspect, a broad classification of unsteady phenomena is in (i) mass oscillations and (ii) elastic oscillations, although both phenomena are generally present at the same time, but with well different time scales.

Mass oscillations can reduce to simple unsteady flows where the inertia itself is negligible and mass conservation is enough to properly describe the problem: a tank with an outflow reduces the fluid level and, if outflow is controlled by gravity (e.g., an orifice), reduces the head and the flow rate in time; it is not necessary to consider inertia. In other situations, like a surge tank or a pressurized tank with the aim of reducing excessive pressure during a decrease of the demand, or to favour water supply if the demand abruptly increases, the inertia of the fluid must be included, although elastic waves are still neglected. They are neglected because their time scale is much lower than mass oscillations time scale.

In other conditions, elastic waves propagation analysis is compulsory, accounting for fluid compressibility and the characteristics of the pipe controlling the celerity of these waves, usually several hundreds meters per second. This last case is commonly defined as waterhammer: elastic waves propagate pressure and velocity variations, and interacts with tanks, junctions, valves, with reflection, transmission and dissipation effects. Waterhammer has been studied in detail theoretically and experimentally, since it is quite common in pumping plants and hydroelectric plants. Several methods of solution are available, most of them amenable of numerical integration.

We propose a series of exercises of increasing complexity which can be solved without requiring a numerical code, with the use of the Allievi interlocking equations and of the method of characteristics for the waterhammer analysis.

---

$C_u$ and $C_{pu}$, that are two integer numbers between 0 and 9, for example, the last and second-last digit of the registration number.

© Springer Nature Switzerland AG 2021
S. Longo et al., *Problems in Hydraulics and Fluid Mechanics*, Springer Tracts
in Civil Engineering, https://doi.org/10.1007/978-3-030-51387-0_8

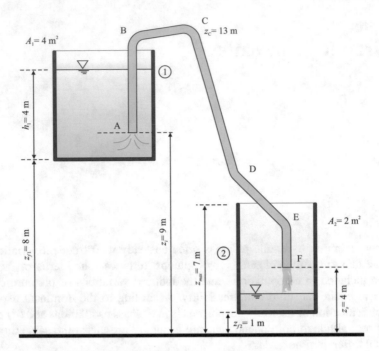

**Fig. 8.1** Siphon between two tanks

**Exercise 8.1**  The siphon in Fig. 8.1 is a series of pipelines with a Gauckler–Strickler roughness $k = 90 \, \mathrm{m}^{1/3} \, \mathrm{s}^{-1}$. The pipelines have length $L_{AB} = 4 \, \mathrm{m}$, $L_{BC} = 3 \, \mathrm{m}$, $L_{CD} = 8 \, \mathrm{m}$, $L_{DE} = 3.5 \, \mathrm{m}$, $L_{EF} = 2 \, \mathrm{m}$ and diameter $D = 300 \, \mathrm{mm}$.

– Calculate the time required for filling tank No 2.

Assume a loss coefficient at the inlet and at the outlet equal to 1.0 and a loss coefficient 0.1 for each curve. The fluid is water at a temperature of $20 \, °\mathrm{C}$, with a vapour pressure of 2314 Pa.

**Solution**  We suppose that the absolute pressure in vertex C is initially greater than the vapour pressure, and the pipe cross-section is fully occupied by the water. Applying the energy balance equation, yields

$$
z_1 - z_s = \frac{Q^2}{2g\Omega^2} \sum \xi_i + \frac{4^{10/3}}{\pi^2 k^2} Q^2 D^{-5.33} \sum L_i,
$$

where $z_1 = z_{f1} + h_1$.
  By introducing the symbol

$$
\beta = \sum \frac{\xi_i}{2g\Omega^2} + \frac{4^{10/3}}{\pi^2 k^2} D^{-5.33} \sum L_i,
$$

results in

$$z_1 - z_s = \beta Q^2,$$

or

$$Q = \sqrt{\frac{z_1 - z_s}{\beta}}.$$

By inserting the numerical values, yields

$$\beta = 2.4 \times \frac{1.0}{2 \times 9.806 \times \left(\frac{\pi \times 0.3^2}{4}\right)^2}$$

$$+ \frac{4^{10/3}}{\pi^2 \times 90^2} \times 0.3^{-5.33} \times 20.5 = \mathbf{40.44 \ s^2 \, m^{-5}}.$$

The initial flow rate is equal to

$$Q = \sqrt{\frac{z_1 - z_s}{\beta}} = \sqrt{\frac{8 + 4 - 4}{40.44}} = \mathbf{0.445 \ m^3 \, s^{-1}},$$

with an average velocity in the pipe equal to

$$V = \frac{4Q}{\pi D^2} = \frac{4 \times 0.445}{\pi \times 0.3^2} = \mathbf{6.3 \ m\,s^{-1}}.$$

The pressure in vertex C is calculated by applying the Bernoulli's theorem for a path that starts from a point where the fluid is at rest in tank No 1:

$$z_1 + \frac{p_1}{\gamma_w} + \frac{V_1^2}{2g} = z_C + \frac{p_C}{\gamma_w} + \frac{V^2}{2g} + (\xi_{in} + \xi_c)\frac{V^2}{2g} + \frac{4^{10/3}}{\pi^2 k^2} Q^2 D^{-5.33} (L_{AB} + L_{BC}),$$

hence

$$p_C = \gamma_w \left[ z_{f1} + h_1 - z_C - \frac{V^2}{2g} \right.$$

$$\left. - (\xi_{in} + \xi_c)\frac{V^2}{2g} - \frac{4^{10/3}}{\pi^2 k^2} Q^2 D^{-5.33} (L_{AB} + L_{BC}) \right] \to$$

$$p_C = 9806 \times \left[ \begin{array}{c} 8 + 4 - 13 - \dfrac{6.3^2}{2 \times 9.806} - (1.0 + 0.1) \times \dfrac{6.3^2}{2 \times 9.806} \\ - \dfrac{4^{10/3}}{\pi^2 \times 90^2} \times \left(\dfrac{\pi \times 0.3^2}{4}\right)^2 \times 6.3^2 \times 0.3^{-5.33} \times (4 + 3) \end{array} \right]$$

$$= \mathbf{-62\,060 \ Pa}.$$

The absolute pressure in C is equal to

$$p_C^* = p_{atm}^* + p_C = 101\,300 - 62\,060 = \mathbf{39\,240\ Pa} > \mathbf{2314\ Pa}.$$

This pressure is greater than the vapour pressure and the initial hypothesis (pipe cross-section fully occupied by the water) is verified.

Applying mass conservation for tank No 1 yields

$$A_1 \frac{dh_1}{dt} = -\sqrt{\frac{z_1 - z_s}{\beta}}.$$

Introducing the variable $\zeta_1 = z_1 - z_s$, yields

$$A_1 \frac{d\zeta_1}{dt} = -\sqrt{\frac{\zeta_1}{\beta}} = -k\zeta_1^{1/2},$$

where

$$k = \frac{1}{\sqrt{\beta}} = \frac{1}{\sqrt{40.44}} = \mathbf{0.157\ m^{5/2}\,s^{-1}},$$

which, upon integration, yields

$$\int_{\zeta_1(t_0)}^{\zeta_1(t)} \frac{d\zeta_1}{\zeta_1^{1/2}} = -\frac{k}{A_1} \int_{t_0}^{t} dt \rightarrow 2\zeta_1^{1/2}\Big|_{\zeta_1(t_0)}^{\zeta_1(t)} = \frac{-k}{A_1} t\Big|_{t_0}^{t} \rightarrow$$

$$t - t_0 = -\frac{2A_1}{k}\left[\sqrt{\zeta_1(t)} - \sqrt{\zeta_1(t_0)}\right]. \quad (8.1)$$

The outflow process changes when the outlet section of the siphon is submerged (submergence limit condition), due to the rising level in tank No 2. This occurs when a volume of water equal to

$$W = A_2\left(z_s - z_{f2}\right) = 2 \times (4 - 1) = \mathbf{6\,m^3}$$

has been transferred from one tank to the other, and corresponds to a lowering of level in tank No 1 equal to

$$\Delta h = \frac{W}{A_1} = \frac{6}{4} = \mathbf{1.5\,m}.$$

In summary, it results

$$\zeta_1(t_0) = z_1 - z_s \equiv z_{f1} + h_1 - z_s = 8 + 4 - 4 = \mathbf{8\,m},$$

$$\zeta_1(t) = z_1' - z_s \equiv z_{f1} + h_1 - \Delta h - z_s = 8 + 4 - 1.5 - 4 = \mathbf{6.5\,m}.$$

By inserting the numerical values in Eq. (8.1), yields

$$t - t_0 = -\frac{2A_1}{k}\left[\sqrt{\varsigma_1(t)} - \sqrt{\varsigma_1(t_0)}\right] = -\frac{2 \times 4}{0.1571} \times \left(\sqrt{6.5} - \sqrt{8}\right) = \mathbf{14.2\,s}.$$

Once the submergence limit is exceeded, the available head is equal to the difference between the levels of the two tanks, and the energy balance equation is

$$z_1 - z_2 = \beta Q^2 \rightarrow Q = \sqrt{\frac{z_1 - z_2}{\beta}}.$$

The mass conservation equation requires that

$$A_1\frac{dh_1}{dt} = -\sqrt{\frac{z_1 - z_2}{\beta}} = -k\sqrt{z_1 - z_2}.$$

The levels of free surface in the tanks are linked by the following relationship:

$$A_1 dz_1 = -A_2 dz_2,$$

and, upon integration

$$z_2 = -\frac{A_1}{A_2}z_1 + \text{const.}$$

The mass conservation equation can be rewritten as

$$A_1\frac{dh_1}{dt} = -k\sqrt{z_1\left(1 + A_1/A_2\right) - \text{const.}} \equiv$$

$$- k\sqrt{\left(z_{f1} + h_1\right)\left(1 + A_1/A_2\right) - \text{const.}}$$

By inserting the auxiliary variable

$$\xi_1 = \left(z_{f1} + h_1\right)\left(1 + A_1/A_2\right) - \text{const.},$$

differentiating, yields

$$\frac{d\xi_1}{dt} = \frac{dh_1}{dt}\left(1 + A_1/A_2\right),$$

or

$$\frac{A_1}{\left(1 + A_1/A_2\right)}\frac{d\xi_1}{dt} = -k\sqrt{\xi_1}.$$

The result of the integration gives, in implicit form, the level change in tank No 1:

$$t - t_0 = -\frac{2A_1}{k\left(1 + A_1/A_2\right)}\left[\sqrt{\xi_1(t)} - \sqrt{\xi_1(t_0)}\right].$$

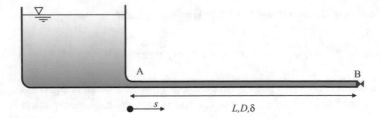

**Fig. 8.2**  Tank and pipeline with gate valve at the outlet section

**Exercise 8.2**  The outlet valve in section B (see Fig. 8.2) of a steel pipe $L = 3000\,\mathrm{m}$ long, diameter $D = 0.1\,\mathrm{m}$ and thickness $\delta = 5\,\mathrm{mm}$, is closed in 2 s, causing a retardation of the flow velocity of the water from $U_0 = 1.5\,\mathrm{m\,s^{-1}}$ to $U_f = 1.0\,\mathrm{m\,s^{-1}}$.

– Calculate the rise in pressure neglecting compressibility of the water and assuming that the pipe is rigid.

**Solution**  The energy balance equation in unsteady flow is

$$\frac{\partial H}{\partial s} + \frac{1}{g}\frac{\partial U}{\partial t} + J = 0.$$

If we assume an instantaneous propagation of the perturbations and neglect the dissipations, it results

$$\frac{\mathrm{d}H}{\mathrm{d}s} = -\frac{1}{g}\frac{\mathrm{d}U}{\mathrm{d}t},$$

where velocity and energy head depend only on time. Integrating, yields

$$\int_{H_0}^{H_f} \mathrm{d}H = -\frac{1}{g}\frac{\mathrm{d}U}{\mathrm{d}t}\int_0^L \mathrm{d}s \rightarrow H_f - H_0 = -\frac{L}{g}\frac{\mathrm{d}U}{\mathrm{d}t} \rightarrow p_f - p_0 \approx -\rho L\frac{\mathrm{d}U}{\mathrm{d}t},$$

where we have neglected the velocity head. Inserting the numerical values, yields

$$\therefore \quad p_f - p_0 = -\rho L\frac{\mathrm{d}U}{\mathrm{d}t} = -1000 \times 3000 \times \frac{1.0 - 1.5}{2} = \mathbf{7.5 \times 10^5\,Pa}.$$

According to this model, the rise in pressure propagates instantaneously in the pipe and the deceleration of the fluid is spatially uniform, with a hydraulic head varying in time but not in space. The rise in pressure is modest, hence the hypotheses of incompressible fluid and rigid pipe are reasonable. With this simplified model, an instantaneous closure results in an infinite pressure rise.

   If we extend the analysis by including the elastic behaviour of the fluid and of the pipe, we find that the celerity of propagation of the perturbation is

$$c = \frac{\sqrt{\dfrac{\varepsilon}{\rho}}}{\sqrt{1 + \dfrac{\varepsilon}{E}\dfrac{D}{\delta}}},$$

where $\varepsilon = 2 \times 10^9$ Pa is the bulk modulus of water, and $E = 2.07 \times 10^{11}$ Pa is the Young modulus of steel. Hence,

$$c = \frac{\sqrt{\dfrac{\varepsilon}{\rho}}}{\sqrt{1 + \dfrac{\varepsilon}{E}\dfrac{D}{\delta}}} = \frac{\sqrt{\dfrac{2 \times 10^9}{1000}}}{\sqrt{1 + \dfrac{2 \times 10^9}{2.07 \times 10^{11}} \times \dfrac{0.1}{0.005}}} = 1295\,\mathrm{m\,s^{-1}},$$

where we have assumed that the pipe is thin-wall (the stresses and the deformations are in membrane regime) and unconstrained. The phase duration, referred to the gate valve section, is equal to

$$\theta = \frac{2L}{c} = \frac{2 \times 3000}{1295} = 4.63\,\mathrm{s}.$$

At the end of the manouvre of 2 s, the reflected wave has not yet reached the gate, hence the pressure rise can be computed as

$$p_f - p_0 = -\rho c (U_f - U_0) = -1000 \times 1295 \times (1.0 - 1.5) = 6.48 \times 10^5\,\mathrm{Pa},$$

a value slightly less than the rise in pressure computed by neglecting the elastic waves. The rise in pressure propagates in the pipe with a celerity $c$, hence the head is space- and time-varying.

---

**Exercise 8.3** We consider a tank with constant level, with a pipe controlled by a valve in section B (see Fig. 8.3). The steel pipe is $L = 150$ m long, with diameter $D = 0.1$ m and relative roughness $\varepsilon/D = 0.002$. The head is $h_0 = 10$ m, the valve is initially closed and then it is instantaneously opened.

– Calculate the time required to reach the steady state.

**Solution** The energy balance equation in unsteady flow is

$$\frac{\partial H}{\partial s} + \frac{1}{g}\frac{\partial U}{\partial t} + J = 0.$$

Assuming an instantaneous propagation of the perturbations (anelastic model), the variables do not depend on $s$, hence, upon integration, it results

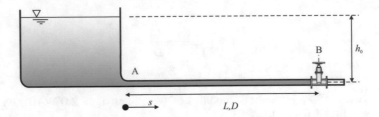

**Fig. 8.3** Tank and pipeline with gate valve initially closed

$$\frac{U^2}{2g} + kU^2 - h_0 = -\frac{L}{g}\frac{dU}{dt},\tag{8.2}$$

where $kU^2$ are the energy losses, concentrated and distributed. Considering $\lambda = \lambda_\infty = 0.023$ and assuming an inlet coefficient of energy loss $\xi_{in} = 0.5$ results in

$$k = \frac{\lambda_\infty L}{2gD} + \frac{\xi_{in}}{2g} = \frac{0.023 \times 150}{2 \times 9.806 \times 0.1} + \frac{0.5}{2 \times 9.806} = 1.83\,\mathrm{m^{-1}\,s^2}.$$

In regime conditions results $dU/dt = 0$ and the fluid velocity is

$$U_0 = \sqrt{\frac{2gh_0}{1 + 2gk}} = \sqrt{\frac{2 \times 9.806 \times 10}{1 + 2 \times 9.806 \times 1.83}} = 2.30\ \mathrm{m\,s^{-1}}.$$

Equation (8.2) can be written as

$$\frac{dU}{dt} = \frac{gh_0}{L}\left(1 - \frac{U^2}{U_0^2}\right),$$

that, upon integration, yields

$$U = U_0 \tanh\left(\frac{gh_0 t}{LU_0}\right).$$

The regime is reached only asymptotically. If we assume that the regime condition corresponds to $U = 0.99U_0$, the required time is

$$\therefore \qquad t = 2.65\frac{LU_0}{gh_0} = 2.65 \times \frac{150 \times 2.30}{9.806 \times 10} = \mathbf{9.3\,s}.$$

If we include the elastic behaviour of the fluid and of the pipe, the regime velocity is reached with a series of oscillations of velocity and pressure, being the first oscillation a pressure reduction. These oscillations are progressively damped, even neglecting dissipations since the system is open. Figure 8.4 shows the time series of velocity

**Fig. 8.4** Velocity in the terminal section. The red dashed line refers to the anelastic model, the blue line includes elastic waves

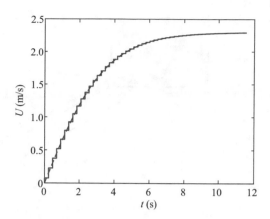

in the terminal section, computed according to an anelastic model (red dashed line) and including elastic waves propagating with a celerity $c = 1296\,\mathrm{m\,s^{-1}}$.

---

**Exercise 8.4**  A steel pipeline is connected to an upstream tank characterized by a constant water level, and to a gate valve, with outflow in atmosphere, see Fig. 8.5. The diameter is $D = (1000 + 10 \times C_{pu})$ mm, the wall thickness is $\delta = 12$ mm, the length is equal to $L = 2000$ m. Starting from a permanent flow condition with a flow rate $Q = 2\,\mathrm{m^3\,s^{-1}}$, a gate valve is closed with a linear variation in a time $\tau = 4\,(2L/c)$ s, where $c$ is the celerity of the perturbations.

Using the equations of Allievi, for the entire duration of the closure manoeuvre and for a subsequent phase calculate:

– the head in the outlet section in the full-phase and half-phase times, collecting the results in a diagram.
– Compare the maximum overpressure with the value given by the Michaud–Allievi formula.
– Draw the head diagram in the cross-section with abscissa $L/2$.

Assume a water isentropic bulk modulus $\varepsilon = 2 \times 10^9$ Pa, and the Young modulus of steel $E = 2.07 \times 10^{11}$ Pa. Neglect energy losses and refer the head to the outflow cross-section, where $h_0 = (400 + 10 \times C_u)$ m.

**Solution**  Before starting computations, we describe the physical process.

• A steady-state velocity $U_0$ exists throughout, the hydraulic grade line is horizontal since we are neglecting losses, and the valve is progressively closed in a finite time $\tau$ starting at time 0.
• As a consequence of valve closure, a wave rises and travels upstream at speed $c$, and behind the wave the velocity is less than $U_0$, the pressure rises, the liquid

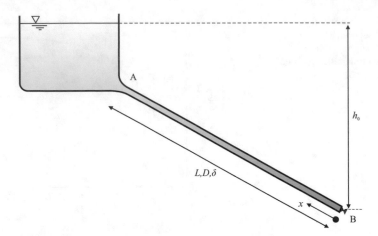

**Fig. 8.5** Tank and pipeline with gate valve at the outlet section

is compressed, and the pipe is slightly expanded radially and possibly longitudinally, depending on the constraints. The wave shape and amplitude is continuously
modulated by the manoeuvre.

- The wave reaches the reservoir at time $L/c$, the pipe pressure reduces to the
  reservoir pressure. A new wave arises propagating downstream to the valve, with
  shape depending on the shape of the manoeuvre at an earlier time of $L/c$.
- The modulated wave reaches the valve at time $\theta = 2L/c$ and determines a velocity with magnitude depending on the degree of closure of the valve and on the
  instantaneous head.
- The process continues for ever since we are neglecting losses and the system is
  closed.

If the pressure behind the wave reduces to vapour pressure, cavitation will occur, a
condition termed "column separation".

Assuming $C_u = C_{pu} = 0$, it results $D = 1000\,\text{mm}$ and $h_0 = 400\,\text{m}$. The average
velocity in the pipeline, in permanent flow condition, is equal to

$$U_0 = \frac{4Q}{\pi D^2} = \frac{4 \times 2}{\pi \times 1.0^2} = 2.55\,\text{m s}^{-1}.$$

Assuming a cylindrical circular pipe made of linear elastic material, behaving like
a membrane, the relative celerity of the perturbations is equal to

$$c = \frac{\sqrt{\dfrac{\varepsilon}{\rho}}}{\sqrt{1 + \varsigma \dfrac{\varepsilon}{E} \dfrac{D}{\delta}}},$$

where $\varsigma$ is a dimensionless parameter that takes into account the type of constraint of the conduct. This parameter has the following expressions:

$\varsigma = 5/4 - \nu \rightarrow$ for upstream anchored pipe and free to deform longitudinally,
$\varsigma = 1 - \nu^2 \rightarrow$ for pipe with precluded longitudinal deformation,
$\varsigma = 1 \rightarrow$ for pipe with free longitudinal deformation,

where $\nu$ is the Poisson coefficient of the pipe material. Assuming that the last condition applies, the celerity is equal to

$$ c = \frac{\sqrt{\dfrac{\varepsilon}{\rho}}}{\sqrt{1 + \varsigma \dfrac{\varepsilon}{E} \dfrac{D}{\delta}}} = \frac{\sqrt{\dfrac{2 \times 10^9}{1000}}}{\sqrt{1 + 1 \times \dfrac{2 \times 10^9}{2.07 \times 10^{11}} \times \dfrac{1.0}{0.012}}} = 1053 \text{ m s}^{-1}. $$

The phase duration, referred to the gate valve section, is equal to

$$ \theta = \frac{2L}{c} = \frac{2 \times 2000}{1053} = 3.8 \text{ s}. $$

Since $\tau > \theta$, the manoeuvre is "slow", that means that the reflected wave reaches the gate before the end of the manouvre. In order to apply the interlocking equations of Allievi, we need a list of the values of the opening of the gate at full-phase and at half-phase times, see Tables 8.1 and 8.2.

**Table 8.1** Ratio of opening of the gate valve at full-phase times

| $t/\theta$ | $\eta$ |
|---|---|
| 0.0 | 1.00 |
| 1.0 | 0.75 |
| 2.0 | 0.50 |
| 3.0 | 0.25 |
| 4.0 | 0.00 |
| 5.0 | 0.00 |

**Table 8.2** Ratio of opening of the gate valve at half-phase times

| $t/\theta$ | $\eta$ |
|---|---|
| 0.5 | 0.875 |
| 1.5 | 0.625 |
| 2.5 | 0.375 |
| 3.5 | 0.125 |
| 4.5 | 0 |

The interlocking equations of Allievi have the following expression:

$$z_i^2 + z_{i-1}^2 - 2 = 2\text{Al}\,(\eta_{i-1}z_{i-1} - \eta_i z_i),$$

where Al is the Allievi number, defined as

$$\text{Al} = \frac{U_0 c}{2gh_0},$$

and

$$z_i = \sqrt{\frac{h_i}{h_0}}, \quad \eta_i = \frac{\omega_i}{\omega_0}, \quad \eta_0 = 1, \quad t_{i+1} = t_i + \theta,$$

where $\omega$ is the gate opening area.

For the present case, it results

$$\text{Al} = \frac{U_0 c}{2gh_0} = \frac{2.55 \times 1053}{2 \times 9.806 \times 400} = \mathbf{0.3418}.$$

We choose the time $0 < t_1 \le \theta$ equal to $t_1 = \theta$; hence, $t_2 = t_1 + \theta = 2\theta$, $t_i = t_{i-1} + \theta = i\theta$. The ratio of opening can be calculated from the linear closure law of the gate valve:

$$\begin{cases} \eta = 1 - \dfrac{t}{\tau} & \text{for } 0 \le t \le \tau, \\[2ex] \eta = 0 & \text{for } t > \tau. \end{cases}$$

Considering the series of full-phase times, at the time $t_1$ the Allievi's equation becomes

$$z_1^2 + z_0^2 - 2 = 2\text{Al}\,(\eta_0 z_0 - \eta_1 z_1) \rightarrow$$

$$z_1^2 + 1 - 2 = 2 \times 0.3418 \times (1 - 0.75 \times z_1) \rightarrow$$

$$z_1^2 + 0.5127 \times z_1 - 1.6836 = 0,$$

which admits two solutions, the physically acceptable one is $z_1 = \mathbf{1.0663}$.

At the time $t_2$ the Allievi's equation becomes

$$z_2^2 + z_1^2 - 2 = 2\text{Al}\,(\eta_1 z_1 - \eta_2 z_2) \rightarrow$$

$$z_2^2 + 1.0663^2 - 2 = 2 \times 0.3418 \times (0.75 \times 1.0663 - 0.5 \times z_2) \rightarrow$$

$$z_2^2 + 0.3418 \times z_2 - 1.4097 = 0,$$

which admits the physically acceptable solution $z_2 = \mathbf{1.0287}$.

A similar procedure applies for the calculation at the half-phase instants.

**Table 8.3** Calculation results in the outlet section

| $t/\theta$ | $z$ | $\dfrac{h\,(t/\theta,0)}{h_0}$ | $h\,(t/\theta,0)$ (m) |
|---|---|---|---|
| 1 | 1.0663 | 1.1369 | 454.8 |
| 2 | 1.0287 | 1.0581 | 423.3 |
| 3 | 1.0550 | 1.1131 | 445.2 |
| 4 | 1.0330 | 1.0672 | 426.9 |
| 5 | 0.9658 | 0.9328 | 373.1 |
| 0.5 | 1.0325 | 1.0660 | 426.4 |
| 1.5 | 1.0502 | 1.1029 | 441.1 |
| 2.5 | 1.0390 | 1.0795 | 431.8 |
| 3.5 | 1.0475 | 1.0973 | 438.9 |
| 4.5 | 0.9961 | 0.9922 | 396.9 |

**Fig. 8.6** Time series of the head in the outlet section. The symbols are the results of the manual calculation, the line is the automatic numerical model output

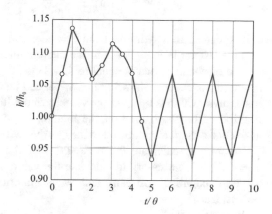

The results are listed in Table 8.3; Fig. 8.6 shows the head (dimensionless) in the outlet section. The continuous line is the numerical computation result, the symbols are the values from the manual calculation.

The maximum head is reached at the end of the first phase, and it is equal to **454.8 m**.

The Michaud–Allievi's formula (also known as Joukowsky's formula) assumes a closure manoeuvre with a linear reduction of fluid velocity, whereas the analysis carried out with the Allievi interlocking equations assumes a closure manoeuvre with a linear reduction of the degree of opening. The Michaud–Allievi's formula gives, at the end of the first phase, an overhead in the outlet section equal to

$$\Delta h = \frac{c}{g} U_0 \frac{\theta}{\tau} = \frac{c}{g} U_0 \frac{1}{4}.$$

By inserting the numerical values yields

**Fig. 8.7** Time series of the non dimensional head in the outlet section. The continuous line is for a gate closing linearly, the dashed line is for a linear variation of the velocity in the outlet section

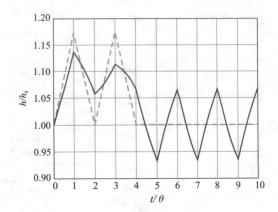

$$\Delta h = \frac{c}{g} U_0 \frac{1}{4} = \frac{1053}{9.806} \times 2.55 \times \frac{1}{4} = \mathbf{68.5\,m},$$

corresponding to a hydraulic head $h = h_0 + \Delta h = \mathbf{468.5\,m}$.

This last value is higher than the one obtained assuming a closure linear in the degree of opening, that is equal to 454.8 m. The comparison between the two different manoeuvres is shown in Fig. 8.7. It can be demonstrated that for each closure manoeuvre which is linear in the velocity and which lasts for a time interval that is a multiple of the phase duration, the combination of direct and reflected waves is such as to cancel them after the end of the closure, with a null residual head overpressure.

In order to calculate the time series of the head in the intermediate section $x = L/2$, it is necessary to calculate the value assumed by the function $F$ (direct wave, propagating upstream) and by the function $f$ (reflected wave, propagating downstream).

We remind that

$$h(t, x) - h_0 = F\left(t - \frac{x}{c}\right) - f\left(t + \frac{x}{c}\right).$$

At the time $t_1, t_2, \ldots, t_i$ and for the outlet section ($x = 0$), it results:

$$\underbrace{h(t_i, 0) - h_0}_{h_i} = \underbrace{F(t_i)}_{F_i} - \underbrace{f(t_i)}_{f_i}.$$

Furthermore, considering the upstream boundary condition, where the head is invariant,

$$h(t, L) - h_0 = F\left(t - \frac{L}{c}\right) - f\left(t + \frac{L}{c}\right) = 0,$$

yields

$$f_i = F_{i-1}.$$

**Table 8.4** Results of the manual calculation for the intermediate section. $h_{1/2}$ is the head at $x = L/2$

| $t/\theta$ | $\dfrac{F(t/\theta)}{h_0}$ | $-\dfrac{f(t/\theta)}{h_0}$ | $\dfrac{F(t/\theta - 0.25)}{h_0}$ | $-\dfrac{f(t/\theta + 0.25)}{h_0}$ | $\dfrac{h_{1/2}(t/\theta) - h_0}{h_0}$ | $h_{1/2}(t/\theta)$ (m) |
|---|---|---|---|---|---|---|
| 1 | 0.1369 | 0 | 0.10145 | −0.033 | 0.06845 | 427.4 |
| 2 | 0.1950 | −0.1369 | 0.18195 | −0.1529 | 0.02905 | 411.6 |
| 3 | 0.3082 | −0.1950 | 0.27825 | −0.22165 | 0.0566 | 422.6 |
| 4 | 0.3753 | −0.3082 | 0.3605 | −0.32695 | 0.03355 | 413.4 |
| 5 | 0.3082 | −0.3753 | 0.323 | | | |
| 0.5 | 0.0660 | 0 | 0.033 | 0 | 0.033 | 413.2 |
| 1.5 | 0.1689 | −0.0660 | 0.1529 | −0.10145 | 0.05145 | 420.6 |
| 2.5 | 0.2483 | −0.1689 | 0.22165 | −0.18195 | 0.0397 | 415.9 |
| 3.5 | 0.3457 | −0.2483 | 0.32695 | −0.27825 | 0.0487 | 419.5 |
| 4.5 | 0.3378 | −0.3457 | 0.35655 | 0.3605 | −0.00395 | 398.4 |

Therefore, it is possible to write the following expression:

$$h_i - h_0 = F_i - F_{i-1} \rightarrow F_i = h_i - h_0 + F_{i-1},$$

where $F_i$ represents the direct wave, propagating upstream, in the section of the valve ($x = 0$) calculated at the time $t_i$; $f_i$ represents the reflected wave, propagating downstream, in the section of the valve ($x = 0$) calculated at the time $t_i$. In the intermediate section, the two functions are out of phase, in advance and late, respectively, by a time interval of a quarter of phase, $t = \dfrac{L/2}{c}$. Hence, the head in the intermediate section has the following expression:

$$h\left(t, \frac{L}{2}\right) - h_0 = F\left(t - \frac{L}{2c}\right) - f\left(t + \frac{L}{2c}\right).$$

By expressing time in dimensionless form with respect to the phase duration, it results:

$$h\left(\frac{t}{\theta}, \frac{L}{2}\right) - h_0 = F\left(\frac{t}{\theta} - 0.25\right) - f\left(\frac{t}{\theta} + 0.25\right).$$

The function $F(t/\theta - 0.25)$ is calculated by linear interpolation of the two closest values, in excess and in defect, to the time $t - 0.25\theta$. A similar calculation is performed for the function $f(t/\theta + 0.25)$.

The results are listed in Table 8.4 and Table 8.5 lists the results of numerical computation. The differences between the values in the two tables are negligible.

Figure 8.8 shows the excess of head (dimensionless) in the intermediate section. The dashed curves represent the two functions $F$ and $f$, that are phase-shifted by a quarter of the phase duration.

**Table 8.5** Results of the numerical computation for the intermediate section. $h_{1/2}$ is the head at $x = L/2$

| $t/\theta$ | $\dfrac{F(t/\theta)}{h_0}$ | $-\dfrac{f(t/\theta)}{h_0}$ | $\dfrac{F(t/\theta - 0.25)}{h_0}$ | $-\dfrac{f(t/\theta + 0.25)}{h_0}$ | $\dfrac{h_{1/2}(t/\theta) - h_0}{h_0}$ | $h_{1/2}(t/\theta)$ (m) |
|---|---|---|---|---|---|---|
| 1 | 0.1369 | 0 | 0.1008 | −0.0324 | 0.0684 | 427.4 |
| 2 | 0.1950 | −0.1369 | 0.1828 | −0.1535 | 0.0293 | 411.7 |
| 3 | 0.3082 | −0.1950 | 0.2773 | −0.2210 | 0.0563 | 422.5 |
| 4 | 0.3753 | −0.3082 | 0.3617 | −0.3277 | 0.0340 | 413.6 |
| 5 | 0.3082 | −0.3753 | 0.3212 | −0.3566 | −0.0354 | 385.8 |
| 0.5 | 0.0660 | 0 | 0.0324 | 0 | 0.0324 | 412.9 |
| 1.5 | 0.1689 | −0.0660 | 0.1535 | −0.1008 | 0.0527 | 421.1 |
| 2.5 | 0.2483 | −0.1689 | 0.2210 | −0.1828 | 0.0382 | 415.2 |
| 3.5 | 0.3457 | −0.2483 | 0.3277 | −0.2773 | 0.0504 | 420.2 |
| 4.5 | 0.3378 | −0.3457 | 0.3566 | −0.3617 | −0.0051 | 397.9 |

**Fig. 8.8** Diagram for the calculation of the head excess in the intermediate section

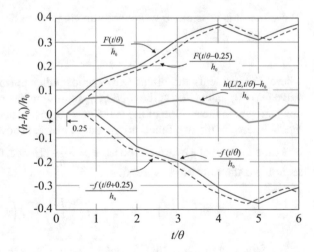

If we are also interested to the velocity evolution in time, we can use the following equation

$$U(t, x) - U_0 = -\frac{g}{c}\left[F\left(t - \frac{x}{c}\right) + f\left(t + \frac{x}{c}\right)\right].$$

Once the two functions $F$ and $f$ have been obtained, the computation of $U(t, x)$ is immediate. Figure 8.9 shows the dimensionless velocity for the inlet, the outlet and the intermediate sections. We notice that the oscillations are not damped, since no dissipation occurs and the system is closed.

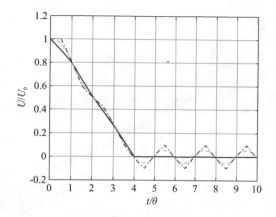

**Fig. 8.9** Diagram for the calculation of the water velocity in the outlet section (bold line), at $x = L/2$ (dashed line), and at the inlet section (dash-dotted line)

**Exercise 8.5** In the plant in Fig. 8.10 the pipe has length $L = (4000 + 100 \times C_u)$ m, diameter $D = (500 + 10 \times C_{pu})$ mm, thickness $\delta = 10$ mm and Gauckler–Strickler coefficient $k = 90 \, \text{m}^{1/3} \, \text{s}^{-1}$. The head is $h_0 = 100$ m.

- Calculate the permanent flow rate.
- Calculate the maximum overpressure in the outlet section for a slow closure manoeuvre which lasts twice the phase duration, neglecting the energy losses.

Assume a water isentropic bulk modulus $\varepsilon = 2 \times 10^9$ Pa, and the Young modulus of steel $E = 2.07 \times 10^{11}$ Pa. Neglect energy losses and refer the head to the outflow cross-section.

**Solution** The permanent flow rate is calculated by applying the energy balance equation:

$$z_A + \frac{p_A}{\gamma_w} + \frac{U_A^2}{2g} = z_B + \frac{p_B}{\gamma_w} + \frac{U_0^2}{2g} + JL + \sum \xi_i \frac{U_0^2}{2g} \rightarrow$$

$$h_0 = \frac{Q^2}{2g\left(\frac{\pi D^2}{4}\right)^2} + \frac{Q^2}{k^2 \left(\frac{D}{4}\right)^{4/3} \left(\frac{\pi D^2}{4}\right)^2} L,$$

where we have neglected the concentrated energy loss at the entrance. The solution for the flow rate is

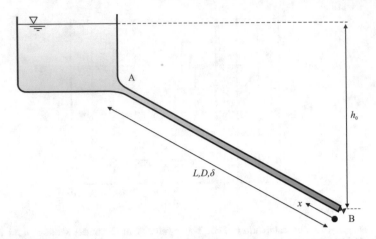

**Fig. 8.10** Tank and pipeline with gate valve at the outlet section

$$Q = \sqrt{\dfrac{h_0}{\dfrac{L}{k^2 \left(\dfrac{D}{4}\right)^{4/3} \left(\dfrac{\pi D^2}{4}\right)^2} + \dfrac{1}{2g \left(\dfrac{\pi D^2}{4}\right)^2}}}.$$

For $C_u = C_{pu} = 0$ it results $L = 4000\,\mathrm{m}$, $D = 500\,\mathrm{mm}$, and

$$\therefore \quad Q = \sqrt{\dfrac{100}{\dfrac{4000}{90^2 \times \left(\dfrac{0.5}{4}\right)^{4/3} \times \left(\dfrac{\pi \times 0.5^2}{4}\right)^2} + \dfrac{1}{2 \times 9.806 \times \left(\dfrac{\pi \times 0.5^2}{4}\right)^2}}}$$

$$= 0.70\,\mathrm{m^3\,s^{-1}},$$

$$U_0 = \dfrac{4Q}{\pi D^2} = \dfrac{4 \times 0.70}{\pi \times 0.5^2} = 3.57\,\mathrm{m\,s^{-1}}.$$

The celerity of the perturbations is equal to

$$c = \sqrt{\dfrac{\varepsilon/\rho}{1 + \dfrac{\varepsilon D}{E\delta}}} = \sqrt{\dfrac{2.14 \times 10^9/1000}{1 + \dfrac{2.14 \times 10^9 \times 0.5}{2.0 \times 10^{11} \times 0.01}}} = 1180\,\mathrm{m\,s^{-1}},$$

and the phase duration is equal to

**Fig. 8.11** Time series of the
opening degree of the valve

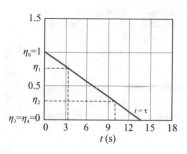

$$\theta = \frac{2L}{c} = \frac{2 \times 4000}{1180} = \textbf{6.78 s}.$$

The duration of the closure manoeuvre is equal to

$$\tau = 2\theta = \textbf{13.56 s}.$$

The Allievi interlocking equations have the following expression

$$z_i^2 + z_{i-1}^2 - 2 = 2\text{Al}\left(\eta_{i-1}z_{i-1} - \eta_i z_i\right),$$

where $\text{Al} = \dfrac{U_0 c}{2gh_0}$ is the Allievi number and

$$z_i = \sqrt{\frac{h_i}{h_0}}, \quad \eta_i = \frac{\omega_i}{\omega_0}.$$

For the present case, it results

$$\text{Al} = \frac{U_0 c}{2gh_0} = \frac{3.57 \times 1180}{2 \times 9.806 \times 100} = \textbf{2.148}.$$

We choose the time $0 < t_1 \leq \theta$ with $t_1 = 0.5 \times \theta = 3.39$ s; hence $t_2 = t_1 + \theta = 10.17$ s, $t_3 = t_2 + \theta = 16.95$ s. The degree of opening can be calculated from the linear closure law of the gate:

$$\begin{cases} \eta = 1 - \dfrac{t}{\tau} & \text{for } 0 \leq t \leq \tau, \\[2mm] \eta = 0 & \text{for } t > \tau, \end{cases}$$

and is shown in Fig. 8.11. The values of the degree of opening at the time of calculation are listed in Table 8.6.

The initial condition is $z_0 = 1$ and

**Table 8.6** Time series of the degree of opening of the gate, half-phase times

| $t_0 = 0\,\text{s}$ | $\eta_0 = 1$ |
|---|---|
| $t_1 = 3.39\,\text{s}$ | $\eta_1 = 0.75$ |
| $t_2 = 10.17\,\text{s}$ | $\eta_2 = 0.25$ |
| $t_3 = 16.95\,\text{s}$ | $\eta_3 = 0$ |

**Table 8.7** Time series of the degree of opening of the gate, full-phase times

| $t'_1 = 6.78\,\text{s}$ | $\eta'_1 = 0.50$ |
|---|---|
| $t'_2 = 13.56\,\text{s}$ | $\eta'_2 = 0$ |
| $t'_3 = 20.34\,\text{s}$ | $\eta'_3 = 0$ |

$$z_1^2 + z_0^2 - 2 = 2\text{Al}\,(\eta_0 z_0 - \eta_1 z_1) \rightarrow z_1^2 + 2\text{Al}\eta_1 z_1 - 1 - 2\text{Al} = 0.$$

By inserting the numerical values yields

$$z_1^2 + 2 \times 2.148 \times 0.75 \times z_1 - 1 - 2 \times 2.148 = 0 \rightarrow z_1^2 + 3.2225 \times z_1 - 5.296 = 0,$$

which leads to the solution $z_1 = \mathbf{1.198}$.

At the time $t_2$, it results:

$$z_2^2 + z_1^2 - 2 = 2\text{Al}\,(\eta_1 z_1 - \eta_2 z_2).$$

By inserting the numerical values, it results:

$$z_2^2 + (1.198)^2 - 2 - 2 \times 2.148 \times 0.75 \times 1.198 + 2 \times 2.148 \times 0.25 \times z_2 = 0 \rightarrow$$

$$z_2^2 + 1.074 \times z_2 - 4.4248 = 0,$$

which leads to the solution $z_2 = \mathbf{1.634}$.

At the time $t_3$, it results:

$$z_3^2 + z_2^2 - 2 = 2\text{Al}\,(\eta_2 z_2 - \eta_3 z_3),$$

and:

$$z_3^2 + (1.634)^2 - 2 - 2 \times 2.148 \times 0.25 \times 1.634 = 0 \rightarrow z_3^2 - 1.085 = 0,$$

which leads to the solution $z_3 = \mathbf{1.0416}$.

We then choose the time $0 < t'_1 \leq \theta$ with $t'_1 = \theta = 6.78\,\text{s}$; hence, $t'_2 = t'_1 + \theta = 13.56\,\text{s}$, $t'_3 = t'_2 + \theta = 20.34\,\text{s}$. The degree of opening can be calculated from the linear closure law. The values are given in Table 8.7. Furthermore, $z'_0 = 1$, $\eta'_0 = 1$. At the time $t'_1$ it is:

$$z'^2_1 + z'^2_0 - 2 = 2\text{Al}\,(\eta'_0 z'_0 - \eta'_1 z'_1) \rightarrow z'^2_1 + 2\text{Al}\eta'_1 z'_1 - 1 - 2\text{Al} = 0,$$

**Table 8.8** Summary table of the results

| $t_0 = 0\,\mathrm{s}$ | $\eta_0 = 1$ | $z_0 = 1$ | $h_0 = 100\,\mathrm{m}$ |
|---|---|---|---|
| $t_1 = 3.39\,\mathrm{s}$ | $\eta_1 = 0.75$ | $z_1 = 1.198$ | $h_1 = 144\,\mathrm{m}$ |
| $t_1' = 6.78\,\mathrm{s}$ | $\eta_1' = 0.5$ | $z_1' = 1.466$ | $h_1' = 215\,\mathrm{m}$ |
| $t_2 = 10.17\,\mathrm{s}$ | $\eta_2 = 0.25$ | $z_2 = 1.634$ | $h_2 = 267\,\mathrm{m}$ |
| $t_2' - 13.56\,\mathrm{s}$ | $\eta_2' = 0$ | $z_2' = 1.732$ | $h_2' = 300\,\mathrm{m}$ |
| $t_3 = 16.95\,\mathrm{s}$ | $\eta_3 = 0$ | $z_3 = 1.042$ | $h_3 = 108\,\mathrm{m}$ |
| $t_3' = 20.34\,\mathrm{s}$ | $\eta_3' = 0$ | | |

and, by inserting the numerical values yields

$$z_1'^2 + 2 \times 2.148 \times 0.50 \times z_1' - 1 - 2 \times 2.148 = 0 \rightarrow z_1'^2 + 2.148 \times z_1' - 5.296 = 0,$$

which leads to the solution $z_1' = \mathbf{1.4656}$.

At the time $t_2'$ it is:

$$z_2'^2 + z_1'^2 - 2 = 2\mathrm{Al}\left(\eta_1' z_1' - \eta_2' z_2'\right).$$

Inserting the numerical values, it results:

$$z_2'^2 + (1.4656)^2 - 2 - 2 \times 2.148 \times 0.5 \times 1.4656 = 0 \rightarrow z_2'^2 - 3.0 = 0,$$

which leads to the solution $z_2' = \mathbf{1.732}$.

At the time $t_3'$ it is:

$$z_3'^2 + z_2'^2 - 2 = 2\mathrm{Al}\left(\eta_2' z_2' - \eta_3' z_3'\right).$$

Inserting the numerical values, it results:

$$z_3'^2 + (1.732)^2 - 2 = 0 \rightarrow z_3'^2 = -1.$$

This last result is equivalent to $h_3' = -h_0$.

The computed values are listed in Table 8.8 and are shown in Fig. 8.12. The last value (and any other head value lower than $-10.33\,\mathrm{m}$) is not admissible: the minimum absolute pressure value is limited to the vapour pressure by the cavitation.

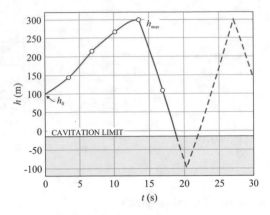

**Fig. 8.12** Hydraulic head in the outlet section. Symbols are manual calculation results, curves are numerical code results. After reaching the cavitation limit, the model is no longer valid (dashed line)

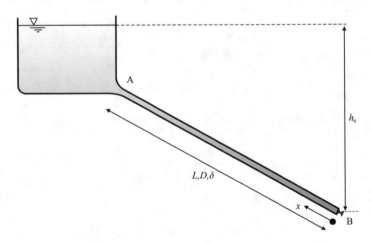

**Fig. 8.13** Tank and pipeline with gate valve at the outlet

**Exercise 8.6** In the system shown in Fig. 8.13, the initial flow rate, when the gate valve is fully open, is $Q_0 = 1 \, \mathrm{m^3 \, s^{-1}}$, with an average velocity $U_0 = 2 \, \mathrm{m \, s^{-1}}$. The pipe is made of steel with a thickness $\delta = 10 \, \mathrm{mm}$, and length $L = 1000 \, \mathrm{m}$. The head is $h_0 = 100 \, \mathrm{m}$.

– Calculate the head rise at the valve, if the gate area is uniformly reduced to zero in 4 s. Use the Allievi interlocking equations, starting at the time $t_1 = 0.5 \, \mathrm{s}$, and for the time interval from 0 to three phases.

   Assume a water isentropic bulk modulus $\varepsilon = 2 \times 10^9 \, \mathrm{Pa}$, a steel Young modulus $E = 2.1 \times 10^{11} \, \mathrm{Pa}$. Neglect pressure losses.

**Solution** The diameter of the pipe is equal to:

$$D = \sqrt{\frac{4Q_0}{\pi U_0}} = \sqrt{\frac{4 \times 1.0}{\pi \times 2.0}} = 0.8 \, \text{m}.$$

The celerity of perturbation is equal to:

$$c = \frac{\sqrt{\frac{\varepsilon}{\rho}}}{\sqrt{1 + \frac{\varepsilon}{E} \frac{D}{\delta}}} = \frac{\sqrt{\frac{2 \times 10^9}{1000}}}{\sqrt{1 + \frac{2 \times 10^9}{2.1 \times 10^{11}} \times \frac{0.8}{0.010}}} = 1065 \, \text{m s}^{-1},$$

and the phase duration is equal to:

$$\theta = \frac{2L}{c} = \frac{2 \times 1000}{1065} = 1.88 \, \text{s}.$$

Since $\tau > \theta$, the closure manoeuvre is slow. The Allievi interlocking equations have the following expression:

$$z_i^2 + z_{i-1}^2 - 2 = 2\text{Al} \left( \eta_{i-1} z_{i-1} - \eta_i z_i \right),$$

where Al is the Allievi number equal to $\text{Al} = \dfrac{U_0 c}{2gh_0}$; furthermore, $z_i = \sqrt{\dfrac{h_i}{h_0}}$, $\eta_i = \dfrac{\omega_i}{\omega_0}$, and $\omega_i$ is the opening area of the gate at the $i$-time. For the present case, results:

$$\text{Al} = \frac{U_0 c}{2gh_0} = \frac{2 \times 1065}{2 \times 9.806 \times 100} = 1.086.$$

We choose the time $0 < t_1 \le \theta$ with $t_1 = 0.5 \, \text{s}$; then $t_2 = t_1 + \theta = 2.38 \, \text{s}$, $t_3 = t_2 + \theta = 4.26 \, \text{s}$ and $t_4 = t_3 + \theta = 6.14 \, \text{s}$. The degree of opening can be calculated from the linear closure equation of the gate valve, is shown in Fig. 8.14 and is listed in Table 8.9:

$$\begin{cases} \eta = 1 - \dfrac{t}{\tau} & \text{for } 0 \le t \le \tau, \\[3mm] \eta = 0 & \text{for } t > \tau, \end{cases}$$

with the additional initial condition

$$z_0 = 1, \quad \eta_0 = 1.$$

At the time $t_1$, results:

$$z_1^2 + z_0^2 - 2 = 2\text{Al} \left( \eta_0 z_0 - \eta_1 z_1 \right) \rightarrow z_1^2 + 2\text{Al}\eta_1 z_1 - 1 - 2\text{Al} = 0.$$

**Fig. 8.14** Time series of the opening degree of the valve

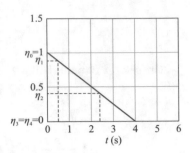

**Table 8.9** Degree of opening of the gate at the calculation instants

| $t_0 = 0\,\text{s}$ | $\eta_0 = 1$ |
|---|---|
| $t_1 = 0.5\,\text{s}$ | $\eta_1 = 0.875$ |
| $t_2 = 2.38\,\text{s}$ | $\eta_2 = 0.405$ |
| $t_3 = 4.26\,\text{s}$ | $\eta_3 = 0$ |
| $t_4 = 6.14\,\text{s}$ | $\eta_4 = 0$ |

Inserting the numerical values yields:

$$z_1^2 + 2 \times 1.086 \times 0.875 \times z_1 - 1 - 2 \times 1.086 = 0 \rightarrow z_1^2 + 1.9005 \times z_1 - 3.172 = 0,$$

which leads to the solution $z_1 = \mathbf{1.069}$.

At the time $t_2$, it results:

$$z_2^2 + z_1^2 - 2 = 2Al\,(\eta_1 z_1 - \eta_2 z_2).$$

Inserting the numerical values yields:

$$z_2^2 + (1.069)^2 - 2 - 2 \times 1.086 \times 0.875 \times 1.069 + 2 \times 1.086 \times 0.405 \times z_2 = 0,$$

that is $z_2^2 + 0.88 \times z_2 - 2.888 = 0$, which leads to the solution $z_2 = \mathbf{1.315}$.

At the time $t_3$, it results:

$$z_3^2 + z_2^2 - 2 = 2Al\,(\eta_2 z_2 - \eta_3 z_3).$$

Inserting the numerical values yields:

$$z_3^2 + (1.315)^2 - 2 - 2 \times 1.086 \times 0.405 \times 1.315 = 0,$$

that is $z_3^2 - 1.428 = 0$, which leads to the solution $z_3 = \mathbf{1.195}$.

At the time $t_4$, it results:

$$z_4^2 + z_3^2 - 2 = 2Al\,(\eta_3 z_3 - \eta_4 z_4).$$

**Table 8.10** Summary table of the results

| $t$ (s) | $\eta$ | $z$ | $h$ (m) |
|---|---|---|---|
| 0 | 1 | 1 | 100 |
| 0.5 | 0.875 | 1.069 | 114 |
| 2.38 | 0.405 | 1.315 | 173 |
| 4.26 | 0 | 1.195 | 143 |
| 6.14 | 0 | 0.756 | 57 |

**Fig. 8.15** Head in the outlet section. The continuous curve is the result of a numerical computation, symbols are the manual calculation results and refer to $t = 0.5, 2.38, 4.26, 6.14$ s

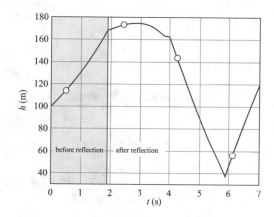

Inserting the numerical values, yields:

$$z_4^2 - 0.572 = 0,$$

which leads to the solution $z_4 = \mathbf{0.756}$.
   The results are listed in Table 8.10 and are plotted in Fig. 8.15.

**Exercise 8.7** In the system in Fig. 8.16, the flow rate in stationary condition is $Q_0 = 1.5\,\mathrm{m^3\,s^{-1}}$ with an average velocity in the pipe $U_0 = 2\,\mathrm{m\,s^{-1}}$. The pipeline is made of steel with a thickness $\delta = 15$ mm, a length $L = 800$ m and $h_0 = 100$ m.

- Calculate the celerity of propagation of the pressure disturbance and the phase duration of the pipeline, $\theta$.
- Calculate the pressure rise at the outlet section for a linear opening manoeuvre of duration $\tau = 3$ s. Use the Allievi interlocking equations, starting from $t_1 = 0.5$ s and for the interval up to three phases.
- Calculate the minimum value of the head at $t = 0$.

   Assume a water isentropic bulk modulus $\varepsilon = 2 \times 10^9$ Pa, a steel Young modulus $E = 2.1 \times 10^{11}$ Pa. Neglect energy losses.

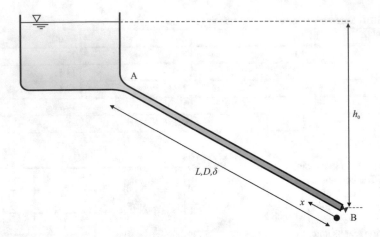

**Fig. 8.16** Tank and pipeline with gate valve at the outlet section

**Solution** The diameter of the pipe is equal to:

$$D = \sqrt{\frac{4Q_0}{\pi U_0}} = \sqrt{\frac{4 \times 1.5}{\pi \times 2}} = 0.98\,\text{m} \approx \mathbf{1.0\,m}.$$

The celerity of the perturbations is equal to:

$$c = \frac{\sqrt{\dfrac{\varepsilon}{\rho}}}{\sqrt{1 + \dfrac{\varepsilon}{E}\dfrac{D}{\delta}}} = \frac{\sqrt{\dfrac{2 \times 10^9}{1000}}}{\sqrt{1 + \dfrac{2 \times 10^9}{2.1 \times 10^{11}} \times \dfrac{1}{0.015}}} = \mathbf{1106\,m\,s^{-1}},$$

and the phase duration of the pipeline is equal to:

$$\theta = \frac{2L}{c} = \frac{2 \times 800}{1106} = \mathbf{1.45\,s}.$$

Since $\tau > \theta$, the manouvre is slow. The Allievi interlocking equations have the following expression:

$$z_i^2 + z_{i-1}^2 - 2 = 2\text{Al}\,(\eta_{i-1}z_{i-1} - \eta_i z_i)\,,$$

where Al is the Allievi number, defined as $\text{Al} = \dfrac{U_0 c}{2gh_0}$; furthermore $z_i = \sqrt{\dfrac{h_i}{h_0}}$, $\eta_i = \dfrac{\omega_i}{\omega_0}$.

For the present case, it results:

**Fig. 8.17** Time series of the opening degree of the valve

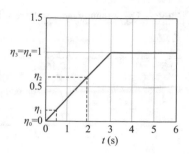

**Table 8.11** Time series of the opening degree of the gate

| | |
|---|---|
| $t_0 = 0$ s | $\eta_0 = 0$ |
| $t_1 = 0.5$ s | $\eta_1 = 0.166$ |
| $t_2 = 1.95$ s | $\eta_2 = 0.65$ |
| $t_3 = 3.4$ s | $\eta_3 = 1$ |
| $t_4 = 4.85$ s | $\eta_4 = 1$ |

$$\text{Al} = \frac{U_0 c}{2gh_0} = \frac{2 \times 1106}{2 \times 9.806 \times 100} = \mathbf{1.128}.$$

We choose the time $0 < t_1 \leq \theta$ with $t_1 = 0.5$ s; then $t_2 = t_1 + \theta = 1.95$ s, $t_3 = t_2 + \theta = 3.4$ s and $t_4 = t_3 + \theta = 4.85$ s. The degree of opening can be calculated from the linear closure equation of the gate valve, is shown in Fig. 8.17 and listed in Table 8.11:

$$\begin{cases} \eta = \dfrac{t}{t} & \text{for } 0 < t \leq \tau, \\[2mm] \eta = 1 & \text{for } t > \tau. \end{cases}$$

The initial condition is $z_0 = 1$. At the time $t_1$, it results:

$$z_1^2 + z_0^2 - 2 = 2\text{Al}\,(\eta_0 z_0 - \eta_1 z_1) \rightarrow z_1^2 + 2\text{Al}\eta_1 z_1 - 1 = 0.$$

Inserting the numerical values, results:

$$z_1^2 + 2 \times 1.128 \times 0.1666 \times z_1 - 1 = 0 \rightarrow z_1^2 + 0.37581 \times z_1 - 1 = 0,$$

which leads to the solution $z_1 = \mathbf{0.829}$.

A the time $t_2$, it results:

$$z_2^2 + z_1^2 - 2 = 2\text{Al}\,(\eta_1 z_1 - \eta_2 z_2).$$

Inserting the numerical values, results:

$$z_2^2 + (0.829)^2 - 2 - 2 \times 1.128 \times 0.1666 \times 0.829 + 2 \times 1.128 \times 0.65 \times z_2 = 0,$$

**Table 8.12**  Summary table of the results

| $t$ (s) | $\eta$ | $z$ | $h$ (m) |
|---------|--------|-------|---------|
| 0 | 0 | 1 | 100 |
| 0.5 | 0.166 | 0.829 | 68.8 |
| 1.95 | 0.65 | 0.737 | 54.3 |
| 3.4 | 1 | 0.824 | 67.9 |
| 4.85 | 1 | 0.982 | 96.4 |

**Fig. 8.18**  Head in the outlet section. The continuous curve is the result of an automatic calculation procedure, symbols are values manually calculated at $t = 0.5, 1.45, 1.95, 3.4, 4.85$ s

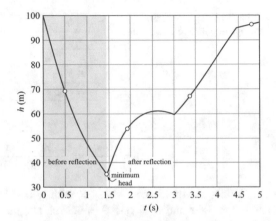

that is $z_2^2 + 1.466 \times z_2 - 1.623 = 0$, which leads to the solution $z_2 = \mathbf{0.737}$.

At the time $t_3$, it results:

$$z_3^2 + z_2^2 - 2 = 2Al\,(\eta_2 z_2 - \eta_3 z_3)\,.$$

Inserting the numerical values, results:

$$z_3^2 + (0.737)^2 - 2 - 2 \times 1.1128 \times 0.65 \times 0.737 + 2 \times 1.1128 \times 1 \times z_3 = 0,$$

that is $z_3^2 - 2.256 \times z_3 - 2.537 = 0$, which leads to the solution $z_3 = \mathbf{0.824}$.

At the time $t_4$, it results:

$$z_4^2 + z_3^2 - 2 = 2Al\,(\eta_3 z_3 - \eta_4 z_4)\,.$$

Inserting the numerical values, results:

$$z_4^2 + 2.256 \times z_4 - 3.18 = 0,$$

which leads to the solution $z_4 = \mathbf{0.982}$.

The calculated values are listed in Table 8.12.

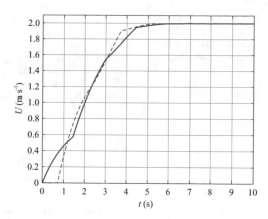

**Fig. 8.19** Velocity in the outlet section (bold line) and in the inlet section (dashed line)

The minimum value of the head occurs at the time $t = \theta = 1.45\,\text{s}$, with $\eta_{min} = 0.4833$. By substituting in the Allievi interlocking equation, results:

$$z_{min}^2 + z_0^2 - 2 = 2Al\,(\eta_0 z_0 - \eta_{min} z_{min}) \rightarrow z_{min}^2 + 2Al\eta_{min} z_{min} - 1 = 0.$$

Inserting the numerical values, results:

$$z_{min}^2 + 2 \times 1.128 \times 0.4833 \times z_{min} - 1 = 0 \rightarrow z_{min}^2 + 1.09 \times z_{min} - 1 = 0,$$

that leads to the solution $z_{min} = 0.593$, corresponding to a head $h_{min} = 35.3\,\text{m}$.

Figures 8.18 and 8.19 show the head in the outlet section and the fluid velocity in the outlet and in the inlet sections, respectively.

---

**Exercise 8.8** In the system in Fig. 8.20, the steel pipeline is made of two trunks with the following characteristics: upstream trunk AC: $D_1 = 950$ mm, $L_1 = 1800$ m, $c_1 = 1037\,\text{m s}^{-1}$; downstream trunk CB: $D_2 = 800$ mm, $L_2 = 2000$ m, $c_2 = 1152\,\text{m s}^{-1}$. The two trunks are free to expand and contract along the axis. The head at the outlet section is $h_0 = 350$ m and the elevation of the sections is $z_A = 300$ m, $z_C = 200$ m, $z_B = 100$ m. Starting from an initial permanent flow rate equal to $Q_0 = 2.1\,\text{m}^3\,\text{s}^{-1}$, a linear closure manoeuvre of duration $\tau = 8\,\text{s}$ is performed.

– Using the method of characteristics, calculate the pressure in section C for two phases.

Neglect the slope of the pipeline axis and the energy losses.

**Solution** The phase duration of the two trunks is equal to:

$$\theta_1 = \frac{2L_1}{c_1} = \frac{2 \times 1800}{1037} = 3.47\,\text{s},$$

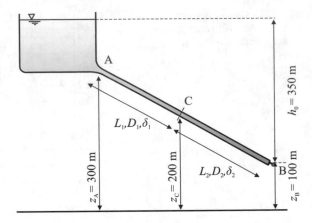

**Fig. 8.20** Tank and pipeline with gate valve at the outlet. The characteristics of the pipes are not uniform

$$\theta_2 = \frac{2L_2}{c_2} = \frac{2 \times 2000}{1152} = \mathbf{3.47\,s},$$

and the phase duration of the pipeline is equal to $\theta = \theta_1 + \theta_2 = \mathbf{6.94\,s}$. Since $\theta < \tau$, the manoeuvre is slow. In addition, section C is reached in equal times by disturbances from the inlet or from the gate valve. The gate closure law is as follows:

$$\begin{cases} \eta = 1 - \dfrac{t}{8} & \text{for } 0 \le t \le \tau, \\ \eta = 0 & \text{for } t > \tau, \end{cases}$$

and, as a function of the time non-dimensional with respect to the phase duration, it results:

$$\begin{cases} \eta = 1 - 0.868\dfrac{t}{\theta} & \text{for } 0 \le \dfrac{t}{\theta} \le 1.152, \\ \eta = 0 & \text{for } \dfrac{t}{\theta} > 1.152. \end{cases}$$

Choosing the coordinate system in Fig. 8.21, it is possible to identify two families of characteristics for each trunk. It is convenient to assume $Q$ and $h$ as the dependent variables, and along these characteristics, it results:

$$\begin{cases} \dfrac{dh}{dt} + \dfrac{c_{1,2}}{g\Omega_{1,2}}\dfrac{dQ}{dt} = -\dfrac{Q}{\Omega_{1,2}}\sin\alpha + c_{1,2}J & \text{if } \dfrac{ds}{dt} = c_{1,2}, \quad \text{downstream, } \lambda^+, \\ \dfrac{dh}{dt} - \dfrac{c_{1,2}}{g\Omega_{1,2}}\dfrac{dQ}{dt} = -\dfrac{Q}{\Omega_{1,2}}\sin\alpha - c_{1,2}J & \text{if } \dfrac{ds}{dt} = -c_{1,2}, \quad \text{upstream, } \lambda^-. \end{cases}$$

Neglecting the slope of the pipeline $\alpha$ and the dissipations, the previous expressions are simplified as follows:

**Fig. 8.21** Schematic adopted to define the propagation celerity of the pressure perturbations

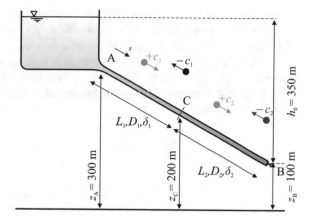

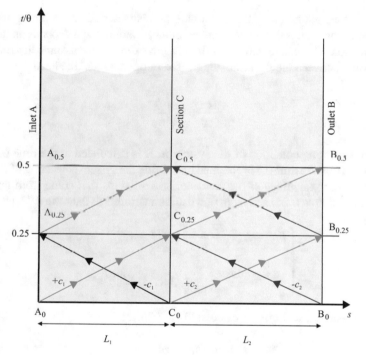

**Fig. 8.22** Characteristic curves in the $s - t$ plane

$$
\begin{cases}
\dfrac{\mathrm{d}h}{\mathrm{d}t} + \dfrac{c_{1,2}}{g\Omega_{1,2}} \dfrac{\mathrm{d}Q}{\mathrm{d}t} = 0 & \text{if } \dfrac{\mathrm{d}s}{\mathrm{d}t} = c_{1,2}, \quad \text{downstream, } \lambda^+, \\[3mm]
\dfrac{\mathrm{d}h}{\mathrm{d}t} - \dfrac{c_{1,2}}{g\Omega_{1,2}} \dfrac{\mathrm{d}Q}{\mathrm{d}t} = 0 & \text{if } \dfrac{\mathrm{d}s}{\mathrm{d}t} = -c_{1,2}, \quad \text{upstream, } \lambda^-.
\end{cases}
$$

Figure 8.22 shows the characteristics in the phase-plane $s - t$. The arrows indicate the direction of propagation of the information. At non dimensional time 0.25 the point C is still undisturbed, since the perturbation triggered at the outlet has not yet arrived. To calculate the condition of the system in C at the time 0.5, it is necessary to preliminary calculate the condition of the system in B at the time 0.25. The operating point at the downstream boundary is obtained at time 0.25 by solving the following system of equations:

$$\begin{cases} h_B^{0.25} - h_C^0 + \dfrac{c_2}{g\Omega_2}\left(Q_B^{0.25} - Q_C^0\right) = 0, \\[3mm] h_B^{0.25} - h_C^0\left(\dfrac{Q_B^{0.25}}{Q_C^0}\dfrac{1}{\eta^{0.25}}\right)^2 = 0. \end{cases}$$

The first equation is the invariant along the characteristic $\lambda^+$, coming from $C_0$. The second equation is the downstream boundary condition and depends on the operating principle of the gate valve. For the present exercise, we assume the following relationship between the flow rate, the degree of opening and the head:

$$\frac{Q}{Q_0} = \eta\sqrt{\frac{h}{h_0}}.$$

Once the operating condition of the system in B is calculated at the time 0.25, the condition in C at the time 0.5 is obtained by solving the system consisting of the finite difference approximations of the characteristic relations that come from upstream (from A at the time 0.25) and from downstream (from B at the time 0.25):

$$\begin{cases} h_C^{0.5} - h_A^{0.25} + \dfrac{c_1}{g\Omega_1}\left(Q_C^{0.5} - Q_A^{0.25}\right) = 0, \\[3mm] h_C^{0.5} - h_B^{0.25} - \dfrac{c_2}{g\Omega_2}\left(Q_C^{0.5} - Q_B^{0.25}\right) = 0. \end{cases}$$

The pressure in section C is calculated as follows:

$$p = \gamma_w\left(h_C + z_B - z_C\right).$$

The results are summarized in Table 8.13 and are shown in Fig. 8.23.

**Table 8.13** Summary table of the results

| $t/\theta$ | $\eta$ | $h_C$ (m) | $p_C$ ($\times 10^5$ Pa) | $Q_C$ (m$^3$ s$^{-1}$) |
|------------|--------|-----------|--------------------------|------------------------|
| 0.25 | 0.783 | 350. | 24.5 | 2.10 |
| 0.5 | 0.566 | 404.5 | 29.8 | 1.73 |
| 0.75 | 0.349 | 472.0 | 36.5 | 1.28 |
| 1.0 | 0.1319 | 480.9 | 37.3 | 0.49 |
| 1.25 | 0 | 485.9 | 37.8 | −0.45 |
| 1.5 | 0 | 410.5 | 30.5 | −0.79 |
| 1.75 | 0 | 238.3 | 13.6 | −0.61 |
| 2.0 | 0 | 192.3 | 9.1 | −0.14 |

**Fig. 8.23** Pressure (gage) diagram in the outlet section (dashed curve) and in the intermediate section C (continuous curve), computed with a numerical code. Symbols are the results of the manual calculations

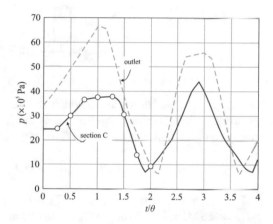

**Exercise 8.9** The pumping plant in Fig. 8.24 conveys, in steady condition, a flow rate $Q_0 = (0.3 + C_u/100)$ m$^3$ s$^{-1}$. The delivery duct is $L = (1500 + C_{pu} \times 10)$ m long, has a diameter $D = 0.4$ m and has a nominal operating pressure of $p_{max} = (25 + 0.5 \times C_{pu}) \times 10^5$ Pa. The Gauckler–Strickler coefficient is $k = 80$ m$^{1/3}$ s$^{-1}$. The static head, measured with respect to a section immediately downstream of the pump, is $H'_s = 210$ m.

– Design an expansion tank (i) without damping restriction, and (ii) with optimal damping restriction, assuming a non-return valve between the tank and the pump.

Assume an instantaneous shut-down of the pump, a water isentropic bulk modulus $\varepsilon = 2 \times 10^9$ Pa, and a steel Young modulus $E = 2 \times 10^{11}$ Pa. The pipe thickness is equal to 8 mm.

**Solution** Consider the plant in regime condition, and abruptly subject to a blackout of the grid supplying the electric motor. After the pump has been switched off, a

**Fig. 8.24** Pumping plant
with expansion tank

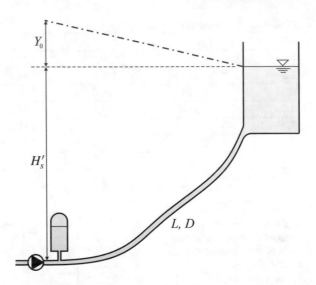

vacuum is generated downstream which can be calculated (for additional safety) by
ignoring the inertia of the impeller and of the fluid and assuming an instantaneous
transition of the flow rate to a zero value, with the following formula:

$$\Delta p = -\rho c V_0.$$

For $C_u = C_{pu} = 0$ it results $Q_0 = 0.3 \, \text{m}^3 \, \text{s}^{-1}$, $L = 1500 \, \text{m}$, $D = 0.4 \, \text{m}$,
$p_{max} = 2.5 \, \text{MPa}$, $k = 80 \, \text{m}^{1/3} \, \text{s}^{-1}$, $H'_s = 210 \, \text{m}$. The initial velocity in the pipeline
is equal to:

$$V_0 = \frac{4Q_0}{\pi D^2} = \frac{4 \times 0.3}{\pi \times 0.4^2} = 2.39 \, \text{m s}^{-1},$$

and the celerity of the perturbation is equal to:

$$c = \frac{\sqrt{\dfrac{\varepsilon}{\rho}}}{\sqrt{1 + \dfrac{\varepsilon}{E}\dfrac{D}{\delta}}} = \frac{\sqrt{\dfrac{2.0 \times 10^9}{1000}}}{\sqrt{1 + \dfrac{2.0 \times 10^9}{2.0 \times 10^{11}} \times \dfrac{0.4}{0.008}}} = 1155 \, \text{m s}^{-1}.$$

The phase duration is equal to:

$$\theta = \frac{2L}{c} = \frac{2 \times 1500}{1155} = 2.60 \, \text{s}.$$

The pressure reduction is equal to $\Delta p = -1000 \times 1155 \times 2.39 = -2.76 \, \text{MPa}$.
The absolute static head is equal to

$$H_s = H'_s + \frac{p^*_{atm}}{\gamma_w} = 210 + 10.33 = 220.33 \, \text{m},$$

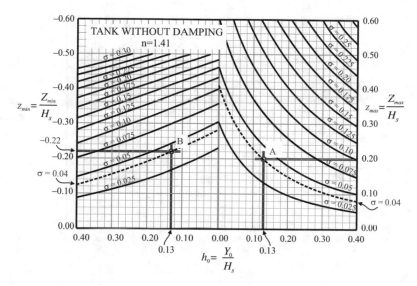

**Fig. 8.25** Diagram for the design of an expansion tank without damping restriction

and the absolute pressure in the section immediately downstream of the pump becomes:

$$p^* = \gamma_w H_s - \Delta p = 9806 \times 220.33 - 2.76 \times 10^6 = -0.59\,\text{MPa} \ll p_{vap}.$$

Since the absolute pressure cannot be lower than the vapour pressure, cavitation occurs and the installation of an expansion tank is required. The design and verification of the tank will be carried out with the help of two graphs shown in Figs. 8.25 and 8.26.

The maximum value of the admissible head depends on the maximum tolerable pressure in the pipeline, and it is equal to:

$$H'_s + Z_{max} = \frac{p_{max}}{\gamma_w} \rightarrow Z_{max} = \frac{p_{max}}{\gamma_w} - H'_s \equiv \frac{2.5 \times 10^6}{9806} - 210 = \textbf{44.9\,m}.$$

*Tank Without Damping Restriction*

The dimensionless parameter to be inserted in the graph in Fig. 8.25 is

$$z_{max} = \frac{Z_{max}}{H_s} = \frac{44.9}{220.33} = \textbf{0.20}.$$

The most critical condition refers to the minimum losses (pipeline new). These losses are equal to:

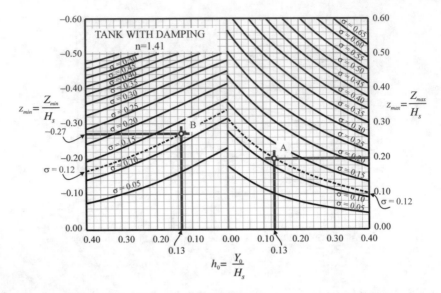

**Fig. 8.26** Diagram for the design of an expansion tank with optimal damping restriction

$$Y_0 = \frac{V_0^2}{k^2 R^{4/3}} L = \frac{2.39^2}{80^2 \times \left(\dfrac{0.4}{4}\right)^{4/3}} \times 1500 = \mathbf{28.8\,m}.$$

The second parameter to be inserted in the diagrams is $h_0 = \dfrac{Y_0}{H_s} = \dfrac{28.8}{220.33} =$ **0.13**.

From the diagram in Fig. 8.25, intersecting $z_{max} = 0.20$ and $h_0 = 0.13$ (point A), a parameter $\sigma$ equal to **0.04** is estimated, where $\sigma$ is a non dimensional group defined as

$$\sigma = \frac{\Omega L}{U_s H_s} \frac{V_0^2}{2g},$$

where $U_s$ is the volume of gas in the tank in static conditions (pump off, fluid at rest, head equal to $H_s$). In the present condition, it is equal to

$$U_s = \frac{\Omega L}{\sigma H_s} \frac{V_0^2}{2g} = \frac{\dfrac{\pi \times 0.4^2}{4} \times 1500}{0.04 \times 220.33} \times \frac{2.39^2}{2 \times 9.806} = \mathbf{6.22\ m^3}.$$

For $h_0 = 0.13$ and $\sigma = 0.04$ (point B), from the left axis of the diagram results $z_{min} = \mathbf{-0.22}$. Using the gas transformation law, yields $H_s U_s^n = H_{min} U_{max}^n$ ($n$ is the exponent of the transformation) and, therefore:

$$U_{max} = \frac{U_s}{(1 + z_{min})^{1/n}} = \frac{6.22}{(1 - 0.22)^{1/1.41}} = \mathbf{7.42\ m^3}.$$

Furthermore,

$$U_{min} = \frac{U_s}{(1 + z_{max})^{1/n}} = \frac{6.22}{(1 + 0.20)^{1/1.41}} = \textbf{5.47 m}^3.$$

When the pump is running, the head is $H_s + Y_0$ and the air volume in the tank is equal to

$$U_0 = \frac{U_s}{(1 + h_0)^{1/n}} = \frac{6.22}{(1 + 0.13)^{1/1.41}} = \textbf{5.70 m}^3.$$

In order to prevent cavitation, it should be:

$$h^*_{min} = H_s (1 + z_{min}) = 220.33 \times (1 - 0.22) \equiv \textbf{171.9 m} > \frac{p^*_{atm}}{\gamma_w}.$$

The minimum value of the relative pressure is equal to:

$$p_{min} = \gamma_w \left(h^*_{min} - 10.33\right) = 9806 \times (171.9 - 10.33) = \textbf{1.58 MPa}.$$

An expansion tank with a total volume of $\textbf{8 m}^3$ is sufficient.

*Tank with Damping Restriction*
We proceed as for the case of a tank without damping restriction, but using the diagram in Fig. 8.26. At point A, it results:

$$\sigma_{opt} = \frac{\Omega L}{U_s H_s} \frac{V_0^2}{2g} = \textbf{0.12}.$$

The gas volume in the tank in static conditions (pump off, fluid at rest, head equal to $H_s$) is equal to:

$$U_{s-opt} = \frac{\Omega L}{\sigma_{opt} H_s} \frac{V_0^2}{2g} = \frac{\frac{\pi \times 0.4^2}{4} \times 1500}{0.12 \times 220.33} \times \frac{2.39^2}{2 \times 9.806} = \textbf{2.07 m}^3.$$

From the left axis of the diagram results $z_{min} = \textbf{-0.27}$. Using the gas transformation law, it results $H_s U_s^n = H_{min} U_{max}^n$, hence

$$U_{max-opt} = \frac{U_{s-opt}}{(1 + z_{min})^{1/n}} = \frac{2.07}{(1 - 0.27)^{1/1.41}} = \textbf{2.58 m}^3.$$

Furthermore:

$$U_{min-opt} = \frac{U_{s-opt}}{(1 + z_{max})^{1/n}} = \frac{2.07}{(1 + 0.20)^{1/1.41}} = \textbf{1.82 m}^3.$$

When the pump is running, the head is $H_s + Y_0$ and the air volume in the tank is equal to

$$U_{0-opt} = \frac{U_{s-opt}}{(1 + h_0)^{1/n}} = \frac{2.07}{(1 + 0.13)^{1/1.41}} = \textbf{1.90 m}^3.$$

To prevent cavitation, it should be:

$$h^*_{min} = H_s (1 + z_{min}) = 220.33 \times (1 - 0.27) \equiv \textbf{160.8 m} > \frac{p^*_{atm}}{\gamma_w}.$$

The minimum value of the relative pressure is equal to: $p_{min} = \gamma_w \left(h^*_{min} - 10.33\right)$ $= 9806 \times (160.8 - 10.33) = \textbf{1.48 MPa}$.

An expansion tank with a total volume of $\textbf{3 m}^3$ is sufficient. Notice that the installation of an optimum restriction allows to reduce considerably the volume of the expansion tank.

To be effective, the restriction should induce a loss, immediately after the pump switch-off, equal to:

$$\Delta H = Y_0 + |Z_{min}| = Y_0 + |H_s z_{min}| = 28.8 + |220.33 \times (-0.27)| = \textbf{88.30 m}.$$

The energy loss coefficient of the restriction is calculated considering that:

$$\beta Q_0^2 = Y_0 + |Z_{min}| = Y_0 + |H_s z_{min}| \rightarrow \beta = \frac{Y_0 + |H_s z_{min}|}{Q_0^2}.$$

Inserting the numerical values, it results:

$$\beta = \frac{Y_0 + |H_s z_{min}|}{Q_0^2} = \frac{28.8 + |220.33 \times (-0.27)|}{0.3^2} = \textbf{981}.$$

The restriction can be realized adding a trunk of smaller diameter in the connection between the box and the pipeline, so as to generate a Borda–Carnot energy loss due to the sudden expansion of the flow. If the connecting trunk has a diameter equal to $D$, the diameter of the restriction is calculated imposing that:

$$\beta Q_0^2 = \underbrace{\frac{Q_0^2}{2g} \left(\frac{4}{\pi d^2} - \frac{4}{\pi D^2}\right)^2}_{\text{Borda-Carnot loss}} \rightarrow \frac{1}{d^2} = \frac{1}{D^2} + \frac{\pi}{4}\sqrt{2g\beta}.$$

Inserting the numerical values, yields

$$\frac{1}{d^2} = \frac{1}{D^2} + \frac{\pi}{4}\sqrt{2g\beta} = \frac{1}{0.4^2} + \frac{\pi}{4}\sqrt{2 \times 9.806 \times 981} = 115.2 \rightarrow d = \textbf{0.093 m}.$$

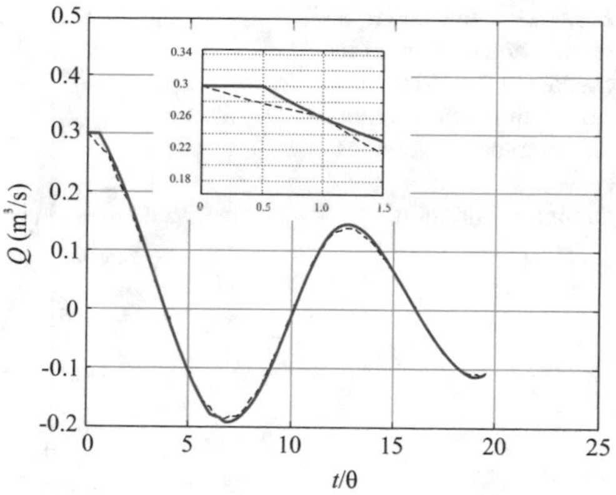

**Fig. 8.27**  Flow rate in the
section downstream of the
expansion tank (dashed
curve) and in the section of
the reservoir (continuous
curve). Tank without
damping restriction

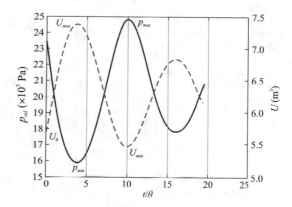

**Fig. 8.28**  Pressure in the
pipeline downstream of the
expansion tank (continuous
line) and air volume (dashed
line) in the expansion tank
without damping restriction

Figure 8.27 shows the flow rate in the section immediately downstream of the expansion tank (dashed curve) and in the reservoir section (continuous curve), as a function of the dimensionless time. The inset is an enlargement of the initial evolution.

The diagrams were obtained with an automatic numerical calculation software that also includes the effect of elastic oscillations. For this reason, the flow rate in the section of the reservoir is affected by the switch-off of the pump only after half of a phase duration. Figure 8.28 shows the pressure in the section immediately downstream of the expansion tank and the volume of air in the expansion tank, respectively.

Figures 8.29 and 8.30 show the same results for an expansion tank with an optimal restriction, with $\beta = 981$. It can be observed that the minimum pressure is slightly higher than 1.2 MPa, while on the basis of the calculations performed using the diagram in Fig. 8.26 it would be equal to 1.48 MPa. The difference between the two results is due to the effects of elastic oscillations which are included in the automatic calculation software. At the end of the first phase, these effects determine a pressure reduction which is combined with the pressure reduction due to the mass oscillation. For the same reason, the maximum pressure in the pipeline (slightly) exceeds 2.5 MPa

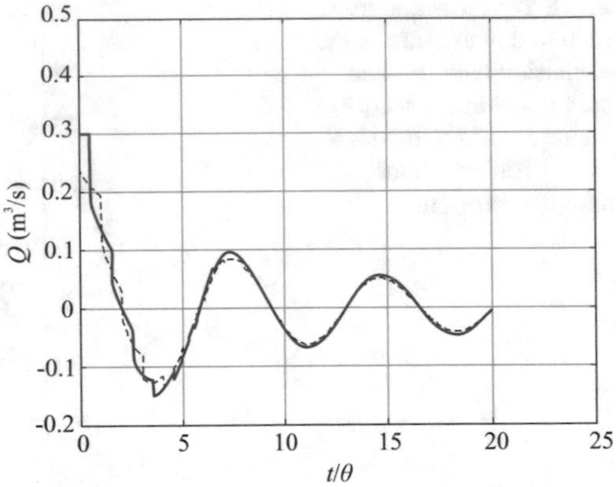

**Fig. 8.29** Flow rate in the section downstream of the expansion tank (dashed curve) and in the section of the reservoir (continuous curve). Tank with optimal damping restriction

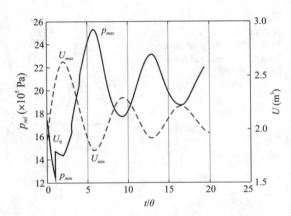

**Fig. 8.30** Pressure in the pipeline downstream of the expansion tank (continuous line) and air volume (dashed line) in the expansion tank with optimal damping restriction

(design value) and the changes of the air volume in the expansion tank are greater than expected. The oscillations have a shorter period than in the case of a expansion tank without restriction, and the damping is faster.

# Chapter 9
# Flow in Open Channels

In an open channel flow, only part of the current is at contact with walls and a free surface is present, usually at atmospheric pressure. In most cases, the curvature of the trajectories is negligible and a hydrostatic pressure distribution is achieved. Gradient pressure in the direction of flow is rather limited and the flow is driven by differences in the elevation head. The source of energy per unit of weight and per unit of length is the slope of the bed, that exactly balances the energy losses in a uniform regime. In other situations, the source of energy is the slope of the free surface, which can drive flow also in the presence of horizontal or counter-sloping bed. The classical terminology defines the area of the cross section of the current; the wetted perimeter, i.e. the length of the contact between the liquid and the walls of the channel; the hydraulic radius, equal to the ratio between the two. The kinematics of the currents is described in terms of Froude number, with distinction of the *subcritical* (Fr < 1) and *supercritical* flows (Fr > 1), also referred to as *tranquil* and *shooting* flow, respectively. If the uniform flow is tranquil, the bed slope is defined *mild*, otherwise it is *steep*. In critical conditions (Fr = 1), the current requires the minimum energy to flow and the critical depth corresponds to a stationary point of energy referred to the local bottom of the channel.

The uniform flow is purely theoretical, as it requires conditions of stationarity in a long infinite channel with homogeneous slope of the bed, roughness, cross-section. Under practical conditions, channel flows are varied and unsteady. For gradually varied steady flows, the hydrostatic pressure distribution assumption still holds, and the evolution of the profile is susceptible to analytical solution for prismatic channels. In the general case, a numerical solution is required.

---

$C_u$ and $C_{pu}$, that are two integer numbers between 0 and 9, for example, the last and second-last digit of the registration number.

© Springer Nature Switzerland AG 2021                                                             319
S. Longo et al., *Problems in Hydraulics and Fluid Mechanics*, Springer Tracts
in Civil Engineering, https://doi.org/10.1007/978-3-030-51387-0_9

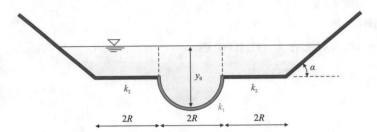

**Fig. 9.1**  Cross-section of the channel

**Exercise 9.1**  In the channel with cross-section shown in Fig. 9.1, the central zone has a Gauckler–Strickler coefficient $k_1 = (40 + C_{pu})$ m$^{1/3}$ s$^{-1}$, and the expansion zone has a Gauckler–Strickler coefficient $k_2 = (35 + C_{pu})$ m$^{1/3}$ s$^{-1}$. The angle of the banks is $\alpha = (30 + C_{pu})°$, the radius is equal to $R = (1 + C_u/100)$ m and the slope of the bed is $i_b = (0.3 + C_{pu}/20)$ %.

– Draw the rating curve every 50 cm, up to $y_0 = 3$ m.

Assume a distribution of the currents in the three bodies delimited by the walls and the dashed verticals as shown in Fig. 9.1.

**Solution**  If the water depth, measured from the lowest point of the cross-section, is less than $R$, the current is contained in the central channel, see Fig. 9.2, and the rating curve is calculated as follows:

$$Q = k_1 R_h^{1/6} \Omega \sqrt{R_h i_b} \equiv k_1 \sqrt{i_b} \frac{\Omega^{5/3}}{P^{2/3}}, \qquad y_0 < R,$$

where $R_h$ is the hydraulic radius, $\Omega$ is the area of the cross-section occupied by the current, $P$ is the wetted perimeter.

From elementary geometry considerations, the area of the cross-section of the current and the wetted perimeter are equal to:

$$\left. \begin{aligned} \Omega &= R^2 \cos^{-1}\left(\frac{R - y_0}{R}\right) - (R - y_0)\sqrt{R^2 - (R - y_0)^2} \\[2ex] P &= 2R \cos^{-1}\left(\frac{R - y_0}{R}\right) \end{aligned} \right\}, \qquad y_0 < R.$$

If the current occupies the expansion zone, the flow rate can be calculated as the sum of the flow rates of the three water bodies separated by the dashed verticals in Fig. 9.3. In the calculation of the wetted perimeters, the contours represented by the the dashed verticals must not be taken into account, since the resistance offered by these to the motion is null by hypothesis.

The central body will contribute with a flow rate equal to:

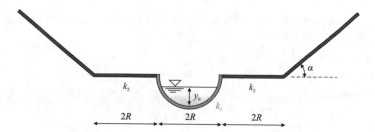

**Fig. 9.2** Cross-section of the channel with the current remaining in the central channel

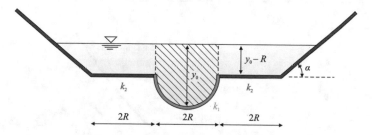

**Fig. 9.3** Separation of the current into water bodies without tangential stress interaction

$$Q_1 = k_1 R_h^{1/6} \Omega \sqrt{R_h i_b} \equiv k_1 \frac{\Omega^{5/3}}{P^{2/3}} \sqrt{i_b}, \qquad y_0 \geq R,$$

with

$$\left. \begin{aligned} \Omega &= \frac{\pi R^2}{2} + 2R\,(y_0 - R) \\ P &= \pi R \end{aligned} \right\}, \qquad y_0 \geq R.$$

The expansion zones will contribute with a flow rate equal to:

$$Q_2 = 2\left(k_2 R_h^{1/6} \Omega \sqrt{R_h i_b}\right) \equiv 2\left(k_2 \frac{\Omega^{5/3}}{P^{2/3}} \sqrt{i_b}\right), \qquad y_0 \geq R,$$

with

$$\left. \begin{aligned} \Omega &= 2R\,(y_0 - R) + \frac{(y_0 - R)^2}{2\tan\alpha} \\ P &= 2R + \frac{(y_0 - R)}{\sin\alpha} \end{aligned} \right\}, \qquad y_0 \geq R.$$

For $C_u = C_{pu} = 0$ it results $k_1 = 40 \text{ m}^{1/3}\,\text{s}^{-1}$, $k_2 = 35 \text{ m}^{1/3}\,\text{s}^{-1}$, $\alpha = 30°$, $R = 1.0$ m, $i_b = 0.3\%$.

The calculated values are given in Table 9.1 and the rating curve is shown in Fig. 9.4.

**Table 9.1** Rating curve values. $Q_1$ is the contribution of the central water body, $Q_2$ is the contribution of the two expansion zones

| $y_0$ (m) | $Q_1$ (m³ s⁻¹) | $Q_2$ (m³ s⁻¹) | $Q$ (m³ s⁻¹) |
|---|---|---|---|
| 0.5 | 0.59 | 0 | 0.59 |
| 1.0 | 2.16 | 0 | 2.16 |
| 1.5 | 4.92 | 2 × 1.27 | 7.46 |
| 2.0 | 8.52 | 2 × 4.39 | 17.30 |
| 2.5 | 12.85 | 2 × 9.42 | 31.69 |
| 3.0 | 17.88 | 2 × 16.55 | 50.98 |

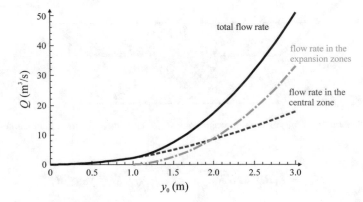

**Fig. 9.4** Rating curve (bold curve) and contribution of the central zone (dashed curve) and of the expansion zones (dash-dotted curve)

**Exercise 9.2** In the rectangular channel with cross-section shown in Fig. 9.5, the flow rate is $Q = (60 + C_u)$ m³ s⁻¹. The bed and the walls have a coefficient of Gauckler–Strickler $k = (40 + C_{pu})$ m¹ᐟ³ s⁻¹. The slope of the bed is $i_b = (0.3 + C_{pu}/20)\%$ and the width is $b = 6$ m.

- Calculate the uniform flow depth.
- Calculate the average wall tangential stress.
- Calculate the critical depth.
- Calculate the uniform flow depth if you insert a vertical septum of negligible thickness and with the same roughness as that of the other walls, as shown in Fig. 9.6.

**Fig. 9.5** Cross-section of
the channel

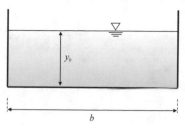

**Fig. 9.6** Cross-section of
the channel after insertion of
the septum

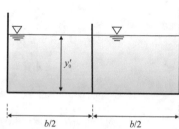

**Solution** Applying Chézy formula, yields

$$Q = k R_h^{1/6} \Omega \sqrt{R_h i_b} \equiv k \left( \frac{b y_0}{b + 2y_0} \right)^{2/3} b y_0 \sqrt{i_b}, \qquad (9.1)$$

where $R_h = b y_0/(b + 2y_0)$ is the hydraulic radius. Equation (9.1) admits only numerical solutions.

The average tangential stress at the wall is

$$\overline{\tau} = \gamma_w R_h i_b.$$

The critical depth is calculated by imposing that the energy head, measured from the bed of a channel, has a stationary point:

$$\frac{\partial E}{\partial y}\bigg|_{y=y_c} = 0 \rightarrow 1 - \frac{Q^2}{g\Omega_c^3} \frac{\partial \Omega}{\partial y}\bigg|_{y=y_c} = 0,$$

equivalent to

$$1 - \frac{Q^2}{g(by_c)^3} b = 0 \rightarrow y_c = \sqrt[3]{\frac{Q^2}{gb^2}}.$$

By inserting a vertical septum in an intermediate position, the flow rate is divided, by symmetry, into equal parts. For each of the two channels, it results:

$$\frac{Q}{2} = k R^{1/6} \Omega \sqrt{R i_b} = k \left( \frac{\frac{b}{2} y_0'}{\frac{b}{2} + 2y_0'} \right)^{2/3} \frac{b}{2} y_0' \sqrt{i_b}.$$

For $C_u = C_{pu} = 0$ it results $Q = 60$ m$^3$ s$^{-1}$, $k = 40$ m$^{1/3}$ s$^{-1}$, $i_b = 0.3\%$, $b = 6$ m,

$$\therefore \quad Q = k \left( \frac{b y_0}{b + 2y_0} \right)^{2/3} b y_0 \sqrt{i_b} \rightarrow$$

$$60 = 40 \left( \frac{6 y_0}{6 + 2y_0} \right)^{2/3} 6 y_0 \sqrt{0.003} \rightarrow y_0 = \mathbf{3.35\ m},$$

$$\therefore \quad \overline{\tau} = \gamma_w R_h i_b = 9806 \times \left( \frac{6 \times 3.35}{6 + 2 \times 3.35} \right) \times 0.003 = \mathbf{46.4\ Pa},$$

$$\therefore \quad y_c = \sqrt[3]{\frac{Q^2}{gb^2}} = \sqrt[3]{\frac{60^2}{9.806 \times 6^2}} = \mathbf{2.17\ m},$$

$$\therefore \quad \frac{Q}{2} = k \left( \frac{\frac{b}{2} y_0'}{\frac{b}{2} + 2y_0'} \right)^{2/3} \frac{b}{2} y_0' \sqrt{i_b} \rightarrow$$

$$\frac{60}{2} = 40 \left( \frac{\frac{6}{2} y_0'}{\frac{6}{2} + 2y_0'} \right)^{2/3} \frac{6}{2} y_0' \sqrt{0.003} \rightarrow y_0' = \mathbf{4.26\ m}.$$

---

**Exercise 9.3** In the system shown in Fig. 9.7, the sluice gate separates the tank from a channel with rectangular cross-section of width $B = 6$ m.

– Calculate the flow rate.
– Determine whether the current downstream is sub- or supercritical.
– Draw qualitatively the flow profile in the downstream channel.

Assume $H = (6 + C_{pu})$ m, $a = (0.6 + C_u/20)$ m and a Gauckler–Strickler coefficient of the downstream channel $k = (40 + 2 \times C_{pu})$ m$^{1/3}$ s$^{-1}$. The bed slope is $i_b = (0.3 + C_{pu}/30)\%$. The contraction coefficient is equal to $C_c = 0.61$ and the outflow coefficient is equal to $\mu = C_c \times C_v = 0.61 \times 0.98 \approx 0.60$, where $C_v$ is the speed correction coefficient.

**Fig. 9.7** Outflow under
vertical flat floodgate

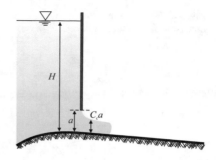

**Solution** Applying Bernoulli's theorem between an upstream section in the tank
and the vena contracta (where we assume that the trajectories are rectilinear and
parallel and the pressure distribution is hydrostatic), and neglecting the energy losses,
yields:

$$z_u + \frac{p_u}{\gamma_w} + \frac{V_u^2}{2g} = z_c + \frac{p_c}{\gamma_w} + \frac{V_c^2}{2g}.$$

Assuming a hydrostatic distribution also in the upstream section (far from the gate),
results in

$$\underbrace{z_u + \frac{p_u}{\gamma_w} + \frac{V_u^2}{2g}}_{H} = \underbrace{z_c + \frac{p_c}{\gamma_w}}_{C_c a} + \frac{V_c^2}{2g} \rightarrow H + \frac{V_u^2}{2g} = C_c a + \frac{V_c^2}{2g}. \qquad (9.2)$$

Defining $q = Q/B$ the flow rate per unit of width of the channel, Eq. (9.2) can be
written as:

$$H + \frac{q^2}{2gH^2} = C_c a + \frac{q^2}{2g(C_c a)^2} \rightarrow q = \sqrt{\frac{2g(H - C_c a)}{\dfrac{1}{(C_c a)^2} - \dfrac{1}{H^2}}}.$$

Since the water comes from a large tank, the speed upstream $V_u$ can be neglected
$(q^2/(2gH^2) \rightarrow 0)$ and it is thus possible to apply the formula of the outflow from a
slot at the bottom of a vertical floodgate:

$$Q = C_c A \sqrt{2g(H - C_c a)},$$

where $A = aB$. If we include the experimental energy losses through the coefficient
$C_v$, it results

$$Q = \mu A \sqrt{2g(H - C_c a)},$$

where $\mu = C_c C_v = 0.61 \times 0.98 \approx 0.60$.

**Fig. 9.8** Flow profile if the
downstream channel is mild

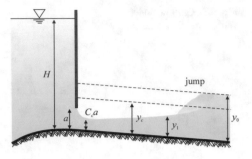

**Fig. 9.9** Flow profile if the
downstream channel is steep
and $C_c a < y_0$

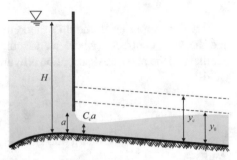

The uniform flow depth downstream is calculated by applying the Chézy formula with Gauckler–Strickler coefficient, $Q = k R_h^{1/6} \Omega \sqrt{R_h i_b}$, which for rectangular channels becomes:

$$Q = k \left( \frac{B y_0}{B + 2 y_0} \right)^{2/3} B y_0 \sqrt{i_b}, \tag{9.3}$$

to be solved numerically with respect to $y_0$.

To determine the state of the current, subcritical or supercritical, we calculate the critical depth:

$$y_c = \sqrt[3]{\frac{q^2}{g}}.$$

For the given data, in the vena contracta the current is supercritical. If the downstream channel is mild, a hydraulic jump will occur downstream of a retarded supercritical flow profile, see Fig. 9.8. The jump connects the uniform flow depth $y_0$ to the sequent flow depth $y_1$.

If the downstream channel is steep, the uniform flow regime is reached asymptotically downstream, with a retarded (Fig. 9.9) or accelerated (Fig. 9.10) supercritical flow profile, according to ratio of water depth in the vena contracta to uniform flow depth.

For $C_u = C_{pu} = 0$ it results $H = 6$ m, $a = 0.6$ m, $k = 40$ m$^{1/3}$ s$^{-1}$, $i_b = 0.3\%$, $C_c = 0.61$, $\mu = 0.60$.

**Fig. 9.10** Flow profile if the
downstream channel is steep
and $C_c a > y_0$

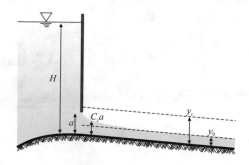

$$\therefore \quad Q = \mu A \sqrt{2g \, (H - C_c a)} =$$
$$0.60 \times 6 \times 0.6 \times \sqrt{2 \times 9.806 \times (6 - 0.61 \times 0.6)} = \mathbf{22.7 \ m^3 \ s^{-1}},$$

$$q = \frac{Q}{B} = \frac{22.7}{6} = \mathbf{3.78 \ m^2 \ s^{-1}}.$$

Inserting numerical values in Eq. (9.3), results

$$22.7 = 40 \left( \frac{6 y_0}{6 + 2 y_0} \right)^{2/3} 6 y_0 \sqrt{0.003},$$

and solving, yields $y_0 = \mathbf{1.65 \ m}$. The critical depth is

$$\therefore \qquad y_c = \sqrt[3]{\frac{q^2}{g}} = \sqrt[3]{\frac{3.78^2}{9.806}} = \mathbf{1.13 \ m}.$$

The uniform downstream flow is subcritical, since $y_0 > y_c$, and the flow profile is as
shown in Fig. 9.8.

---

**Exercise 9.4** The jet of water flowing out of a tap, impacting on a flat horizontal
plane forms a hydraulic jump at the distance $r = (5 + C_{pu}/2)$ cm from the axis,
see Fig. 9.11. The depth of the radial current upstream of the jump is equal to
$y_1 = (0.1 + C_u)$ cm, downstream of the jump is equal to $y_2 = (0.5 + C_u)$ cm.

– Calculate the flow rate.

Consider the jump as if it were in a rectangular prismatic channel.

**Solution** After choosing the control volume shown in Fig. 9.12, we write the
momentum balance and mass conservation. $\Pi_1$, $\Pi_2$ and $\Pi_{lat}$ represent the forces

**Fig. 9.11** Radial jump

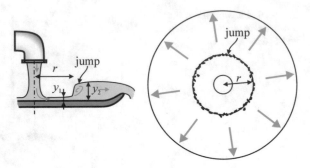

**Fig. 9.12** Schematic for computing the characteristics of the radial jump

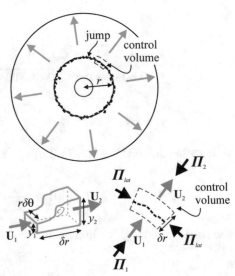

due to pressure and we are neglecting the resistance at the bottom. If we analyze the jump as if it happened in a prismatic rectangular channel, we can neglect the components of the lateral force $\boldsymbol{\Pi}_{lat}$ in radial direction, and the relationship between the sequent flow depths is the classic one for a prismatic rectangular cross-section channel:

$$\frac{y_2}{y_1} = \frac{1}{2}\left(-1 + \sqrt{1 + 8\mathrm{Fr}_1^2}\right), \quad \mathrm{Fr}_1 = \frac{U_1}{\sqrt{gy_1}},$$

where $\mathrm{Fr}_1$ is the Froude number of the upstream current (supercritical). Since the sequent depths are known, we can calculate $\mathrm{Fr}_1$:

$$\mathrm{Fr}_1 = \sqrt{\frac{1}{8}\left[\left(2\frac{y_2}{y_1} + 1\right)^2 - 1\right]},$$

and the velocity $U_1$. The flow rate is $Q = 2\pi r y_1 U_1$.

For $C_u = C_{pu} = 0$ it results $r = 5$ cm, $y_1 = 0.1$ cm, $y_2 = 0.5$ cm,

$$\mathrm{Fr}_1 = \sqrt{\frac{1}{8}\left[\left(2\frac{y_2}{y_1}+1\right)^2 - 1\right]} = \sqrt{\frac{1}{8} \times \left[\left(2 \times \frac{0.5}{0.1}+1\right)^2 - 1\right]} = \mathbf{3.87},$$

$$U_1 = \mathrm{Fr}_1 \sqrt{gy_1} = 3.87 \times \sqrt{9.806 \times 0.001} = \mathbf{0.38\ m\ s^{-1}},$$

$$\therefore \qquad Q = 2\pi r y_1 U_1 = 2 \times \pi \times 0.05 \times 0.001 \times 0.38 = \mathbf{0.12\ l\ s^{-1}}.$$

**Exercise 9.5** Figure 9.13 shows the cross-section of a channel with geometry listed in Table 9.2. The bed slope is $i_b = (2 + C_{pu}/30)\%$ and the Gauckler–Strickler coefficient is $k = (40 + 2C_{pu})$ m$^{1/3}$ s$^{-1}$. The flow rate is $Q = 50$ m$^3$ s$^{-1}$.

- Calculate the uniform flow depth.
- Calculate the critical depth.

Hint: linearly interpolate the geometric data of the table.

**Solution** The solution of the problem requires the calculation of the rating curve. At a uniform flow depth value $y_0 = 1.0$ m, the cross-section has a surface area equal to

**Fig. 9.13** Cross-section of the channel

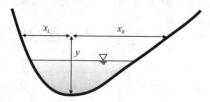

**Table 9.2** Geometric characteristics of the cross-section of the channel

| $y$ (m) | $x_L$ (m) | $x_R$ (m) |
|---------|-----------|-----------|
| 0       | 0         | 0         |
| 1.0     | 1.50      | 1.10      |
| 2.0     | 2.60      | 2.00      |
| 3.0     | 3.25      | 3.30      |
| 4.0     | 3.60      | 5.50      |
| 5.0     | 3.85      | 8.20      |
| 6.0     | 4.00      | 10.0      |

**Table 9.3** Characteristic quantities of the current in the channel for increasing water depth. Uniform flow conditions

| $y_0$ (m) | $x_L$ (m) | $x_R$ (m) | $\Omega$ (m$^2$) | $P$ (m) | $R_h$ (m) | $Q_0$ (m$^3$ s$^{-1}$) |
|---|---|---|---|---|---|---|
| 1.0 | 1.50 | 1.10 | 1.30 | 3.29 | 0.40 | 3.96 |
| **2.0** | 2.60 | 2.00 | 4.90 | 6.12 | 0.80 | **23.90** |
| **3.0** | 3.25 | 3.30 | 10.48 | 8.95 | 1.17 | **65.79** |
| 4.0 | 3.60 | 5.50 | 18.30 | 12.43 | 1.47 | 133.97 |
| 5.0 | 3.85 | 8.20 | 28.88 | 16.34 | 1.77 | 238.75 |
| 6.0 | 4.00 | 10.00 | 41.90 | 19.41 | 2.16 | 395.89 |

$$\Omega = \frac{x_L + x_R}{2} y_0 = \frac{1.5 + 1.1}{2} \times 1.0 = \mathbf{1.3 \ m^2},$$

a wetted perimeter

$$P = \sqrt{x_L^2 + y_0^2} + \sqrt{x_R^2 + y_0^2} = \sqrt{1.5^2 + 1.0^2} + \sqrt{1.1^2 + 1.0^2} = \mathbf{3.29 \ m},$$

a hydraulic radius

$$R_h = \frac{\Omega}{P} = \frac{1.3}{3.29} = \mathbf{0.40 \ m}.$$

The flow rate is

$$Q = k R_h^{1/6} \Omega \sqrt{R_h i_b} \equiv k R_h^{1/6} \Omega \sqrt{R_h i_b}.$$

For water depth greater than 1.0 m it is advisable to add the new contributions of wetted area and perimeter to the values previously calculated. These contributions are considered as trapezoidal surfaces. For example, for $y = 2.0$ m:

$$\Delta\Omega = \frac{[x_L(2.0) + x_R(2.0)] + [x_L(1.0) + x_R(1.0)]}{2} \times 1.0,$$

$$\Delta P = \sqrt{[x_L(2.0) - x_L(1.0)]^2 + 1.0^2} + \sqrt{[x_R(2.0) - x_R(1.0)]^2 + 1.0^2}.$$

For $C_u = C_{pu} = 0$ it results $i_b = 2\%$, $k = 40$ m$^{1/3}$ s$^{-1}$.
For $y_0 = 1.0$ m, it results

$$Q = k R^{1/6} \Omega \sqrt{R i_b} = 40 \times 0.40^{1/6} \times 1.3 \times \sqrt{0.40 \times 0.02} = \mathbf{3.96 \ m^3 \ s^{-1}}.$$

Performing the calculations for increasing water depth, yields the results listed in Table 9.3.

For flow rate $Q = 50$ m$^3$ s$^{-1}$, the uniform flow depth is calculated by interpolating between $y_0 = 2.0$ m and $y_0 = 3.0$ m:

**Table 9.4** Characteristic quantities of the current in the channel for the increasing water depth. Critical flow conditions

| $y_c$ (m) | $x_L$ (m) | $x_R$ (m) | $\Omega_c$ (m) | $B_c$ (m) | $Q_c$ (m$^3$ s$^{-1}$) |
|---|---|---|---|---|---|
| 1.0 | 1.50 | 1.10 | 1.30 | 2.60 | 2.88 |
| 2.0 | 2.60 | 2.00 | 4.90 | 4.60 | 15.84 |
| 3.0 | 3.25 | 3.30 | 10.48 | 6.55 | **41.48** |
| 4.0 | 3.60 | 5.50 | 18.30 | 9.10 | **81.26** |
| 5.0 | 3.85 | 8.20 | 28.88 | 12.05 | 139.97 |
| 6.0 | 4.00 | 10.00 | 41.90 | 14.00 | 226.99 |

**Fig. 9.14** Rating curve and critical flow rate

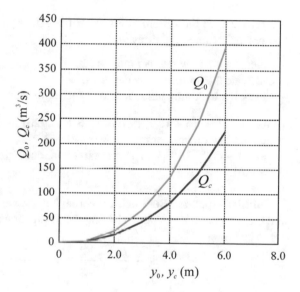

$$\therefore \quad \frac{y_0 - 2.0}{3.0 - 2.0} = \frac{Q - 23.90}{65.79 - 23.90} \rightarrow y_0|_{Q=50.0} = 2.0 + \frac{50.00 - 23.90}{65.79 - 23.90} = \mathbf{2.62 \text{ m.}}$$

For the calculation of the critical depth, it is convenient to calculate the value of the critical flow rate for the assigned geometry of the current. The critical condition for the current is as follows:

$$1 - \frac{Q_c^2}{g\Omega_c^3} B_c = 0 \rightarrow Q_c = \sqrt{\frac{g\Omega_c^3}{B_c}},$$

where the subscript "$c$" indicates that the value refers to the critical condition. $B_c$ is the top width in critical conditions.

By performing the calculations for increasing critical depth, the results listed in Table 9.4 are obtained.

For the flow rate $Q = 50.0$ m$^3$ s$^{-1}$, the critical depth is calculated by interpolating between $y_c = 3.0$ m and $y_c = 4.0$ m:

$$\therefore \quad \frac{y_c - 3.0}{4.0 - 3.0} = \frac{Q - 41.48}{81.26 - 41.28} \rightarrow y_c|_{Q=50.0} = 3.0 + \frac{50.00 - 41.48}{81.26 - 41.28} = \mathbf{3.21\ m}.$$

The current is supercritical, since $y_0 < y_c$. Figure 9.14 shows the rating curve and the critical flow rate.

---

**Exercise 9.6** In the system in Fig. 9.15, the gate separates the tank from a channel that can be assumed infinitely wide.

– Calculate the flow rate per unit width.
– Calculate the uniform flow depth $y_0$.
– Check if a hydraulic jump develops and, if so, calculate the sequent depth $y_1$.

Assume $H = (6 + C_{pu})$ m, $a = (0.6 + C_u/20)$ m and a Gauckler–Strickler coefficient of the downstream bed $k = (40 + 2 \times C_{pu})$ m$^{1/3}$ s$^{-1}$. The bed slope is $i_b = (0.3 + C_{pu}/30)\%$ and the contraction coefficient (only in the vertical) is $C_c = 0.61$.

**Solution** Applying Bernoulli's theorem between an upstream section in the tank and the vena contracta, and neglecting the dissipations, yields

$$z_u + \frac{p_u}{\gamma_w} + \frac{V_u^2}{2g} = z_c + \frac{p_c}{\gamma_w} + \frac{V_c^2}{2g},$$

and, due to the hydrostatic distribution in the two sections,

$$\underbrace{z_u + \frac{p_u}{\gamma_w} + \frac{V_u^2}{2g}}_{H} = \underbrace{z_c + \frac{p_c}{\gamma_w}}_{C_c a} + \frac{V_c^2}{2g} \rightarrow H + \frac{V_u^2}{2g} = C_c a + \frac{V_c^2}{2g}.$$

**Fig. 9.15** Flow profile downstream of a flat vertical floodgate

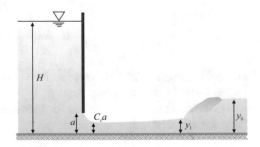

Defining $q$ the flow rate per unit width, it results

$$H + \frac{q^2}{2gH^2} = C_c a + \frac{q^2}{2g(C_c a)^2} \rightarrow q = \sqrt{\frac{2g\,(H - C_c a)}{\frac{1}{(C_c a)^2} - \frac{1}{H^2}}}.$$

The uniform flow depth downstream is calculated by applying Chézy formula with the Gauckler–Strickler coefficient, $Q = kR_h^{1/6}\Omega\sqrt{R_h i_b}$, which, for a very wide channel, reduces to $q = ky_0^{5/3}\sqrt{i_b}$, hence

$$y_0 = \frac{q^{3/5}}{k^{3/5} i_b^{3/10}}.$$

To check if the downstream current is sub- or supercritical, we calculate the critical depth:

$$y_c = \sqrt[3]{\frac{q^2}{g}}.$$

If the current below the gate is supercritical and the uniform flow in the downstream channel is subcritical, a hydraulic jump occurs. To calculate the sequent depth of $y_0$, we impose the balance of momentum in integral form and per unit width:

$$\frac{1}{2}\gamma_w y_1^2 + \rho\frac{q^2}{y_1} = \frac{1}{2}\gamma_w y_0^2 + \rho\frac{q^2}{y_0},$$

which can be written as follows:

$$\frac{y_1}{y_0} = \frac{1}{2}\left(-1 + \sqrt{1 + 8\mathrm{Fr}_0^2}\right),$$

where $\mathrm{Fr}_0 = V_0/\sqrt{gy_0}$ is the Froude number of the subcritical uniform flow depth in the channel downstream.

For $C_u = C_{pu} = 0$ it results $H = 6$ m, $a = 0.6$ m, $k = 40$ m$^{1/3}$ s$^{-1}$, $i_b = 0.3\%$, $C_c = 0.61$,

$$\therefore \quad q = \sqrt{\frac{2g\,(H - C_c a)}{\frac{1}{(C_c a)^2} - \frac{1}{H^2}}} = \sqrt{\frac{2 \times 9.806 \times (6 - 0.61 \times 0.6)}{\frac{1}{(0.61 \times 0.6)^2} - \frac{1}{6^2}}} = 3.85 \text{ m}^2\,\text{s}^{-1},$$

$$\therefore \quad y_0 = \frac{q^{3/5}}{k^{3/5} i_b^{3/10}} = \frac{3.85^{3/5}}{40^{3/5} \times 0.003^{3/10}} = 1.40 \text{ m},$$

$$y_c = \sqrt[3]{\frac{q^2}{g}} = \sqrt[3]{\frac{3.85^2}{9.806}} = \mathbf{1.14\ m}.$$

The uniform downstream flow current is subcritical, since $y_0 > y_c$. The uniform downstream flow velocity is

$$V_0 = \frac{q}{y_0} = \frac{3.85}{1.40} = \mathbf{2.74\ m\ s^{-1}},$$

and the downstream current Froude number is

$$\mathrm{Fr}_0 = \frac{V_0}{\sqrt{g\,y_0}} = \frac{2.75}{\sqrt{9.806 \times 1.40}} = \mathbf{0.74}.$$

The sequent depth of the hydraulic jump is

$$\therefore \quad y_1 = \frac{y_0}{2}\left(-1 + \sqrt{1 + 8\mathrm{Fr}_0^2}\right) = \frac{1.40}{2} \times \left(-1 + \sqrt{1 + 8 \times 0.74^2}\right) = \mathbf{0.93\ m}.$$

---

**Exercise 9.7** The prismatic channel with cross-section represented in Fig. 9.16 (scale drawing), conveys the flow rate $Q = (300 + 0.1 \times C_u)$ m$^3$ s$^{-1}$, has a uniform bed slope $i_b = (0.05 + 0.005 \times C_u)\%$ and roughness uniform on the walls and on the bed defined by the Gauckler–Strickler coefficient $k = (50 + 0.5 \times C_{pu})$ m$^{1/3}$ s$^{-1}$.

– Calculate the uniform flow depth and the critical depth.

**Solution** We adopt Chézy formula $Q = k R_h^{2/3} \Omega \sqrt{i_b}$. For the cross-section in Fig. 9.16, it results

**Fig. 9.16** Cross-section of the channel, scale drawing

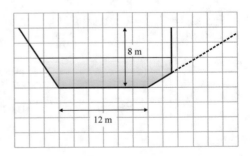

$$Q = k \frac{\left[\left(2b_0 + \frac{7}{3}y_0\right)\frac{y_0}{2}\right]^{5/3}}{\left[b_0 + \frac{y_0}{3}\left(\sqrt{13} + \sqrt{34}\right)\right]^{2/3}} \sqrt{i_b}, \quad 0 < y_0 \leq 2 \text{ m},$$

$$Q = k \frac{\left[\left(2b_0 + \frac{14}{3}\right) + (y_0 - 2)\left(b_0 + \frac{14}{3}\right) + \frac{(y_0 - 2)^2}{3}\right]^{5/3}}{\left[b_0 + \frac{y_0}{3}\sqrt{13} + \frac{2}{3}\sqrt{34} + (y_0 - 2)\right]^{2/3}} \sqrt{i_b}, \quad y_0 \geq 2 \text{ m}.$$

The specific energy is

$$E = y + \frac{Q^2}{2g\Omega^2},$$

and, in critical condition, it results

$$\left.\frac{\partial E}{\partial y}\right|_{y=y_c} \equiv 1 - \left.\frac{Q^2}{g\Omega_c{}^3}\frac{\partial \Omega}{\partial y}\right|_{y=y_c} = 0.$$

For the cross-section in Fig. 9.16, it results

$$1 - \left.\frac{Q^2}{g\Omega_c{}^3}\frac{\partial \Omega}{\partial y}\right|_{y=y_c} \equiv 1 - \frac{72Q^2}{gy_c{}^3}\frac{(3b_0 + 7y_c)}{(6b_0 + 7y_c)^3} = 0, \quad 0 < y_c \leq 2 \text{ m},$$

$$1 - \left.\frac{Q^2}{g\Omega_c{}^3}\frac{\partial \Omega}{\partial y}\right|_{y=y_c} \equiv 1 - \frac{9Q^2}{g}\frac{(3b_0 + 2y_c + 10)}{\left(y_c^2 + 10y_c + 3b_0y_c - 10\right)^3} = 0, \quad y_c \geq 2 \text{ m}.$$

For $C_u = C_{pu} = 0$ it results $Q = 300$ m$^3$ s$^{-1}$, $k = 50$ m$^{1/3}$ s$^{-1}$, $i_b = 0.05\%$.
The uniform flow depth is equal to $y_0 = $ **6.41 m**, the critical depth is equal to $y_c = $ **3.51 m**. The flow is tranquil since $y_c < y_0$.

---

**Exercise 9.8** In the channel with cross-section shown in Fig. 9.17, with width $b = 12$ m, the flow rate is $Q = (60 + C_u)$ m$^3$ s$^{-1}$. The walls and the bed have a coefficient of Gauckler–Strickler $k = (40 + C_{pu})$ m$^{1/3}$ s$^{-1}$ and the bed slope is $i_b = (0.3 + C_{pu}/20)\%$.

- Calculate the uniform flow depth.
- Calculate the critical depth.
- Calculate the sequent depth of the uniform flow depth $y_0$ in the hydraulic jump.

**Fig. 9.17** Cross-section of
the channel

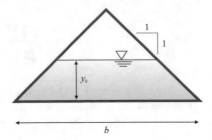

**Solution** The uniform flow depth is calculated by applying the Chézy formula:

$$Q = kR_h^{1/6}\Omega\sqrt{R_h i_b} \equiv k\frac{\left(by_0 - y_0^2\right)^{5/3}}{\left(b + 2\sqrt{2}y_0\right)^{2/3}}\sqrt{i_b}.$$

The critical depth corresponds to a stationary point of the specific energy function:

$$\left.\frac{\partial E}{\partial y}\right|_{y=y_c} \equiv 1 - \frac{Q^2}{g\Omega_c^3}\left.\frac{\partial \Omega}{\partial y}\right|_c = 0,$$

where $\partial\Omega/\partial y|_c \equiv b'_c$ is the top width in critical conditions. For the triangular cross-section channel in Fig. 9.17, it results

$$1 - \frac{Q^2}{g\left(by_c - y_c^2\right)^3}\,(b - 2y_c) = 0.$$

The sequent depth of the hydraulic jump is obtained by imposing the momentum balance in integral form,

**Fig. 9.18** Rating curve

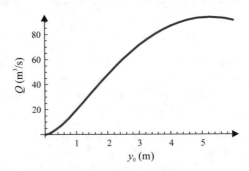

$$\underbrace{\frac{1}{2}\gamma_w y_0^2 (b - 2y_0)}_{\substack{\text{pressure force} \\ \text{on the rectangle}}} + \underbrace{\frac{2}{3}\gamma_w y_0^3}_{\substack{\text{pressure force} \\ \text{on the triangles}}} + \underbrace{\rho\frac{Q^2}{(by_0 - y_0^2)}}_{\text{momentum flux}} =$$

$$\underbrace{\frac{1}{2}\gamma_w y_1^2 (b - 2y_1)}_{\substack{\text{pressure force} \\ \text{on the rectangle}}} + \underbrace{\frac{2}{3}\gamma_w y_1^3}_{\substack{\text{pressure force} \\ \text{on the triangles}}} + \underbrace{\rho\frac{Q^2}{(by_1 - y_1^2)}}_{\text{momentum flux}}, \quad (9.4)$$

where, for convenience, we have computed the normal forces separating the trapezoidal section into a rectangle plus two triangles.

For $C_u = C_{pu} = 0$ it results $Q = 60 \text{ m}^3 \text{ s}^{-1}$, $k = 40 \text{ m}^{1/3} \text{ s}^{-1}$, $i_b = 0.3\%$.
The rating curve is shown in Fig. 9.18, with a maximum for $y_0 = 5.25$ m, corresponding to 87.5% of the maximum water depth.
The uniform flow depth is equal to $y_0 = \textbf{2.49 m}$, the critical depth is equal to $y_c = \textbf{1.41 m}$ and the flow is tranquil, since $y_0 > y_c$. The sequent depth (shooting flow) of the hydraulic jump is obtained by solving Eq. (9.4) and is equal to $y_1 = \textbf{0.73 m}$.

---

**Exercise 9.9** A flow rate $Q = (60 + C_u) \text{ m}^3 \text{ s}^{-1}$ flows in a channel with cross-section shown in Fig. 9.19. The walls have a coefficient of Gaucklcr–Strickler $k = (40 + C_{pu}) \text{ m}^{1/3} \text{ s}^{-1}$ and the bed slope is $i_b = (0.3 + C_{pu}/20)\%$.

– Calculate the uniform flow depth.
– Calculate the critical depth.
– Calculate the sequent depth of the uniform flow depth in the hydraulic jump.

**Solution** The uniform flow depth is calculated by applying Chézy formula:

$$Q = k R_h^{1/6} \Omega \sqrt{R_h i_b},$$

which, for the cross-section in Fig. 9.19, yields

**Fig. 9.19** Cross-section of
the channel

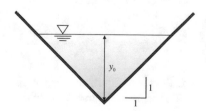

$$Q = k \frac{y_0^{8/3}}{\left(2\sqrt{2}\right)^{2/3}} \sqrt{i_b} \rightarrow y_0 = \left[\frac{Q\left(2\sqrt{2}\right)^{2/3}}{k\sqrt{i_b}}\right]^{3/8}.$$

The specific energy is

$$E = y + \frac{Q^2}{2g\Omega^2},$$

and, in critical conditions, results

$$\left.\frac{\partial E}{\partial y}\right|_{y=y_c} \equiv 1 - \frac{Q^2}{g\Omega_c{}^3}\left.\frac{\partial \Omega}{\partial y}\right|_{y=y_c} = 0.$$

For the cross-section in Fig. 9.19, it results

$$1 - \frac{Q^2}{g\Omega_c{}^3}\left.\frac{\partial \Omega}{\partial y}\right|_{y=y_c} \equiv 1 - \frac{Q^2}{g y_c{}^6}2y_c = 0 \rightarrow y_c = \left(\frac{2Q^2}{g}\right)^{1/5}.$$

The sequent depths of the hydraulic jump result from the linear momentum balance equation in integral form:

$$S = p_G\Omega + \rho\frac{Q^2}{\Omega} = \text{constant} \rightarrow \gamma_w\frac{y_0^3}{3} + \rho\frac{Q^2}{y_0^2} = \gamma_w\frac{y_1^3}{3} + \rho\frac{Q^2}{y_1^2},$$

where $S$ is the total force, sum of the hydrostatic force and of the flux of momentum, and $p_G$ is the pressure in the centroid.

For $C_u = C_{pu} = 0$ it results $Q = 60 \text{ m}^3 \text{ s}^{-1}$, $k = 40 \text{ m}^{1/3} \text{ s}^{-1}$, $i_b = 0.3\%$.
The uniform flow depth is

$$\therefore \quad y_0 = \left[\frac{Q\left(2\sqrt{2}\right)^{2/3}}{k\sqrt{i_b}}\right]^{3/8} = \left[\frac{60 \times \left(2 \times \sqrt{2}\right)^{2/3}}{40 \times \sqrt{\frac{0.3}{100}}}\right]^{3/8} = 4.49 \text{ m},$$

and the critical depth is

$$\therefore \quad y_c = \left(\frac{2Q^2}{g}\right)^{1/5} = \left(\frac{2 \times 60^2}{9.806}\right)^{1/5} = 3.74 \text{ m}.$$

The flow is tranquil, since $y_c < y_0$.
The sequent depths of the hydraulic jump satisfy the following equation:

**Fig. 9.20** Diagram of
specific energy and total
force

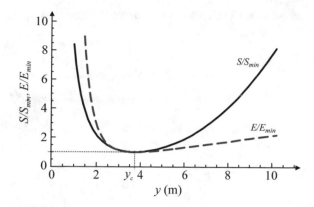

$$\gamma_w \frac{y_0^3}{3} + \rho \frac{Q^2}{y_0^2} = \gamma_w \frac{y_1^3}{3} + \rho \frac{Q^2}{y_1^2} \rightarrow$$

$$9806 \times \frac{4.49^3}{3} + 1000 \times \frac{60^2}{4.49^2} = 9806 \times \frac{y_1^3}{3} + 1000 \times \frac{60^2}{y_1^2},$$

which admits the solution $y_1 = \mathbf{3.08\ m}$. The flow is supercritical.

Figure 9.20 shows the diagrams of the total force $S$ and the specific energy $E$ computed for the triangular cross-section, dimensionless with respect to their minimum value. Under the assumption of a unitary value for the correction coefficients of momentum flux and of kinetic power, the two functions have the same minimum corresponding to the critical condition.

---

**Exercise 9.10**  A flow rate $Q = (60 + C_u)\ \mathrm{m^3\,s^{-1}}$ flows in the channel shown in Fig. 9.21. The base is $b = (6 + C_{pu})\ \mathrm{m}$, the walls and the bed have a coefficient of Gauckler–Strickler $k = (40 + C_{pu})\ \mathrm{m^{1/3}\,s^{-1}}$ and the bed slope is $i_b = (0.3 + C_{pu}/20)\%$.

- Calculate the uniform flow depth.
- Calculate the critical depth.
- Calculate the sequent depth of the uniform flow depth in the hydraulic jump.

**Fig. 9.21** Cross-section of
the channel

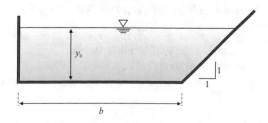

**Solution** The uniform flow depth is calculated by applying the Chézy formula:

$$Q = kR_h^{1/6}\Omega\sqrt{R_h i_b} \equiv k\,\frac{\left(b y_0 + \dfrac{y_0^2}{2}\right)^{5/3}}{\left(b + y_0 + \sqrt{2}\,y_0\right)^{2/3}}\sqrt{i_b},$$

and the critical depth is calculated by imposing a stationary point of the specific energy

$$1 - \frac{Q^2}{g\Omega_c^3}\frac{d\Omega}{dy}\bigg|_{y=y_c} = 0 \rightarrow 1 - \frac{Q^2}{g\left(b y_c + \dfrac{y_c^2}{2}\right)^3}(b + y_c) = 0,$$

to be solved numerically.

The sequent depth of the hydraulic jump is computed by imposing the linear momentum balance in integral form. For computing hydrostatic forces, it is convenient to separate the trapezoidal cross-section into a rectangle and a triangle:

$$\underbrace{\frac{1}{2}\gamma_w y_0^2 b}_{\substack{\text{hydrostatic force} \\ \text{on rectangle}}} + \underbrace{\frac{1}{6}\gamma_w y_0^3}_{\substack{\text{hydrostatic force} \\ \text{on triangle}}} + \underbrace{\rho\,\frac{Q^2}{\left(b y_0 + \dfrac{y_0^2}{2}\right)}}_{\text{flux of momentum}} =$$

$$\underbrace{\frac{1}{2}\gamma_w y_1^2 b}_{\substack{\text{hydrostatic force} \\ \text{on rectangle}}} + \underbrace{\frac{1}{6}\gamma_w y_1^3}_{\substack{\text{hydrostatic force} \\ \text{on triangle}}} + \underbrace{\rho\,\frac{Q^2}{\left(b y_1 + \dfrac{y_1^2}{2}\right)}}_{\text{flux of momentum}}.$$

For $C_u = C_{pu} = 0$ it results $Q = 60$ m$^3$ s$^{-1}$, $b = 6$ m, $k = 40$ m$^{1/3}$ s$^{-1}$, $i_b = 0.3\%$, and $y_0 = \mathbf{2.72}$ **m**, $y_c = \mathbf{2.04}$ **m**, $y_1 = \mathbf{1.48}$ **m**.

---

**Exercise 9.11** A flow rate $Q = (60 + C_u)$ m$^3$ s$^{-1}$ flows in the channel with the cross-section shown in Fig. 9.22. The base is $b = 6$ m, the side walls and the bed have a coefficient of Gauckler–Strickler $k = (40 + C_{pu})$ m$^{1/3}$ s$^{-1}$ and the bed slope is $i_b = (0.3 + C_{pu}/20)\%$.

– Calculate the uniform flow depth.
– Calculate the critical depth.
– Calculate the sequent depth of the uniform flow depth in the hydraulic jump.

**Fig. 9.22** Cross-section of the channel

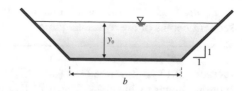

**Fig. 9.23** Total force diagram

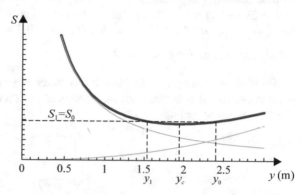

**Solution** The uniform flow depth is calculated by applying the Chézy formula:

$$Q = kR^{1/6}\Omega\sqrt{R_h i_b} \equiv k\frac{\left(by_0 + y_0^2\right)^{5/3}}{\left(b + 2\sqrt{2}y_0\right)^{2/3}}\sqrt{i_b},$$

and the critical depth is calculated by imposing a stationary point to the specific energy

$$1 - \frac{Q^2}{g\Omega_c^3}\frac{d\Omega}{dy}\bigg|_{y=y_c} = 0 \rightarrow 1 - \frac{Q^2}{g\left(by_c + y_c^2\right)^3}(b + 2y_c) = 0,$$

which requires a numerical solution.

The sequent depth is obtained by solving the linear momentum balance in integral form

$$\underbrace{\frac{1}{2}\gamma_w y_0^2 b}_{\substack{\text{hydrostatic force} \\ \text{on rectangle}}} + \underbrace{\frac{1}{3}\gamma_w y_0^3}_{\substack{\text{hydrostatic force} \\ \text{on triangles}}} + \underbrace{\rho\frac{Q^2}{\left(by_0 + y_0^2\right)}}_{\text{flux of momentum}} =$$

$$\underbrace{\frac{1}{2}\gamma_w y_1^2 b}_{\substack{\text{hydrostatic force} \\ \text{on rectangle}}} + \underbrace{\frac{1}{3}\gamma_w y_1^3}_{\substack{\text{hydrostatic force} \\ \text{on triangles}}} + \underbrace{\rho\frac{Q^2}{\left(by_1 + y_1^2\right)}}_{\text{flux of momentum}} .$$

For $C_u = C_{pu} = 0$ it results $Q = 60 \text{ m}^3 \text{ s}^{-1}$, $k = 40 \text{ m}^{1/3} \text{ s}^{-1}$, $i_b = 0.3\%$, $y_0 = \mathbf{2.40}$ **m**, $y_c = \mathbf{1.94}$ **m**, $y_1 = \mathbf{1.52}$ **m**.

Figure 9.23 shows the total force as a function of the water depth, and the two sequent depths of the hydraulic jump.

---

**Exercise 9.12** A flow rate $Q = (60 + C_u)$ m$^3$ s$^{-1}$ flows in the channel having cross-section shown in Fig. 9.24. The wall $\overline{BD}$ has a coefficient of Gauckler–Strickler $k_{\overline{BD}} = (30 + C_u)$ m$^{1/3}$ s$^{-1}$, all other walls have a coefficient of Gauckler–Strickler $k = (40 + C_{pu})$ m$^{1/3}$ s$^{-1}$. The bed slope is $i_b = (0.3 + C_{pu}/20)\%$.

– Calculate the uniform flow depth.

**Solution** Assuming that the overall resistance is given by the sum of the contributions due to the different parts of the wetted perimeter, an equivalent roughness coefficient can be estimated as:

$$k_{s,eq} = \sqrt{\frac{P}{\sum \frac{P_i}{k_{s,i}^2}}},$$

where $P_i$ indicates the length of the wetted perimeter with a $k_{s,i}$ roughness index. Notice that this is not the only approach and in literature there are other methods for calculating $k_{s,eq}$.

In the present case, it results:

$$k_{s,eq} = \left( \frac{y_0 + 6 + y_0\sqrt{101}}{\frac{y_0 + 6}{k^2} + \frac{y_0\sqrt{101}}{k_{\overline{BD}}^2}} \right)^{1/2} \quad \text{for } y_0 < 2 \text{ m,}$$

and

$$k_{s,eq} = \left( \frac{2y_0 + 4 + 2\sqrt{101}}{\frac{2y_0 + 4}{k^2} + \frac{2\sqrt{101}}{k_{\overline{BD}}^2}} \right)^{1/2} \quad \text{for } y_0 \geq 2 \text{ m.}$$

No assumption is made about the subdivision of the section.

**Fig. 9.24** Cross-section of the channel with variable roughness of the walls

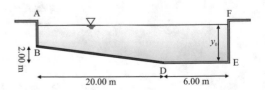

The flow rate is obtained from Chézy formula, $Q = k_{s,eq} R_h^{1/6} \Omega \sqrt{R_h i_b}$:

$$Q = \left( \frac{y_0 + 6 + y_0\sqrt{101}}{\dfrac{y_0 + 6}{k^2} + \dfrac{y_0\sqrt{101}}{k_{\overline{BD}}^2}} \right)^{1/2} \frac{(6y_0 + 5y_0^2)^{5/3}}{\left(y_0 + 6 + y_0\sqrt{101}\right)^{2/3}} \sqrt{i_b} \quad \text{for } y_0 < 2 \text{ m,}$$

and

$$Q = \left( \frac{2y_0 + 4 + 2\sqrt{101}}{\dfrac{2y_0 + 4}{k^2} + \dfrac{2\sqrt{101}}{k_{\overline{BD}}^2}} \right)^{1/2} \frac{(26y_0 - 20)^{5/3}}{\left(2y_0 + 4 + 2\sqrt{101}\right)^{2/3}} \sqrt{i_b} \quad \text{for } y_0 \geq 2 \text{ m,}$$

which require a numerical solution.

For $C_u = C_{pu} = 0$ it results $Q = 60 \text{ m}^3 \text{ s}^{-1}, k_{\overline{BD}} = 30 \text{ m}^{1/3} \text{ s}^{-1}, k = 40 \text{ m}^{1/3} \text{ s}^{-1}$, $i_b = 0.3\%$, and $y_0 = \mathbf{1.98}$ **m.**

---

**Exercise 9.13** The channel with cross-section shown in Fig. 9.25 has the spanwise bed inclination $1 : n_s$, with $n_s = (4 + C_{pu}/10)$, and vertical walls. The bed slope is $i_b = (0.5 + 0.5 \times C_u)\%$, and the coefficient of Gauckler–Strickler is $k = (35 + C_{pu} \times 5) \text{ m}^{1/3} \text{ s}^{-1}$. Calculate, with a step of 1.0 m and up to 4.0 m:

- the rating curve,
- the critical depth scale.

Assume $B = (8 + C_u \times 0.1)$ m.

**Fig. 9.25** Cross-section of the channel

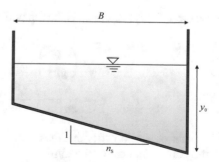

**Solution** The cross-section is triangular for $0 < y_0 \le B/n_s$, with a top width of the current $b = y_0 n_s$, an area $\Omega = y_0^2 n_s / 2$, a wetted perimeter equal to:

$$P = y_0 + y_0 \sqrt{1 + n_s^2}.$$

The hydraulic radius is

$$R_h = \frac{y_0 n_s}{2 + 2\sqrt{1 + n_s^2}}.$$

The cross-section is trapezoidal for $y_0 > B/n_s$, with a top width $B$, an area equal to:

$$\Omega = B y_0 - \frac{B^2}{2 n_s},$$

and a wetted perimeter equal to:

$$P = 2 y_0 - \frac{B}{n_s} + B \sqrt{1 + \frac{1}{n_s^2}}.$$

The hydraulic radius is

$$R_h = \frac{B y_0 - \dfrac{B^2}{2 n_s}}{2 y_0 - \dfrac{B}{n_s} + B \sqrt{1 + \dfrac{1}{n_s^2}}}.$$

The flow rate in uniform regime is calculated using the Chézy formula, $Q = k R_h^{2/3} \Omega \sqrt{i_b}$.

The critical depth is computed by imposing a stationary point of the energy:

$$1 - \frac{Q^2}{g \Omega^3} \frac{\partial \Omega}{\partial y} \bigg|_{y = y_c} = 0,$$

where $\partial \Omega / \partial y |_{y = y_c} \equiv b_c$ is the top width of the current in critical conditions. Hence

$$\left.
\begin{aligned}
y_c &= \left( \frac{8 Q^2}{g n_s^2} \right)^{1/5} && \text{if } \ 0 < y_c \le B/n_s, \\
y_c &= \frac{B}{2 n_s} + \left( \frac{Q^2}{g B^2} \right)^{1/3} && \text{if } \ y_c > B/n_s.
\end{aligned}
\right\}$$

It is necessary to verify that the calculated critical depth is actually contained within the correct domain of the formula.

**Table 9.5** Characteristics of the current for increasing flow rate

| $y_0$ (m) | $y_c$ (m) | $Q$ (m$^3$ s$^{-1}$) | $\Omega$ (m$^2$) | $P$ (m) | $R_h$ (m) | $k$ (m$^{1/3}$ s$^{-1}$) | $U$ (m s$^{-1}$) |
|---|---|---|---|---|---|---|---|
| 0.50 | 0.38 | 0.42 | 0.50 | 2.56 | 0.20 | 35.00 | 0.83 |
| 1.00 | 0.81 | 2.64 | 2.00 | 5.12 | 0.39 | 35.00 | 1.32 |
| 1.50 | 1.25 | 7.80 | 4.50 | 7.68 | 0.59 | 35.00 | 1.73 |
| 2.00 | 1.70 | 16.79 | 8.00 | 10.25 | 0.78 | 35.00 | 2.10 |
| 2.50 | 2.15 | 31.01 | 12.00 | 11.25 | 1.07 | 35.00 | 2.58 |
| 3.00 | 2.52 | 47.32 | 16.00 | 12.25 | 1.31 | 35.00 | 2.96 |
| 3.50 | 2.89 | 65.14 | 20.00 | 13.25 | 1.51 | 35.00 | 3.26 |
| 4.00 | 3.24 | 84.10 | 24.00 | 14.25 | 1.68 | 35.00 | 3.50 |

**Fig. 9.26** Rating curve and critical depth scale

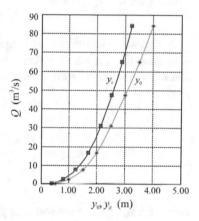

It is also possible to calculate the critical flow rate corresponding to an imposed critical depth. In this case, it results:

$$Q_c = \sqrt{\frac{g\Omega_c^3}{b_c}},$$

expressing $\Omega_c$ and $b_c$ with the relationship appropriate to the numerical value of the imposed water depth.

For $C_u = C_{pu} = 0$ it results $n_s = (4 + C_{pu}/10), i_b = 0.5\%, k = 35$ m$^{1/3}$ s$^{-1}$. The numerical results are listed in Table 9.5. The rating curve and the critical depth scale are shown in Fig. 9.26.

**Fig. 9.27** Cross-section of
the channel with varying
roughness of the walls

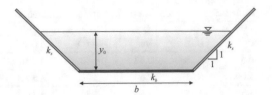

**Exercise 9.14**  The channel with cross-section shown in Fig. 9.27, has a bottom width
$b = 6$ m and a bed slope $i_b = (0.5 + C_{pu}/10)\%$. The banks are made of concrete
with a Gauckler–Strickler coefficient $k_s = 60$ m$^{1/3}$ s$^{-1}$, the bed is made of gravel
with a Gauckler–Strickler coefficient $k_b = 45$ m$^{1/3}$ s$^{-1}$.

– Draw the rating curve up to $y_0 = 3$ m.

**Solution** We will calculate the equivalent roughness with two different approaches.
Assuming that the total resistance exerted on the current is given by the sum of the
contributions due to the different parts of the wetted perimeter, results in an equivalent
roughness:

$$k_{s,eq} = \sqrt{\frac{P}{\sum \dfrac{P_i}{k_{s,i}^2}}} = \left(\frac{b + 2\sqrt{2}y_0}{\dfrac{2\sqrt{2}y_0}{k_s^2} + \dfrac{b}{k_b^2}}\right)^{1/2},$$

where the resistances are acting in parallel.

With a different approach, we assume that the average flow velocity is the same
in each arbitrary part in which the section can be divided, hence

$$\frac{U}{\sqrt{i_b}} = k_{s,eq}\left(\frac{\Omega}{P}\right)^{2/3} \equiv k_{s,i}\left(\frac{\Omega_i}{P_i}\right)^{2/3}, \quad \text{with} \quad \Omega = \sum \Omega_i, \quad P = \sum P_i,$$

and, substituting $\Omega_i$ in the summation, yields

$$k'_{s,eq} = \left(\frac{P}{\sum \dfrac{P_i}{k_{s,i}^{3/2}}}\right)^{2/3} \equiv \left(\frac{b + 2\sqrt{2}y_0}{\dfrac{2\sqrt{2}y_0}{k_s^{3/2}} + \dfrac{b}{k_b^{3/2}}}\right)^{2/3}.$$

Also with the second approach the resistances are in parallel, but with a different
characteristic exponent.

The flow rate is obtained from Chézy formula:

**Table 9.6** Rating curves values computed with two different expressions for the equivalent roughness

| $y_0$ (m) | $k_{s,eq}$ $Q$ (m³ s⁻¹) | $k'_{s,eq}$ $Q$ (m³ s⁻¹) |
|---|---|---|
| 0.5 | 6.23 | 6.25 |
| 1.0 | 20.58 | 20.66 |
| 1.5 | 42.11 | 42.31 |
| 2.0 | 70.85 | 71.20 |
| 2.5 | 107.00 | 107.55 |
| 3.0 | 150.89 | 151.67 |
| 3.5 | 202.84 | 203.88 |

**Fig. 9.28** Rating curves computed with two different expressions for the equivalent roughness

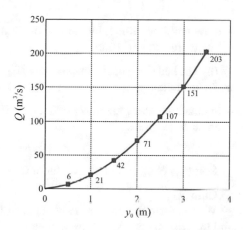

$$Q = k_{s,eq} R_h^{1/6} \Omega \sqrt{R_h i_b} \equiv \left( \frac{b + 2\sqrt{2} y_0}{\frac{2\sqrt{2} y_0}{k_s^2} + \frac{b}{k_b^2}} \right)^{1/2} \frac{\left( b y_0 + y_0^2 \right)^{5/3}}{\left( b + 2\sqrt{2} y_0 \right)^{2/3}} \sqrt{i_b},$$

or

$$Q = k'_{s,eq} R_h^{1/6} \Omega \sqrt{R_h i_b} \equiv \left( \frac{b + 2\sqrt{2} y_0}{\frac{2\sqrt{2} y_0}{k_s^{3/2}} + \frac{b}{k_b^{3/2}}} \right)^{2/3} \frac{\left( b y_0 + y_0^2 \right)^{5/3}}{\left( b + 2\sqrt{2} y_0 \right)^{2/3}} \sqrt{i_b}.$$

For $C_{pu} = 0$ it results $i_b = 0.5\%$. The rating curves computed with the two different expressions for the equivalent roughness are listed in Table 9.6 and are shown in Fig. 9.28. The results obtained by applying the two formulas are practically coincident.

---

**Exercise 9.15** In the channel in Fig. 9.29 the upstream section is gently connected to the downstream section with an expansion. The channel Gauckler–Strickler coefficient is $k_{s1} = k_{s3} = 50 \text{ m}^{1/3} \text{ s}^{-1}$, the expansion Gauckler–Strickler coefficient is $k_{s2} = 60 \text{ m}^{1/3} \text{ s}^{-1}$. The length of the expansion is $L = 50$ m, the upstream section width is $b = 1.5$ m and the downstream section width is $B = 2$ m. The flow rate is $Q = (3.2 + 0.1 \times C_{pu}) \text{ m}^3 \text{ s}^{-1}$.

(a) The bed slope upstream is $i_{b1} = 0.1\%$, while the bed slope downstream and in the expansion is $i_{b2} = i_{b3} = (0.3 + 0.01 \times C_u)\%$:

– draw the flow profile, the normal depth and the critical depth. Assume a spatial integration step of 10 m.

(b) The bed slope upstream is equal to $i_{b1} = 2\%$, see Fig. 9.30, while the bed slope downstream and in the expansion is $i_{b2} = i_{b3} = (0.3 + 0.01 \times C_u)\%$:

– qualitatively draw the flow profile, describing the critical aspects of a perspective quantitative draw.

**Solution** Numerical results refer to $C_u = C_{pu} = 0$.

Case (a)
We calculate the uniform flow depth of the upstream and downstream sections, and the critical depths. For the upstream section the flow rate is:

**Fig. 9.29** Geometry of the channel, case (a)

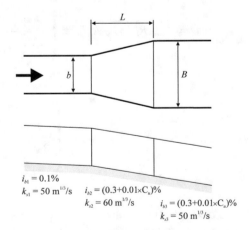

$i_{b1} = 0.1\%$
$k_{s1} = 50 \text{ m}^{1/3}/\text{s}$     $i_{b2} = (0.3+0.01 \times C_u)\%$
$k_{s2} = 60 \text{ m}^{1/3}/\text{s}$     $i_{b3} = (0.3+0.01 \times C_u)\%$
$k_{s3} = 50 \text{ m}^{1/3}/\text{s}$

**Fig. 9.30**  Geometry of the
channel, case (b)

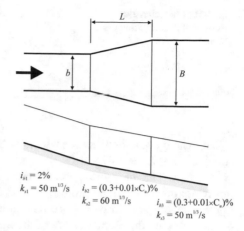

$i_{b1} = 2\%$
$k_{s1} = 50 \text{ m}^{1/3}/\text{s}$   $i_{b2} = (0.3+0.01 \times C_u)\%$
$k_{s2} = 60 \text{ m}^{1/3}/\text{s}$   $i_{b3} = (0.3+0.01 \times C_u)\%$
$k_{s3} = 50 \text{ m}^{1/3}/\text{s}$

$$Q = k_{s1} R_{h1}{}^{2/3} b y_{01} \sqrt{i_{b1}} \equiv k_{s1} \frac{(b y_{01})^{5/3}}{(b + 2 y_{01})^{2/3}} \sqrt{i_{b1}},$$

and solving, yields $y_{01} = \mathbf{2.02 \; m}$. The specific energy is

$$E_1 = y_{01} + \frac{Q^2}{2g (b y_{01})^2} = 2.02 + \frac{3.2^2}{2 \times 9.806 \times (1.5 \times 2.02)^2} = \mathbf{2.07 \; m},$$

and the critical depth is

$$y_{c1} = \sqrt[3]{\frac{Q^2}{g b^2}} = \mathbf{0.77 \; m}.$$

The uniform flow is tranquil.
    For the downstream section, it results

$$Q = k_{s3} R_{h3}{}^{2/3} B y_{03} \sqrt{i_{b3}} = k_{s3} \frac{(B y_{03})^{5/3}}{(B + 2 y_{03})^{2/3}} \sqrt{i_{b3}}.$$

Solving, yields $y_{03} = \mathbf{0.95 \; m}$. The energy is

$$E_3 = y_{03} + \frac{Q^2}{2g (B y_{03})^2} = 0.95 + \frac{3.2^2}{2 \times 9.806 \times (2 \times 0.95)^2} = \mathbf{1.09 \; m},$$

and the critical depth is

$$y_{c3} = \sqrt[3]{\frac{Q^2}{g B^2}} = \mathbf{0.64 \; m}.$$

The uniform flow is tranquil.

**Fig. 9.31** Characteristic
variables of the current in the
channel

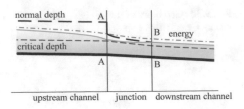

The energy variation between section B-B and section A-A, see Fig. 9.31, is
equal to $E_1 - E_3 = 2.07 - 1.09 = \mathbf{0.98\ m}$, hence the current must reduce the specific energy. The uniform regime for tranquil flow is only reached asymptotically
upstream: downstream the profile must be connected with the uniform flow depth
immediately in section B-B, since there are no flow profiles that are energetically
compatible with the characteristics of the flow. Hence, any change in the profile can
only occur upstream of this section. The profile can be drawn for discrete values on
the basis of the specific energy balance equation:

$$\frac{\mathrm{d}E}{\mathrm{d}s} = i_b - J,$$

where $J$ is the friction loss per unit length.

Actually, the expansion of the current generates an additional dissipation, tradi-
tionally assumed to be proportional to the lost velocity head (Borda–Carnot loss).
For the purposes of the calculation, this loss could be concentrated at the beginning
of the expansion, or evenly distributed along the entire expansion.

We divide the expansion into 5 steps of length $\Delta s = 10$ m. The variation of energy
for each step is equal to:

$$\Delta E = \left( i_b - \bar{J} \right) \Delta s,$$

where $\bar{J}$ is the average loss per unit length. In performing integration, we assume
that the average characteristics of each step coincide with the characteristics of the
upstream section of the step. The profile is shown in Fig. 9.31. The normal depth (by
definition, it is the uniform flow depth for a prismatic channel with a uniform bed
slope having the same geometric and roughness characteristics as the calculation
section) is calculated, in the junction, for discrete values in a certain number of
sections. The critical depth can be calculated analytically in explicit form and has
the following expression:

$$y_c(s) = \sqrt[3]{\frac{Q^2}{g(b + ns)^2}} \quad \text{for } 0 < s < L, \quad \text{with } n = \frac{B - b}{L}.$$

In the upstream channel, the profile is of subcritical accelerated current. The depth
at the intersection section between the expansion and the upstream section (section
A-A) cannot be less than the critical depth (if this were not the case, the upstream

section would have a supercritical current profile influenced by downstream, which is kinematically not admitted). In the downstream channel, the profile is immediately of uniform flow starting from the intersection section with the expansion (B-B), since there is no energetic compatible profile reaching asymptotically downstream the uniform tranquil flow. Spatial integration proceeds from downstream to upstream, starting from the intersection between the expansion and the downstream section (section B-B), where the depth must be equal to the uniform flow depth in the downstream section. The flow profile in the expansion cannot be drawn using the results valid for the prismatic channel, since the junction is not a prismatic channel. Actually, it would be necessary to analyze an equation of the profile wherein an additional term appears that depends on the spatial variation of the cross-section of the channel. For the geometry of this exercise, this equation develops as follows.

The specific energy in the junction is a function of the abscissa and of the depth of the current:

$$E(s, y(s)) = y(s) + \frac{Q^2}{2g[b(s)]^2[y(s)]^2} \equiv y + \frac{Q^2}{2g(b + ns)^2 y^2}, \quad \text{with } n = \frac{B - b}{L},$$

whereas for a prismatic channel $E = E(\cancel{s}, y(s))$. Since

$$\frac{dH}{ds} = -J; \quad H = E + z_b \rightarrow \frac{dE}{ds} = i_b - J, \quad \text{with } \frac{dE}{ds} = \frac{\partial E}{\partial s} + \frac{\partial E}{\partial y}\frac{dy}{ds},$$

we can write

$$\frac{dy}{ds} = \frac{i_b - J - \frac{\partial E}{\partial s}}{\frac{\partial E}{\partial y}} \rightarrow \frac{dy}{ds} = \frac{i_b - J + n\frac{Q^2}{g(b + ns)^3 y^2}}{1 - \frac{Q^2}{g(b + ns)^2 y^3}}. \tag{9.5}$$

The new term in the numerator of Eq. (9.5),

$$n\frac{Q^2}{g(b + ns)^3 y^2}$$

is always positive for a diverging channel, and is equivalent to an increase in the bed slope. The denominator of Eq. (9.5) is positive or negative for tranquil and shooting flow, respectively.

Assuming that the Gauckler–Stricler coefficient $\chi = k_s R_h^{1/6}$ is uniform in the diverging channel (a reasonable hypothesis if the depth of the current is subject to a modest variation), there are nine possible profiles shown in Fig. 9.32 for varying bed slope and roughness (Noseda 1965).

The parameters have the following definition:

**Fig. 9.32** Expansion mild slope channel: possible flow profiles. (From G. Noseda, 1965. Current profiles in divergent and converging rectangular channels. Acts of the IX Conference on Hydraulics and Hydraulic Constructions, Trieste, 1965-in Italian)

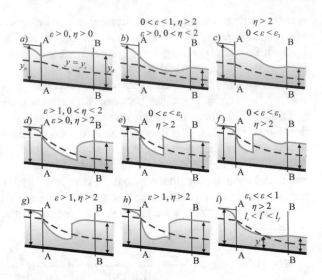

$$\varepsilon = \frac{i_b \chi^2}{g}, \quad \eta = n\frac{\chi^2}{g}, \quad \varepsilon_1 = 1 + \frac{3(\eta - 2)}{4 - 2\sqrt{5\eta}\,(\eta - 2)},$$

and $l^*$ is the width of the section in which the current reaches the critical depth:

$$l^* = \left(\frac{Q}{\sqrt{g}}\right)^{2/5} \left(\frac{\varepsilon - 1}{2 - \eta}\right)^{3/5}.$$

$l_i = b$ and $l_f = B$ are the initial and final widths of the expansion. In the present case, assuming an average depth in the expansion of 1.0 m, the hydraulic radius is on average equal to 0.46 m and results:

$$\varepsilon = 0.85 < 1, \quad \eta = 2.85 > 2, \quad \varepsilon_1 = 0.14, \quad l^* = 0.35 \text{ m}.$$

Therefore, profiles $(a)$, $(b)$ and $(d)$ can develop, but not profile $(i)$ because $l^* < l_i \equiv b$.

Case (b)

For the upstream channel the flow rate is:

$$Q = k_{s1} R_{h1}^{2/3} b y_{01} \sqrt{i_{b1}} \equiv k_{s1} \frac{(b y_{01})^{5/3}}{(b + 2y_{01})^{2/3}} \sqrt{i_{b1}},$$

and solving, yields $y_{01} = \mathbf{0.62}$ **m**. The specific energy is

$$E_1 = y_{01} + \frac{Q^2}{2g(b y_{01})^2} = 0.62 + \frac{3.2^2}{2 \times 9.806(1.5 \times 0.62)^2} = \mathbf{1.22} \text{ m},$$

and the critical depth upstream is

$$y_{c1} = \sqrt[3]{\frac{Q^2}{gb^2}} = 0.77 \text{ m}.$$

The flow is shooting since $y_{01} < y_c$.

For the downstream channel it is still $y_{03} = 0.95$ m, the energy is $E_3 = 1.09$ m and the critical depth is

$$y_{c3} = \sqrt[3]{\frac{Q^2}{gB^2}} = 0.64 \text{ m}.$$

The flow is tranquil.

The current must reduce the specific energy and a hydraulic jump can develop, unless the conditions for a type ($i$) profile in Fig. 9.32 are met.

In the expansion channel, the profile must immediately connect with the uniform flow depth from the intersection section between the expansion and the downstream channel (unless the jump is pushed downstream). The uniform flow depth in the upstream section is maintained up to the intersection with the expansion (unless a jump develops in the upstream section).

---

**Exercise 9.16** A channel with bed slope $i_b = 0.0025$, and a rectangular cross-section of width $b = (3 + 0.1 \times C_u)$ m, is fed by a tank of infinite capacity. The flow in the channel is regulated by the gates P1 and P2 positioned at a very large distance, see Fig. 9.33. In the channel, which has a coefficient of Gauckler–Strickler $k = 70$ m$^{1/3}$ s$^{-1}$, the flow rate is $Q = (20 + C_{pu})$ m$^3$ s$^{-1}$. The water depth in the vena contracta downstream of the floodgate P1 is equal to $C_c a = 0.70$ m, while the water depth upstream of the floodgate P2 is equal to $y_{u2} = 3.50$ m.

- Determine the water level $y_{u1}$ in the tank that feeds the channel.
- Determine the uniform flow depth of the current.
- Determine the character of the bed slope in relation to the flow rate.
- Draw qualitatively the flow profile.
- Calculate the sequent depths of any hydraulic jump that may occur.

**Solution** To solve the problem, it is necessary to calculate the uniform flow depth $y_0$ of the current, which is essential for determining the character of the bed slope in relation to the flow rate $Q$, and for drawing up the permanent flow profile.

To this aim, we apply the Chézy formula:

$$Q = k R_h^{1/6} \Omega \sqrt{R_h i_b} \equiv k \left( \frac{b y_0}{b + 2 y_0} \right)^{2/3} b y_0 \sqrt{i_b}.$$

Since it is not possible to render explicit the unknown $y_0$, the calculation must be performed numerically. Once the uniform flow depth $y_0$ of the current has been calculated, a comparison is made with the critical depth of the current in the rectangular bed, equal to:

$$y_c = \sqrt[3]{\frac{Q^2}{gb^2}}.$$

In this exercise, always results $y_c < y_0$ and the bed slope is mild for the given flow rate $Q$. The qualitative profile of the current is shown in Fig. 9.34. In the hypothesis of free outflow, at short distance from the sluice gate P1 a vena contracta is established (it is a section with rectilinear and parallel trajectories; there the current is therefore rectilinear with consequent hydrostatic distribution of the pressure) and the current is supercritical if $a < y_c/C_c$.

For the calculation of $y_{u1}$ we impose the energy balance between the upstream section of the sluice gate and the vena contracta, neglecting losses:

$$y_{u1} + \frac{Q^2}{2gb^2y_{u1}^2} = C_c a + \frac{Q^2}{2gb^2(C_c a)^2}.$$

The resulting equation is cubic in the unknown $y_{u1}$ and can be solved analytically or numerically. The flow profile downstream of the sluice gate, between section A and section B, is a decelerated supercritical current profile in a mild slope channel, and is drawn downstream starting from the water depth $C_c a$ in the vena contracta.

**Fig. 9.33** Channel with two control gates

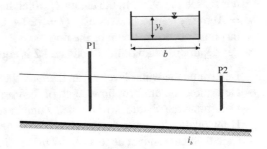

**Fig. 9.34** Profile of the current. The surge immediately upstream of the gates is equal to the velocity head and is due to the almost complete stop of the flow, with consequent transformation of kinetic energy into piezometric head

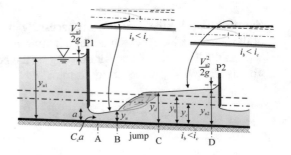

**Fig. 9.35** Current profile
without hydraulic jump
between the two gates

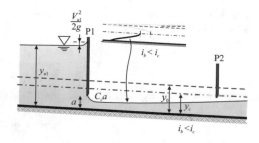

This profile, if it could develop completely, would reach with (theoretical) vertical tangent the critical depth $y_c$ at a finite distance from the vena contracta.

Upstream of the gate P2 the flow is tranquil, if $y_{u2} > y_0$. The profile is of a subcritical decelerated current, tends asymptotically upstream to uniform flow depth and must be drawn from downstream to upstream starting from the water depth $y_{u2}$.

Between the two sluice gates, the transition between the supercritical upstream current and the subcritical downstream current (governed by the presence of sluice gate P2) must take place. If the channel is prismatic, the transition occurs abruptly with a hydraulic jump. The sequent depths of the hydraulic jump, $y_u$ and $y_d$, satisfy the linear momentum balance in integral form:

$$\underbrace{\frac{\gamma_w b y_u^2}{2} + \frac{\rho Q^2}{b y_u}}_{S_u} = \underbrace{\frac{\gamma_w b y_d^2}{2} + \frac{\rho Q^2}{b y_d}}_{S_d}.$$

Moreover, given the long distance between the two gates, it can be reasonably assumed that the downstream depth of the jump, $y_d$, is practically coincident with the uniform flow depth of the current $y_0$. If this is not the case, the position of the jump can be detected by drawing the diagrams of the total force of the two currents (starting from the vena contracta downstream, for the supercritical current, and from the floodgate P2 upstream, for the subcritical current) and identifying the intersection. The resulting equation is cubic in the unknown $y_u$ and can be solved analytically, dividing the third degree polynomial in $y_u$ by the binomial $(y_u - y_0)$, since the uniform flow depth $y_0$ is known and is a root of the equation, or numerically.

From an energy balance point of view, the current must recover energy in order to cross the slot underneath the P1 sluice gate. In the profile between sections A and B and in the jump, the current dissipates part of the gained energy. Finally, to cross the slot underneath gate P2, it must recover energy again between sections C and D. As a special case, it may happen that if the two floodgates are relatively close and if $a$ is sufficiently small, the supercritical current coming out of the floodgate P1 flows under the floodgate P2, see Fig. 9.35. In these conditions, the presence of the second floodgate is irrelevant and the hydraulic jump develops downstream of the P2 floodgate.

**Fig. 9.36** Current profile
with a submerged hydraulic
jump

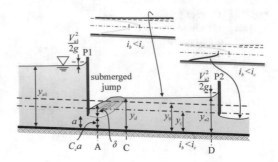

It can also happen that the hydraulic jump is pushed towards the gate P1 becoming
a submerged jump, see Fig. 9.36. In this case, the calculation can be carried out taking
into account that the roller of the hydraulic jump in the vena contracta section does
not contribute to the flow, but contributes with its weight to the pressure. The water
depth upstream of the gate P1 is calculated by imposing the following energy balance:

$$y_{u1} + \frac{Q^2}{2gb^2 y_{u1}^2} = C_c a + \delta + \frac{Q^2}{2gb^2(C_c a)^2},$$

where $\delta$ is the thickness of the roller of the jump in correspondence of the vena
contracta. The water depth $C_c a + \delta$ is calculated imposing the balance of the total
force between the section A and the section C:

$$\underbrace{\frac{\gamma_w b(C_c a + \delta)^2}{2} + \frac{\rho Q^2}{b C_c a}}_{S_u} = \underbrace{\frac{\gamma_w b y_d^2}{2} + \frac{\rho Q^2}{b y_d}}_{S_d}.$$

The presence of the submerged jump implies the loss of hydraulic disconnection,
when the flow rate out of the sluice gate is a function only of the depth level upstream.
When the gate is no longer disconnected, the flow rate is a function of both the
upstream and downstream water depths. The gate is defined as drowned.

For $\ C_u = C_{pu} = 0 \ $ it results $\ i_b = 0.0025, \quad b = 3 \text{ m}, \quad k = 70 \text{ m}^{1/3}\text{ s}^{-1},$
$Q = 20 \text{ m}^3 \text{ s}^{-1}, C_c a = 0.70 \text{ m}, y_{u2} = 3.50 \text{ m}.$

$$y_{u1} + \frac{Q^2}{2gb^2 y_{u1}^2} = C_c a + \frac{Q^2}{2gb^2(C_c a)^2} \rightarrow$$

$$y_{u1} + \frac{20^2}{2 \times 9.806 \times 3^2 \times y_{u1}^2} = 0.70 + \frac{20^2}{2 \times 9.806 \times 3^2 \times 0.70^2} \rightarrow$$

$$y_{u1}^3 - 5.325 y_{u1}^2 + 2.266 = 0. \quad (9.6)$$

Equation (9.6) admits the three solutions $y_{u1} = -0.62$ m, $y_{u1} = 0.70$ m, $y_{u1} = \textbf{5.24 m}$, of which the first is physically meaningless, and the second coincides with the depth in the vena contracta. The third solution is the correct value.

$$Q = k\left(\frac{by_0}{b+2y_0}\right)^{2/3}by_0\sqrt{i_b} \rightarrow$$

$$20 = 70 \times \left(\frac{3 \times y_0}{3+2y_0}\right)^{2/3} \times 3 \times y_0 \times \sqrt{0.0025} \rightarrow \qquad y_0 = \textbf{2.09 m.}$$

$$y_c = \sqrt[3]{\frac{Q^2}{gb^2}} = \sqrt[3]{\frac{20^2}{9.806 \times 3^2}} = \textbf{1.65 m.}$$

As $y_c < y_0$ the bed slope is mild.

To calculate the sequent depth upstream of the hydraulic jump:

$$\underbrace{\frac{\gamma_w by_u^2}{2} + \frac{\rho Q^2}{by_u}}_{S_u} = \underbrace{\frac{\gamma_w by_d^2}{2} + \frac{\rho Q^2}{by_d}}_{S_d} \rightarrow$$

$$\frac{9806 \times 3 \times y_u^2}{2} + \frac{1000 \times 20^2}{3 \times y_u} = \frac{9806 \times 3 \times 2.09^2}{2} + \frac{1000 \times 20^2}{3 \times 2.09} \rightarrow$$

$$y_u^3 - 8.706y_u + 9.065 = 0,$$

which admits the three solutions $y_u = -3.38$ m, $y_u = \textbf{1.29 m}$, $y_u = 2.09$ m. The correct solution is the second one.

Between section A and section B, the water depth increases from 0.70 m to 1.29 m. For the hypotheses, in section C, immediately downstream of the jump, the water depth assumes a value equal to the uniform flow depth $y_0 = 2.09$ m. Between section C and section D, the water depth further increases up to the value of 3.50 m.

The profile of the current is qualitatively shown in Fig. 9.36.

---

**Exercise 9.17** A channel with Gauckler–Strickler roughness $k = (60 + 2 \times C_{pu})$ m$^{1/3}$ s$^{-1}$ and bed slope $i_b = 0.02$, derives water from a large quiet basin by means of a well-connected inlet with negligible losses, see Fig. 9.37. The cross-section of the channel is triangular with a slope of the banks of $\alpha = 45°$. The water level in the basin has level $H = (4 + C_u/5)$ m above the threshold at the inlet section.

– Determine the flow rate.
– Determine if the channel has a steep or mild slope.
– Qualitatively draw the profile of the current.

Suppose the channel of unlimited length.

**Fig. 9.37** Channel of
derivation from a reservoir

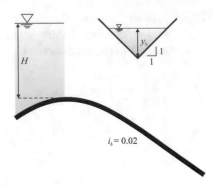

$i_b = 0.02$

**Solution** First of all, it is necessary to determine the value of the flow rate $Q$
derived from the still basin by making appropriate assumptions. Given the high value
of the slope of the channel, it is reasonable to assume that it is steep for the derived
flow $Q$. This hypothesis implies that the uniform flow current is supercritical and
that will tend to reach the depth $y_0$ asymptotically downstream. In correspondence
of the crest there is a spontaneous transition through the critical state. Moreover,
there is no energy dissipation because the entrance is well connected, and the energy
balance equation and the critical state condition can be set up as follows:

$$
\begin{cases}
H = y_c + \dfrac{Q^2}{2g\Omega_c^2}, \\[3mm]
\dfrac{\Omega_c^3}{B_c} = \dfrac{Q^2}{g},
\end{cases}
$$

where $B_c$ is the top width of the current in critical condition. For the triangular
cross-section with $\alpha = 45°$ results $\Omega_c = y_c^2$, $B_c = 2y_c$, hence

$$
\begin{cases}
\dfrac{\Omega_c^3}{B_c} = \dfrac{Q^2}{g} \;\rightarrow\; \dfrac{Q^2}{g\Omega_c^2} = \dfrac{y_c}{2}, \\[3mm]
H = y_c + \dfrac{Q^2}{2g\Omega_c^2} \;\rightarrow\; H = \dfrac{5y_c}{4}.
\end{cases}
$$

Solving the system of equations, results

$$
y_c = \frac{4H}{5}, \qquad Q = \sqrt{\frac{g\Omega_c^2 y_c}{2}} \equiv \sqrt{\frac{gy_c^5}{2}}.
$$

Once the value of the flow rate $Q$ has been determined, it is necessary to calculate
the uniform flow depth $y_0$ to ensure the character of the bed slope and to draw the

**Fig. 9.38**  Profile of the
current

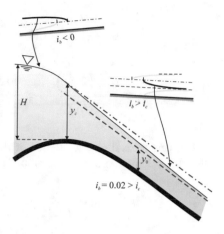

profile of the current. Applying Chézy formula yields

$$Q = kR_h^{1/6}\Omega\sqrt{R_h i_b}, \quad \text{with } \Omega = y_0^2, \quad P = 2\sqrt{2}y_0, \quad R_h = \frac{\Omega}{P} \equiv \frac{y_0}{2\sqrt{2}}.$$

Hence, it results

$$Q = k\left(\frac{y_0}{2\sqrt{2}}\right)^{2/3} y_0^2\sqrt{i_b} \equiv \frac{k\sqrt{i_b}}{\left(2\sqrt{2}\right)^{2/3}} y_0^{8/3},$$

which admits the analytical solution

$$y_0 = \frac{Q^{3/8}\left(2\sqrt{2}\right)^{1/4}}{\left(k\sqrt{i_b}\right)^{3/8}}.$$

For the channel in Fig. 9.37, the relation $y_0 < y_c$ always holds and confirms the initial hypothesis of a steep channel for the flow rate $Q$ derived from the basin. The resulting profile is qualitatively shown in Fig. 9.38.

Starting from the critical state, in correspondence of the crest, an accelerated supercritical current profile develops which tends asymptotically downstream to the uniform flow depth $y_0$. In the basin, upstream of the crest, a profile of subcritical accelerated current develops.

For $C_u = C_{pu} = 0$ it results $k = 60$ m$^{1/3}$ s$^{-1}$, $i_b = 0.02$, $H = 4$ m,

$$y_c = \frac{4H}{5} = \frac{4 \times 4}{5} = \textbf{3.20 m},$$

$$Q = \sqrt{\frac{gy_c^5}{2}} = \sqrt{\frac{9.806 \times 3.2^5}{2}} = 40.56 \text{ m}^3 \text{ s}^{-1},$$

$$y_0 = \frac{Q^{3/8}\left(2\sqrt{2}\right)^{1/4}}{\left(k\sqrt{i_b}\right)^{3/8}} = \frac{40.56^{3/8} \times \left(2 \times \sqrt{2}\right)^{1/4}}{\left(60 \times \sqrt{0.02}\right)^{3/8}} = 2.33 \text{ m}.$$

The geometrical characteristics of the channel and the roughness of the walls, allow the calculation of the critical bed slope, equal to

$$i_c = \frac{2g}{k^2 y_c^{1/3}} = \frac{2 \times 9.806}{60^2 \times 3.20^{1/3}} = 0.0037.$$

Since $i_b > i_c$, the flow rate is independent of the slope of the channel and reaches the maximum value. The channel is hydraulically disconnected.

If $i_b < i_c$, the channel is a mild slope one and the uniform flow depth is reached immediately downstream of the intake. In this case, the flow rate is a function of the bed slope and of the walls roughness, and decreases for decreasing bed slope or for increasing roughness. The channel is hydraulically connected.

---

**Exercise 9.18**  A channel with rectangular cross-section of width $b = 8$ m is divided into two sections with a bed slope $i_{b1} = 0.01$ and $i_{b2} = 0.002$, with Gauckler–Strickler coefficient $k_1 = 70 \text{ m}^{1/3} \text{ s}^{-1}$ and $k_2 = 60 \text{ m}^{1/3} \text{ s}^{-1}$, respectively, see Fig. 9.39. The flow rate is $Q = (80 + 5 \times C_u) \text{ m}^3 \text{ s}^{-1}$.

– Draw qualitatively the profile of the current.
– Determine the values of the sequent depths if there a hydraulic jump develops.

**Solution**  It is preliminary necessary to establish the character of the two channels, with bed slope and roughness respectively $i_{b1}, k_1$ and $i_{b2}, k_2$, for the given flow rate $Q$. It is necessary to calculate the value of the critical depth $y_c$, unique for the two sections (being a function of the quantities $Q$, $g$ and $b$ that do not vary in the

**Fig. 9.39**  Rectangular cross-section channel with bed slope variation

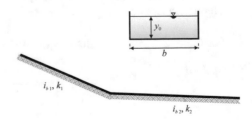

two channels), and the uniform flow depths $y_{01}$ and $y_{02}$ of the current through the following relations:

$$y_c = \sqrt[3]{\frac{Q^2}{gb^2}},$$

$$Q = k_1 R_{h1}^{1/6} \Omega_1 \sqrt{R_{h1} i_{b1}}, \quad \text{with } \Omega_1 = b y_{01}, \quad R_{h1} = \frac{\Omega_1}{P_1} \equiv \frac{b y_{01}}{b + 2 y_{01}},$$

and

$$Q = k_2 R_{h2}^{1/6} \Omega_2 \sqrt{R_{h2} i_{b2}}, \quad \text{with } \Omega_2 = b y_{02}, \quad R_{h2} = \frac{\Omega_2}{P_2} \equiv \frac{b y_{02}}{b + 2 y_{02}}.$$

Since for a rectangular cross-section no analytical solution is available for the unknown $y_0$, the calculation must be performed numerically.

For $C_u = C_{pu} = 0$ it results $b = 8$ m, $i_{b1} = 0.01$, $i_{b2} = 0.002$, $k_1 = 70$ m$^{1/3}$ s$^{-1}$, $k_2 = 60$ m$^{1/3}$ s$^{-1}$, $Q = 80$ m$^3$ s$^{-1}$, and we calculate $y_{01} = \mathbf{1.39}$ m and $y_{02} = \mathbf{2.71}$ m, $y_c = \mathbf{2.17}$ m.

Since $y_{01} < y_c < y_{02}$, the upstream and downstream sections are steep and mild slope, respectively, for the assigned flow rate $Q$. The transition from steep to mild bed slope, if the channel is prismatic, requires a hydraulic jump. The spatial position of the jump derives from consideration on the total force of the currents with depth $y_{01}$ and $y_{02}$.

The total forces are

$$S_{01} \equiv \frac{\gamma_w b y_{01}^2}{2} + \frac{\rho Q^2}{b y_{01}} = \frac{9806 \times 8 \times 1.39^2}{2} + \frac{1000 \times 80^2}{8 \times 1.39} = \mathbf{651.3 \ kN},$$

and

$$S_{02} \equiv \frac{\gamma_w b y_{02}^2}{2} + \frac{\rho Q^2}{b y_{02}} = \frac{9806 \times 8 \times 2.71^2}{2} + \frac{1000 \times 80^2}{8 \times 2.71} = \mathbf{583.3 \ kN},$$

hence it results

$$S_{01} > S_{02}.$$

This condition requires that the hydraulic jump develops in the mild slope section of the channel. In fact, if the jump developed upstream, in the steep slope channel, the only possible profile starting from the sequent depth of the jump would be of subcritical decelerated current, which increases the total force in the direction of the motion and, therefore, could never drop to assume the value $S_{02} < S_{01}$ corresponding to the uniform flow downstream.

The qualitative flow profile is shown in Fig. 9.40.

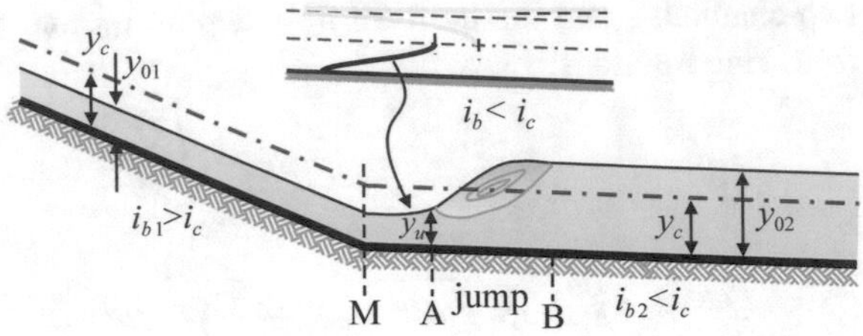

**Fig. 9.40** Current profile if the jump is in the downstream mild slope channel

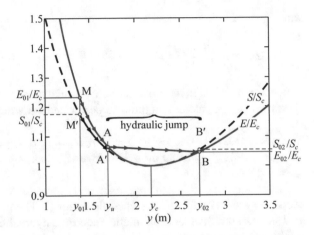

**Fig. 9.41** Evolution of specific energy $E/E_c$ and total force $S/S_c$ of the current

The uniform flow current in the upstream channel has a specific energy greater than that of the uniform flow current in the downstream channel. Starting from section M, the only possible supercritical current profile develops on a mild slope bed. This profile dissipates specific energy (the current dissipates more energy than that supplied by gravity) and stops at $y_u$, the sequent depth of the uniform flow depth $y_{02}$. The determination of $y_u$ is obtained by imposing an equal total force downstream and upstream of the jump:

$$\underbrace{\frac{\gamma_w b y_{02}^2}{2} + \frac{\rho Q^2}{b y_{02}}}_{S_{y_{02}}} = \underbrace{\frac{\gamma_w b y_u^2}{2} + \frac{\rho Q^2}{b y_u}}_{S_{y_u}}.$$

The resulting equation is cubic in the unknown $y_u$ and can be solved analytically or numerically. It is also possible to reduce the degree of the third degree polynomial dividing it by the binomial $(y_u - y_{02})$, since the uniform flow depth $y_{02}$ is a root of the equation.

The solution is $y_u = \mathbf{1.70\ m}$. The evolution of total force and energy (dimensionless with respect to the values assumed in critical condition) is shown in Fig. 9.41. The continuous curve represents the specific energy for the assigned flow rate $Q$, the dashed curve represents the total force.

**Fig. 9.42** Current profile if
the jump is formed in the
steep slope channel

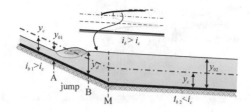

Starting from the change of slope (section M), along the decelerated supercritical current profile the total force drops until it reaches the total force value associated to $y_u$ (section A), the sequent depth of $y_{02}$. From an energy point of view, the specific energy is also reduced to the value corresponding to $y_u$. The hydraulic jump dissipates further energy, equal to:

$$\Delta E_{jump} = \frac{(y_u - y_{02})^3}{4 y_u y_{02}}.$$

This dissipation occurs, for real fluids, with the transfer of energy toward smaller vortices up to the geometric scale controlled by viscosity, with final conversion of mechanical energy to thermal energy.

It is a question of considerable importance how the dissipation in the jump can be foreseen in a calculation scheme valid for ideal fluids (the viscosity of the fluid, in fact, does not appear in the equation of conservation of mass and balance of linear momentum, used to describe the jump). A now accepted interpretation, originally due to Rayleigh (Rayleigh, Lord, 1910. Aerial plane waves of finite amplitude. Proc. R. Soc. Lond. A84, 247–284) attributes this inconsistency to the inadequacy of the differential equations adopted in the presence of a discontinuity. By their nature the mathematical equations used would not, therefore, be suitable to describe the behavior of functions (such as speed, pressure, etc.) which, in the jump, are locally without derivative. A coherent model, able to explain the variation of energy even for ideal fluids (which can be described with Euler's equations), postulates that the subtraction of energy in the jump is the transfer of mechanical energy from the water current to eddies of virtually null geometric scale, conventionally described as "heat" or "internal energy" (Broadwell, J.E., 1997. Shocks and energy dissipation in inviscid fluids: a question proposed by Lord Rayleigh. Journal of Fluid Mechanics 347, 375–380).

If it were $S_{01} < S_{02}$, the jump should develop in the steep slope channel, see Fig. 9.42. In fact, if it were to occur in the mild slope channel, the sequent depth downstream would necessarily be equal to $y_{02}$ and the sequent depth upstream would reach $y_{01}$, starting from section M, with a profile of decelerated supercritical current. This profile reduces the total force in the direction of motion and could never reach the value $S_{02}$.

Finally, if it were $S_{01} = S_{02}$ the jump would be formed exactly in section M, with the sequent depths equal to $y_{01}$ and $y_{02}$, respectively.

**Fig. 9.43** Schematic of the
channel with flat sluice gate
for control of the flow rate

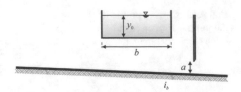

**Exercise 9.19** In the channel shown in Fig. 9.43, with rectangular cross-section of
width $b = (15 + C_u)$ m, bed slope $i_b = 0.01$ and a coefficient of Gauckler–Strickler
$k = 70$ m$^{1/3}$ s$^{-1}$, the flow rate is $Q = 100$ m$^3$ s$^{-1}$. In an intermediate section there is
a sluice gate with a rectangular slot near the bottom of height $a = (0.9 + C_{pu}/20)$ m.

- Determine the water depth upstream of the floodgate.
- Draw qualitatively the profile of the current.
- If a hydraulic jump develops, determine the value of sequent depths.

The outflow occurs without dissipation and the contraction coefficient is equal to
$C_c = 0.61$. Consider the channel of indefinite length.

**Solution** It is preliminary necessary to determine whether the bed has a steep or
a mild slope for the given flow rate $Q$. It is necessary to calculate the critical depth
$y_c$ and the uniform flow depth $y_0$ through the relationships:

$$y_c = \sqrt[3]{\frac{Q^2}{gb^2}},$$

and

$$Q = kR_h^{1/6}\Omega\sqrt{R_h i_b}, \quad \text{with } \Omega = by_0, \quad R_h = \frac{\Omega}{P} \equiv \frac{by_0}{b + 2y_0}.$$

For a rectangular cross-section it is not possible to analytically express the
unknown $y_0$, and the calculation must be performed numerically.

For $C_u = C_{pu} = 0$ it results $b = 15$ m, $i_b = 0.01$, $k = 70$ m$^{1/3}$ s$^{-1}$,
$Q = 100$ m$^3$ s$^{-1}$, $a = 0.9$ m, and we calculate $y_0 = \mathbf{1.02}$ **m** and $y_c = \mathbf{1.65}$ **m**. Since
$y_0 < y_c$ the bed is steep for the given flow rate $Q$.

The qualitative profile of the current is shown in Fig. 9.44.

*Upstream of the Floodgate*

Upstream of the floodgate, if the slot height $a$ is less than the uniform flow depth $y_0$,
the current must gain specific energy. This energy recovery can only be achieved in
subcritical current regime, since the cause of disturbance (the floodgate) could not
affect upstream in the presence of a supercritical current. The determination of $y_{u1}$

**Fig. 9.44** Qualitative profile
of the current

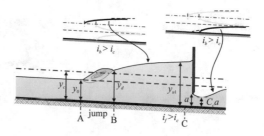

requires the energy balance between the upstream section of the sluice gate and the vena contracta, neglecting energy losses:

$$y_{u1} + \frac{Q^2}{2gb^2 y_{u1}^2} = C_c a + \frac{Q^2}{2gb^2(C_c a)^2}.$$

The resulting equation is cubic in the unknown $y_{u1}$ and can be solved analytically or numerically selecting solutions greater than $y_c$. The result is $y_{u1} = \textbf{8.03 m}$.

The profile that is established is a slow decelerated current in a steep sloping bed, which from the critical condition tends to a horizontal asymptote downstream. The transition from a uniform supercritical flow to a decelerated subcritical flow requires a hydraulic jump. This jump connects the uniform flow depth $y_0$ and the sequent depth $y_d$, computed by imposing the linear momentum balance in integral form:

$$\underbrace{\frac{\gamma_w b y_0^2}{2} + \frac{\rho Q^2}{b y_0}}_{S_{y_0}} = \underbrace{\frac{\gamma_w b y_d^2}{2} + \frac{\rho Q^2}{b y_d}}_{S_{y_d}}.$$

The resulting equation is cubic in the unknown $y_d$ and can be solved analytically or numerically, or dividing the third degree polynomial by the binomial $(y_d - y_0)$, the uniform flow depth $y_0$ also being one of the root of the equation. The result is $y_d = \textbf{2.51 m}$.

*Downstream of the Floodgate*

The current through the slot is supercritical. Starting from the vena contracta (with a water depth $C_c a$) the profile is of a supercritical decelerated current and tends asymptotically downstream to the uniform flow depth $y_0$.

---

**Exercise 9.20** A channel with Gauckler–Strickler coefficient $k = (60 + 2 \times C_{pu})$ m$^{1/3}$ s$^{-1}$ derives water from a large quiet reservoir by means of a well-connected inlet with negligible losses. The channel has a bed slope $i_{b1} = 0.02$ in a first part,

**Fig. 9.45** Intake channel
with slope variation

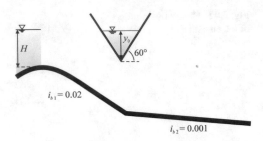

and $i_{b2} = 0.001$ in a second part. The change of slope is very far from the crest of
the inlet and the $i_{b2}$ slope channel is homogeneous indefinitely downstream. The
cross-section of the channel is triangular with a slope of the banks $\alpha = 60°$ as shown
in Fig. 9.45. The level in the reservoir is at $H = (4 + C_u/5)$ m above the entrance
crest.

- Determine the flow rate into the channel.
- Qualitatively draw the profile of the current.
- Quantitatively draw the profile starting from the section of the change of slope.

  Assume a number of steps for spatial integration $n = 10$.

  **Solution** It is preliminary necessary to determine the value of the flow rate $Q$
derived from the reservoir with some simplifying hypotheses. It is reasonable to
assume that the initial channel has a steep slope for the flow $Q$ derived from the
reservoir. This implies a spontaneous transition of the current through the critical
state in correspondence of the crest. We can also neglect the dissipations at the inlet
because the intake is well-connected. We can write a system of equations of energy
balance of the current between the reservoir and the crest, and the condition of critical
state:

$$
\begin{cases}
H = y_c + \dfrac{Q^2}{2g\Omega_c^2}, \\[2mm]
\dfrac{\Omega_c^3}{B_c} = \dfrac{Q^2}{g},
\end{cases}
$$

with the solution

$$
H = y_c + \frac{\Omega_c}{2B_c}, \quad \text{with} \quad \Omega_c = \frac{y_c^2}{\sqrt{3}}, \quad B_c = \frac{2y_c}{\sqrt{3}}.
$$

Substituting, it results

$$
y_c = \frac{4H}{5},
$$

and

$$
Q = \sqrt{\frac{g\Omega_c^2 y_c}{2}} \equiv \sqrt{\frac{g y_c^5}{6}}.
$$

Applying Chézy formula

$$Q = k\left(\frac{y_0}{4}\right)^{1/6} \frac{y_0^2}{\sqrt{3}} \sqrt{\frac{y_0}{4} i_b} \equiv \frac{k\sqrt{i_b} y_0^{8/3}}{4^{2/3}\sqrt{3}},$$

gives the uniform flow depth

$$y_0 = \left(\frac{4^{2/3}\sqrt{3} Q}{k\sqrt{i_b}}\right)^{3/8}. \tag{9.7}$$

Substituting $i_{b1}$ and $i_{b2}$ into Eq. (9.7), we calculate the uniform flow depths $y_{01}$ and $y_{02}$.

In the present case the relation $y_{01} < y_c < y_{02}$ always holds: the first channel is steep while the second one is mild. At the crest there is the spontaneous transition through the critical state. The profile that develops in the first channel is an accelerated supercritical current profile that tends to reconstitute the uniform flow asymptotically downstream. Given the long distance between the crest and the section where the change of slope takes place, it can be confidently assumed that the uniform flow depth $y_{01}$ is reached before the change of slope. The absence of disturbing causes at the far end of downstream of the mild slope channel guarantees the condition of uniform flow depth equal to $y_{02}$.

The supercritical-subcritical current transition requires a hydraulic jump, whose spatial position must be determined through the comparison of the total forces relative to the currents of depth $y_u$ and $y_d$. In the particular case under examination, it can be assumed that (i) $y_u \approx y_{01}$, given the long distance between the crest (where the flow is critical) and the section of change of bed slope; (ii) the sequent depth $y_d$ coincides with the uniform flow depth $y_{02}$. The total forces are:

$$S_u = \gamma_w z_{Gu} \Omega_u + \frac{\rho Q^2}{\Omega_u} \equiv S_{y_{01}} = \gamma_w \frac{y_{01}^3}{3\sqrt{3}} + \frac{\sqrt{3}\rho Q^2}{y_{01}^2},$$

and

$$S_d = \gamma_w z_{Gd} \Omega_d + \frac{\rho Q^2}{\Omega_d} \equiv S_{y_{02}} = \gamma_w \frac{y_{02}^3}{3\sqrt{3}} + \frac{\sqrt{3}\rho Q^2}{y_{02}^2}.$$

$z_{Gu}$ and $z_{Gd}$ are the depth of the centres of pressure of the cross-sections of the current with respect to the free surface. In the present case results $S_u < S_d$ and the hydraulic jump develops in the steep channel. In fact, the difference in total force can only be recovered with a subcritical decelerated current profile downstream of the jump. If the jump were located in the mild slope channel, it would be directly connected with the uniform flow depth $y_{02}$, preceded by a profile of supercritical decelerated current that further reduces the total force in the direction of motion. In the steep channel, between the downstream depth of the jump and the section where the slope changes, a subcritical decelerated current profile develops, the only possible

**Table 9.7** Experimental length of the hydraulic jump

| $\mathrm{Fr}_u$ | $L/(y_d - y_u)$ |
| --- | --- |
| 2 | 7.6 |
| 3 | 7.2 |
| 5 | 7.0 |
| 10 | 6.6 |
| 15 | 6.2 |
| 20 | 5.7 |

subcritical current profile in a steep slope bed (profile between sections M and B). If this profile could develop entirely, it would reach upstream the critical state with a vertical tangent at a finite distance from the change of slope (actually, the strong curvature of the trajectories would make no longer valid the hypothesis of hydrostatic distribution of pressure, contained in the equations adopted to describe the profile of gradually varied currents). Given the presence of the uniform supercritical current from upstream, this profile (we have assumed that the upstream trunk is long enough to allow the current to reach the uniform flow regime) will stop at a value $y_d$ (section A), the sequent depth of $y_{01}$. The sequent depth $y_d$ is calculated by imposing the balance of the total force:

$$\underbrace{\gamma_w \frac{y_{01}^3}{3\sqrt{3}} + \frac{\sqrt{3}\rho Q^2}{y_{01}^2}}_{S_u \equiv S_{y01}} = \underbrace{\gamma_w \frac{y_d^3}{3\sqrt{3}} + \frac{\sqrt{3}\rho Q^2}{y_d^2}}_{S_d}.$$

The hydraulic jump has a finite length that, on the basis of experimental data (see Table 9.7), is a function of difference between the sequent depths and of the Froude number of the upstream current:

$$\mathrm{Fr}_u = \frac{V}{\sqrt{g y_h}},$$

where $y_h = \Omega/B$ is the hydraulic depth, equal to the ratio between the area of the cross-section of the current and the top width.

The qualitative profile is shown in Fig. 9.46.

Quantitative profile drawing from section M (change of slope) can be done numerically. Assume, for example, a step of variation of the water depth of the current equal to:

$$\Delta y = \frac{(y_{02} - y_d)}{n}.$$

The origin of the coordinate system with a positive downstream abscissa is fixed in section M. The spatial interval $\Delta s_r$ corresponding to the depth variation of the current $\Delta y_r$ can be calculated using the discrete form of the profile equation:

**Fig. 9.46** Profile of the current

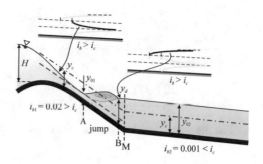

$$\Delta s_r = \frac{\Delta E_r}{\left(i_b - J_{ave,r}\right)},$$

where $\Delta s_r$ is the difference in coordinate between the sections where the current has depths $y_{r+1}$ and $y_r$, and $\Delta E_r = E_{r+1} - E_r$, $J_{ave,r} = 0.5(J_{r+1} + J_r)$.

For $C_u = C_{pu} = 0$ it results $k = 60$ m$^{1/3}$ s$^{-1}$, $i_{b1} = 0.02$, $i_{b2} = 0.001$, $H = 4$ m, hence

$$y_c = \frac{4H}{5} = \frac{4 \times 4}{5} = \mathbf{3.2\ m},$$

$$Q = \sqrt{\frac{gy_c^5}{6}} = \sqrt{\frac{9.806 \times 3.2^5}{6}} = \mathbf{23.42\ m^3\,s^{-1}},$$

$$y_{01} = \left(\frac{\sqrt{3} \times 4^{2/3} Q}{k\sqrt{i_{b1}}}\right)^{3/8} = \left(\frac{23.42 \times \sqrt{3} \times 4^{2/3}}{60 \times \sqrt{0.02}}\right)^{3/8} = \mathbf{2.54\ m},$$

$$y_{02} = \left(\frac{\sqrt{3} \times 4^{2/3} Q}{k\sqrt{i_{b2}}}\right)^{3/8} = \left(\frac{23.42 \times \sqrt{3} \times 4^{2/3}}{60 \times \sqrt{0.001}}\right)^{3/8} = \mathbf{4.46\ m},$$

$$S_u = \gamma_w \frac{y_{01}^3}{3\sqrt{3}} + \frac{\sqrt{3}\rho Q^2}{y_{01}^2} \rightarrow S_u = 9806 \times \frac{2.54^3}{3 \times \sqrt{3}} + \frac{\sqrt{3} \times 1000 \times 23.42^2}{2.54^2}$$
$$= \mathbf{177\ 932\ N},$$

$$S_d = \gamma_w \frac{y_{02}^3}{3\sqrt{3}} + \frac{\sqrt{3}\rho Q^2}{y_{02}^2} \rightarrow S_v = 9806 \times \frac{4.46^3}{3 \times \sqrt{3}} + \frac{\sqrt{3} \times 1000 \times 23.42^2}{4.46^2}$$
$$= \mathbf{215\ 092\ N}.$$

The result is $S_u < S_d$. We calculate the sequent depth of the hydraulic jump:

$$\underbrace{\gamma_w \frac{y_{01}^3}{3\sqrt{3}} + \frac{\sqrt{3}\rho Q^2}{y_{01}^2}}_{S_u \equiv S_{y01}} = \underbrace{\gamma_w \frac{y_d^3}{3\sqrt{3}} + \frac{\sqrt{3}\rho Q^2}{y_d^2}}_{S_d} \rightarrow$$

$$177\,932 = 9806 \times \frac{y_d^3}{3 \times \sqrt{3}} + \frac{\sqrt{3} \times 1000 \times 23.42^2}{y_d^2} \rightarrow y_d = \mathbf{3.96\ m.}$$

For the numerical integration, we consider the following depth step

$$\Delta y = \frac{(y_{02} - y_d)}{n} = \frac{(4.46 - 3.96)}{10} = \mathbf{0.05\ m.}$$

From the downstream section ($r = 1$), with $y_1 = y_{02}$, the cross-section of the current is calculated:

$$\Omega_1 = \frac{y_1^2}{\sqrt{3}} = \frac{4.46^2}{\sqrt{3}} = \mathbf{11.48\ m^2,}$$

the top width

$$B_1 = \frac{2y_1}{\sqrt{3}} = \frac{2 \times 4.46}{\sqrt{3}} = \mathbf{5.15\ m,}$$

the wetted perimeter

$$P_1 = \frac{4y_0}{\sqrt{3}} = \frac{4 \times 4.46}{\sqrt{3}} = \mathbf{10.30\ m,}$$

the hydraulic radius

$$R_{h1} = \frac{y_1}{4} = \frac{4.46}{4} = \mathbf{1.12\ m.}$$

Then, the energy gradient $J_1$ and the energy $E_1$ are calculated on the basis of the data:

$$J_1 = \frac{3 \times 4^{4/3} Q^2}{k^2 y_1^{16/3}} = \frac{3 \times 4^{4/3} \times 23.42^2}{60^2 \times 4.46^{16/3}} = \mathbf{0.000\,999,}$$

$$E_1 = y_1 + \frac{Q^2}{2g\Omega_1^2} = 4.46 + \frac{23.42^2}{2 \times 9.806 \times 11.48^2} = \mathbf{4.672\ m.}$$

The calculation is repeated in section 2 with the following results:

$$y_2 = \mathbf{4.41\ m,} \quad \Omega_2 = \mathbf{11.23\ m^2,} \quad B_2 = \mathbf{5.09\ m,}$$
$$P_2 = \mathbf{10.18\ m,} \quad R_{h2} = \mathbf{1.10\ m,} \quad J_2 = \mathbf{0.001\,061,} \quad E_2 = \mathbf{4.632\ m.}$$

For the current between sections 1 and 2, an energy gradient equal to the average of $J_1$ and $J_2$ is calculated:

$$J_{ave,1} = \frac{J_1 + J_2}{2} = \frac{0.000\,999 + 0.001\,061}{2} = \textbf{0.001\,03},$$

and the energy drop

$$\Delta E_1 = E_2 - E_1 = 4.63 - 4.67 = \textbf{-0.040 m}.$$

The profile equation allows the calculation of the distance between the two sections:

$$\Delta s_1 = \frac{\Delta E_1}{(i_{b1} - J_{ave,1})} = \frac{-0.040}{(0.02 - 0.001\,03)} = \textbf{-2.12 m}.$$

The coordinate of section 2 will be equal to:

$$s_2 = s_1 + \Delta s_1 = 0 - 2.12 = \textbf{-2.12 m}.$$

The calculation is repeated for all subsequent sections, until we reach the sequent depth of the hydraulic jump, $y_d$. The results are listed in Table 9.8.

Section B is located 19.91 m upstream of section M.

To estimate the length of the jump, we calculate the Froude number of the upstream supercritical current:

$$\Omega_{01} = \frac{y_{01}^2}{\sqrt{3}} = \frac{2.54^2}{\sqrt{3}} = \textbf{3.72 m}^2,$$

$$B_{01} = \frac{2y_{01}}{\sqrt{3}} = \frac{2 \times 2.54}{\sqrt{3}} = \textbf{2.93 m},$$

$$y_{h,u} \equiv y_{h,01} = \frac{\Omega_{01}}{B_{01}} = \frac{3.72}{2.93} = \textbf{1.27 m},$$

$$V_u \equiv V_{01} = \frac{Q}{\Omega_{01}} = \frac{23.42}{3.72} = \textbf{6.30 m s}^{-1},$$

$$Fr_u = \frac{V_u}{\sqrt{g\,y_{h,u}}} = \frac{6.30}{\sqrt{9.806 \times 1.27}} = \textbf{1.78}.$$

Using the experimental data in Table 9.7, yields:

$$L \approx 7.6\,(y_d - y_u) = 7.6 \times (3.96 - 2.54) = \textbf{10.8 m}.$$

Section A (toe of the jump) is approximately 10.8 m upstream of section B.

**Table 9.8** Characteristic values of the current in the channel from section M to section A

| r | y (m) | Ω (m²) | B (m) | P (m) | $R_h$ (m) | J | $J_{ave}$ | E (m) | ΔE (m) | Δs (m) | $s_r$ (m) | S (kN) |
|---|---|---|---|---|---|---|---|---|---|---|---|---|
| $y_1 \equiv y_{02}$ | 4.46 | 11.48 | 5.15 | 10.30 | 1.12 | 0.000999 | | 4.672 | | | 0.00 | 215.2 |
| | | | | | | | 0.001030 | | −0.040 | −2.12 | | |
| 2 | 4.41 | 11.23 | 5.09 | 10.18 | 1.10 | 0.001061 | | 4.632 | | | −2.12 | 210.7 |
| | | | | | | | 0.001094 | | −0.040 | −2.10 | | |
| 3 | 4.36 | 10.98 | 5.03 | 10.07 | 1.09 | 0.001128 | | 4.592 | | | −4.22 | 206.4 |
| | | | | | | | 0.001163 | | −0.039 | −2.07 | | |
| 4 | 4.31 | 10.72 | 4.98 | 9.95 | 1.08 | 0.001199 | | 4.553 | | | −6.29 | 202.2 |
| | | | | | | | 0.001238 | | −0.038 | −2.05 | | |
| 5 | 4.26 | 10.48 | 4.92 | 9.84 | 1.07 | 0.001276 | | 4.515 | | | −8.34 | 198.2 |
| | | | | | | | 0.001318 | | −0.038 | −2.02 | | |
| 6 | 4.21 | 10.23 | 4.86 | 9.72 | 1.05 | 0.001359 | | 4.477 | | | −10.35 | 194.4 |
| | | | | | | | 0.001404 | | −0.037 | −1.99 | | |
| 7 | 4.16 | 9.99 | 4.80 | 9.61 | 1.04 | 0.001448 | | 4.440 | | | −12.34 | 190.8 |
| | | | | | | | 0.001497 | | −0.036 | −1.95 | | |
| 8 | 4.11 | 9.75 | 4.75 | 9.49 | 1.03 | 0.001545 | | 4.404 | | | −14.29 | 187.3 |
| | | | | | | | 0.001597 | | −0.035 | −1.92 | | |
| 9 | 4.06 | 9.52 | 4.69 | 9.38 | 1.02 | 0.001649 | | 4.369 | | | −16.21 | 183.9 |
| | | | | | | | 0.001706 | | −0.034 | −1.88 | | |
| 10 | 4.01 | 9.28 | 4.63 | 9.26 | 1.00 | 0.001762 | | 4.334 | | | −18.08 | 180.8 |
| | | | | | | | 0.001823 | | −0.033 | −1.83 | | |
| $y_d$ | 3.96 | 9.05 | 4.57 | 9.15 | 0.99 | 0.001884 | | 4.301 | | | −19.91 | 177.9 |

**Fig. 9.47** Schematic of the channel with change of bed slope and with a control floodgate

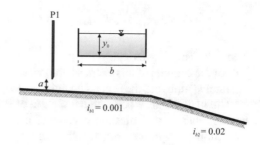

**Exercise 9.21**  In a channel with a rectangular cross-section, $b = (3.5 + C_{pu}/5)$ m wide, with roughness of the bed and of the walls equal to $k = 70$ m$^{1/3}$ s$^{-1}$, a floodgate is inserted with a rectangular slot of height $a = 0.8$ m at the bottom ($C_c = 0.61$), see Fig. 9.47. Immediately upstream of the floodgate the water depth is $y_u = (2 + C_u/10)$ m. The bed slope is $i_{b1} = 0.001$ in one section and $i_{b2} = 0.02$ in a second section.

- Determine the flow rate.
- Qualitatively draw the profile of the current.
- Check whether there is a hydraulic jump and, if so, calculate the values of the sequent depths and the dissipated energy.
- Quantitatively draw the profile, starting from the section of the change of bed slope.

The section of change of slope is very far from the flood gate. Assume $n = 10$ steps for numerical integration of the profile.

**Solution** The flow rate is computed on the basis of energy balance between the upstream section of the sluice gate and the vena contracta, neglecting dissipations:

$$\underbrace{z_u + \frac{p_u}{\gamma_w} + \frac{V_u^2}{2g}}_{H_u} = \underbrace{z_c + \frac{p_c}{\gamma_w} + \frac{V_c^2}{2g}}_{H_c} \rightarrow y_{u1} + \frac{Q^2}{2gb^2 y_{u1}^2} = C_c a + \frac{Q^2}{2gb^2 (C_c a)^2},$$

hence

$$Q = \sqrt{\frac{2gb^2 (y_{u1} - C_c a)}{\left[\dfrac{1}{(C_c a)^2} - \dfrac{1}{y_{u1}^2}\right]}}.$$

The critical depth is

$$y_c = \sqrt[3]{\frac{Q^2}{gb^2}}.$$

We use Chézy formula to determine the value of the uniform flow depth for the two channels with bed slope $i_{b1}$ and $i_{b2}$, respectively:

$$Q = kR_h^{1/6}\Omega\sqrt{R_h i_b}, \quad \text{with } \Omega = by_0, \quad R_h = \frac{\Omega}{P} \equiv \frac{by_0}{b + 2y_0}.$$

The solution is numerical.

For the channel of this exercise, the relation $y_{02} < y_c < y_{01}$ always holds; the first channel is mild and the second channel is steep slope for the computed flow rate $Q$.

The long distance between the section of the sluice gate and the change of slope ensures that in the mild slope channel a subcritical current can develop, with a transition through the critical state and a hydraulic jump. In a section in the mild slope channel, the shooting flow through the slot and the tranquil flow governed by the presence of the critical state at the change of slope must meet, in a section where a balance holds between the total forces $S_u$ and $S_d$:

$$\underbrace{\frac{\gamma_w by_u^2}{2} + \frac{\rho Q^2}{by_u}}_{S_u} = \underbrace{\frac{\gamma_w by_d^2}{2} + \frac{\rho Q^2}{by_d}}_{S_d}.$$

$y_u$ and $y_d$ are the sequent depths of the hydraulic jump.

Moreover, given the long distance between the change of slope and the sluice gate, it can be reasonably assumed that the sequent depth $y_d$ (subcritical current) is practically coincident with the uniform flow depth $y_{01}$. Therefore, it results:

$$\underbrace{\frac{\gamma_w by_u^2}{2} + \frac{\rho Q^2}{by_u}}_{S_u} = \underbrace{\frac{\gamma_w by_{01}^2}{2} + \frac{\rho Q^2}{by_{01}}}_{S_{01}}.$$

The resulting equation is cubic in the unknown $y_u$ and can be solved numerically starting from values lower than $y_c$ or dividing the third degree polynomial by the binomial $(y_u - y_{01})$, since the uniform flow depth $y_{01}$ is also a root of the equation. The energy dissipated in the hydraulic jump is:

$$\Delta E_{jump} = -\frac{(y_d - y_u)^3}{4 y_d y_u} \equiv -\frac{(y_{01} - y_u)^3}{4 y_{01} y_u}.$$

The profile of the current is qualitatively shown in Fig. 9.48.

The profile of the current downstream of the sluice gate is a decelerated supercritical profile in a mild slope, and is drawn downwards from the vena contracta. If this profile could develop completely, it would reach with a vertical tangent the critical height $y_c$ at a finite distance from the vena contracta. The presence of a spontaneous passage through the critical state at the change of slope guarantees the presence of a tranquil flow current upstream of the change of slope itself. The profile is a tranquil accelerated current that is traced upstream from the change of slope, and is interrupted by the hydraulic jump. Downstream of the change of slope, the current profile is supercritical accelerated current and must be traced from upstream to downstream

**Fig. 9.48** Profile of the current

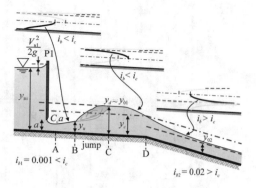

from the critical depth $y_c$. The current asymptotically tends to the uniform flow depth downstream.

The profile from section D (change of slope) downstream can be traced numerically. Assume, for example, a step of variation of the water depth equal to:

$$\Delta y = \frac{(y_c - y_{02})}{n}.$$

The origin of the coordinate system, positive downstream, is at the change of slope. The spatial interval $\Delta s_r$ corresponding to the depth variation $\Delta y_r$ can be calculated by means of the discrete form of the profile equation:

$$\Delta s_r = \frac{\Delta E_r}{(i_b - J_{ave,r})},$$

where $\Delta s_r$ is the difference in coordinate between the sections where the current has depth $y_{r+1}$ and $y_r$, and $\Delta E_r = E_{r+1} - E_r$, $J_{ave,r} = 0.5(J_{r+1} + J_r)$.

For $C_u = C_{pu} = 0$ it results $b = 3.5$ m, $k = 70$ m$^{1/3}$ s$^{-1}$, $a = 0.8$ m, $C_c = 0.61$, $y_u = 2$ m, $i_{b1} = 0.001$, $i_{b2} = 0.02$,

$$Q = \sqrt{\frac{2gb^2 (y_{u1} - C_c a)}{\left(\dfrac{1}{(C_c a)^2} - \dfrac{1}{y_{u1}^2}\right)}} = \sqrt{\frac{2 \times 9.806 \times 3.5^2 \times (2 - 0.61 \times 0.8)}{\left(\dfrac{1}{0.61^2 \times 0.8^2} - \dfrac{1}{2^2}\right)}}$$

$$= 9.59 \text{ m}^3 \text{ s}^{-1}.$$

$$y_c = \sqrt[3]{\frac{Q^2}{gb^2}} = \sqrt[3]{\frac{9.59^2}{9.806 \times 3.5^2}} = \textbf{0.91 m},$$

$$y_{01} = \textbf{1.45 m}, \quad y_{02} = \textbf{0.51 m},$$

**Table 9.9** Characteristic values of the current in the channel from section D downstream

| $r$ | $y$ (m) | $\Omega$ (m²) | $P$ (m) | $R_h$ (m) | $J$ | $J_{ave}$ | $E$ (m) | $\Delta E$ (m) | $\Delta s$ (m) | $s_r$ (m) |
|---|---|---|---|---|---|---|---|---|---|---|
| $y_1 \equiv y_c$ | 0.91 | 3.19 | 5.32 | 0.6 | 0.003667 | | 1.372 | | | 0.00 |
| | | | | | | 0.003921 | | 0.003 | 0.22 | |
| 2 | 0.87 | 3.05 | 5.24 | 0.58 | 0.004174 | | 1.376 | | | 0.22 |
| | | | | | | 0.004479 | | 0.010 | 0.64 | |
| 3 | 0.83 | 2.91 | 5.16 | 0.56 | 0.004784 | | 1.386 | | | 0.86 |
| | | | | | | 0.005154 | | 0.018 | 1.19 | |
| 4 | 0.79 | 2.77 | 5.08 | 0.54 | 0.005524 | | 1.403 | | | 2.05 |
| | | | | | | 0.005978 | | 0.027 | 1.94 | |
| 5 | 0.75 | 2.63 | 5.00 | 0.53 | 0.006431 | | 1.431 | | | 3.99 |
| | | | | | | 0.006994 | | 0.039 | 2.99 | |
| 6 | 0.71 | 2.49 | 4.92 | 0.51 | 0.007556 | | 1.469 | | | 6.97 |
| | | | | | | 0.008263 | | 0.053 | 4.55 | |
| 7 | 0.67 | 2.35 | 4.84 | 0.48 | 0.008969 | | 1.523 | | | 11.52 |
| | | | | | | 0.009870 | | 0.072 | 7.08 | |
| 8 | 0.63 | 2.21 | 4.76 | 0.46 | 0.010770 | | 1.594 | | | 18.60 |
| | | | | | | 0.011937 | | 0.095 | 11.81 | |
| 9 | 0.59 | 2.07 | 4.68 | 0.44 | 0.013103 | | 1.690 | | | 30.41 |
| | | | | | | 0.014642 | | 0.126 | 23.47 | |
| 10 | 0.55 | 1.93 | 4.60 | 0.42 | 0.016181 | | 1.815 | | | 53.88 |
| | | | | | | 0.018256 | | 0.166 | $\infty$ | |
| $y_{02}$ | 0.51 | 1.79 | 4.52 | 0.39 | 0.020000 | | 1.982 | | | $\infty$ |

$$\Delta y = \frac{(y_c - y_{02})}{n} = \frac{(0.91 - 0.51)}{10} = \textbf{0.04 m}.$$

The results are listed in Table 9.9. The sequent depth of the hydraulic jump is computed by imposing the equilibrium of total forces:

$$\underbrace{\frac{\gamma_w b y_u^2}{2} + \frac{\rho Q^2}{b y_u}}_{S_u} = \underbrace{\frac{\gamma_w b y_{01}^2}{2} + \frac{\rho Q^2}{b y_{01}}}_{S_{01}} \rightarrow$$

$$\frac{9806 \times 3.5 \times y_u^2}{2} + \frac{1000 \times 9.59^2}{3.5 \times y_u} = \frac{9806 \times 3.5 \times 1.45^2}{2} + \frac{1000 \times 9.59^2}{3.5 \times 1.45} \rightarrow$$

$$y_u = \textbf{0.53 m}.$$

$$\Delta E_{jump} = -\frac{(y_{01} - y_u)^3}{4 y_{01} y_u} = -\frac{(1.45 - 0.53)^3}{4 \times 1.45 \times 0.53} = \textbf{-0.25 m}.$$

As a special case, if the channel downstream of the floodgate before the change of slope is not long enough, it can happen that the jump does not develop and the current remains supercritical. The profile in the second channel can be accelerated or decelerated depending on whether the depth reached in the section D is greater or less than $y_{02}$.

# Appendix A
# Geometry Properties of Common Plane Shapes

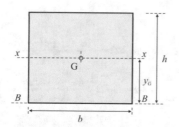

$$y_G = \frac{h}{2} \qquad A = bh$$

$$I_{Gxx} = \frac{1}{12}bh^3 \qquad I_{Bxx} = \frac{1}{3}bh^3$$

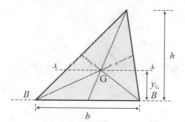

$$y_G = \frac{h}{3} \qquad A = \frac{bh}{2}$$

$$I_{Gxx} = \frac{1}{36}bh^3 \qquad I_{Bxx} = \frac{1}{12}bh^3$$

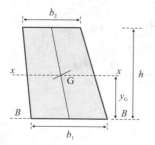

$$y_G = \frac{h}{3}\frac{b_1 + 2b_2}{b_1 + b_2} \qquad A = \frac{b_1 + b_2}{2}h$$

$$I_{Gxx} = \frac{b_1^2 + 4b_1b_2 + b_2^2}{36(b_1 + b_2)}h^3 \qquad I_{Bxx} = \frac{1}{12}(b_1 + 3b_2)h^3$$

© Springer Nature Switzerland AG 2021
S. Longo et al., *Problems in Hydraulics and Fluid Mechanics*, Springer Tracts in Civil Engineering, https://doi.org/10.1007/978-3-030-51387-0

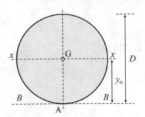

$$y_G = \frac{D}{2} \qquad A = \pi \frac{D^2}{4}$$

$$I_{Gxx} = \pi \frac{D^4}{64} \qquad I_{Bxx} = 5\pi \frac{D^4}{64}$$

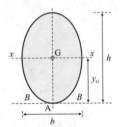

$$y_G = \frac{h}{2} \qquad A = \pi \frac{bh}{4}$$

$$I_{Gxx} = \pi \frac{bh^3}{64} \qquad I_{Bxx} = 5\pi \frac{bh^3}{64}$$

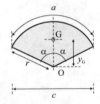

$$y_G = \frac{c^3}{6(ar - hc)} \qquad A = \frac{r^2}{2}(2\alpha - \sin 2\alpha) = \frac{r(a-c) + c(r-h)}{2}$$

$$I_{Gxx} = \frac{1}{16} r^4 (4\alpha - \sin 4\alpha) \qquad I_{Oxx} = \frac{1}{16} r^4 \left(4\alpha - \sin 4\alpha + \frac{64}{9} \frac{\sin^6 \alpha}{\sin \alpha \cos \alpha - \alpha}\right)$$

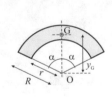

$$y_G = \frac{2rc}{3a} = \frac{2}{3} \frac{\sin \alpha}{\alpha} \qquad\qquad A = \alpha r^2$$

$$I_{Gxx} = \frac{r^4}{4}\alpha + \frac{r^4}{8}\sin 2\alpha - \frac{4r^4}{9}\frac{\sin^2 \alpha}{\alpha} \qquad I_{Oxx} = \frac{r^4}{4}\alpha + \frac{r^4}{8}\sin 2\alpha$$

$$y_G = \frac{2}{3}\frac{(R^3 - r^3)}{(R^2 - r^2)}\frac{\sin \alpha}{\alpha} \qquad\qquad A = \alpha (R^2 - r^2)$$

$$I_{Gxx} = \frac{(R^4 - r^4)}{4}\alpha + \frac{(R^4 - r^4)}{8}\sin 2\alpha$$

$$I_{Oxx} = \frac{R^4 - r^4}{4}\alpha + \frac{R^4 - r^4}{8}\sin 2\alpha - \frac{4}{9}\frac{(R^3 - r^3)^2}{(R^2 - r^2)}\frac{\sin^2 \alpha}{\alpha}$$

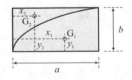

$$y_1 = \frac{3}{8}b, x_1 = \frac{3}{5}a \qquad A_1 = \frac{2}{3}ab$$

$$y_2 = \frac{3}{4}b, x_2 = \frac{3}{10}a \qquad A_2 = \frac{1}{3}ab$$

# Appendix B
# Volume and Surface Area of Solid Figures

$$V = \frac{4}{3}\pi R^3 \qquad A_{lat} = 4\pi R^2$$

$$V = \frac{\pi}{3}h^2(3R - h) = \frac{\pi}{2}h\left(r^2 + \frac{h^2}{3}\right) \qquad A_{lat} = 2\pi Rh = \pi\left(r^2 + h^2\right)$$

$$V = \frac{2\pi}{3}R^2 h \qquad A_{lat} = \pi R(r + 2h)$$

$$V = \frac{\pi h}{2}\left(r_1^2 + r_2^2\right) + \frac{\pi h^3}{6} \qquad A_{lat} = 2\pi Rh$$

© Springer Nature Switzerland AG 2021
S. Longo et al., *Problems in Hydraulics and Fluid Mechanics*, Springer Tracts
in Civil Engineering, https://doi.org/10.1007/978-3-030-51387-0

$$V = \frac{2}{3}R^3\alpha \quad A_{lat} = 2R^2\alpha$$

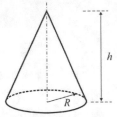

$$V = \frac{1}{3}\pi R^2 h \quad A_{lat} = \pi R\sqrt{R^2 + h^2}$$

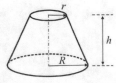

$$V = \frac{1}{3}\pi h\left(R^2 + r^2 + rR\right) \quad A_{lat} = \pi\left(R+r\right)\sqrt{\left(R-r\right)^2 + h^2}$$

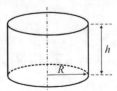

$$V = \pi R^2 h \quad A_{lat} = 2\pi R h$$

# Appendix C
# Physical Properties of the Fluids

**Fig. C.1** Water density as a function of temperature

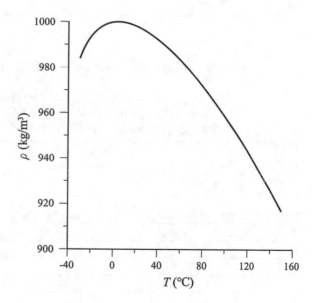

© Springer Nature Switzerland AG 2021
S. Longo et al., *Problems in Hydraulics and Fluid Mechanics*, Springer Tracts
in Civil Engineering, https://doi.org/10.1007/978-3-030-51387-0

**Table C.1** Density of water as a function of temperature

| $T(^\circ C)$ | $\rho$ (kg m$^{-3}$) | $T(^\circ C)$ | $\rho$ (kg m$^{-3}$) |
|---|---|---|---|
| −20 | 993.5500 | 40 | 992.2187 |
| −15 | 996.2860 | 45 | 990.2162 |
| −10 | 998.1200 | 50 | 988.0393 |
| −5 | 999.2590 | 55 | 985.6982 |
| 0 | 999.8425 | 60 | 983.2018 |
| 5 | 999.9668 | 65 | 980.5578 |
| 10 | 999.7026 | 70 | 977.7726 |
| 15 | 999.1026 | 75 | 974.8519 |
| 18 | 998.5986 | 80 | 971.8007 |
| 20 | 998.2071 | 85 | 968.6232 |
| 25 | 997.0479 | 90 | 965.3230 |
| 30 | 995.6502 | 95 | 961.9033 |
| 35 | 994.0349 | 100 | 958.3665 |

Interpolating equation between $-30\,^\circ C$ and $+150\,^\circ C$:

$$\rho\,(\text{kg m}^{-3}) = \frac{\left(\begin{array}{l} 999.839\,52 + 16.945\,176T - 7.987\,040\,1 \cdot 10^{-3}T^2 \\ -46.170\,461 \cdot 10^{-6}T^3 + 105.563\,02 \cdot 10^{-9}T^4 \\ -280.542\,53 \cdot 10^{-12}T^5 \end{array}\right)}{1 + 16.879\,850 \cdot 10^{-3}T}, \quad T \text{ in } ^\circ C$$

**Table C.2** Density of degassed pure water as a function of temperature

| $T(^\circ C)$ | $\rho$ (kg m$^{-3}$) | $T(^\circ C)$ | $\rho$ (kg m$^{-3}$) |
|---|---|---|---|
| 0 | 999.87 | 45 | 990.25 |
| 3.98 | 1000.00 | 50 | 988.07 |
| 5 | 999.99 | 55 | 985.73 |
| 10 | 999.73 | 60 | 983.24 |
| 15 | 999.13 | 65 | 980.59 |
| 18 | 998.62 | 70 | 977.81 |
| 20 | 998.23 | 75 | 974.89 |
| 25 | 997.07 | 80 | 971.83 |
| 30 | 995.67 | 85 | 968.65 |
| 35 | 994.06 | 90 | 965.34 |
| 38 | 992.99 | 95 | 961.92 |
| 40 | 992.24 | 100 | 958.38 |

Interpolating equation for air density:

$$\rho \left(\text{kg m}^{-3}\right) = \frac{1.293}{(1 + 0.003\,67T) \cdot 10^5}\,p, \quad T \text{ in } °C, p \text{ in pascal.}$$

**Table C.3** Density of air at $10^5$ Pa

| $T(°C)$ | $\rho$ (kg m$^{-3}$) | $T(°C)$ | $\rho$ (kg m$^{-3}$) |
|---------|----------------------|---------|----------------------|
| 10 | 1.247 | 21 | 1.201 |
| 11 | 1.243 | 22 | 1.197 |
| 12 | 1.239 | 23 | 1.193 |
| 13 | 1.234 | 24 | 1.189 |
| 14 | 1.230 | 25 | 1.185 |
| 15 | 1.226 | 26 | 1.181 |
| 16 | 1.221 | 27 | 1.177 |
| 17 | 1.217 | 28 | 1.173 |
| 18 | 1.213 | 29 | 1.169 |
| 19 | 1.209 | 30 | 1.165 |
| 20 | 1.205 | | |

**Table C.4** Surface tension for air–water interface

| $T(°C)$ | $\sigma$ ($\times 10^{-3}$ N m$^{-1}$) | $T(°C)$ | $\sigma$ ($\times 10^{-3}$ N m$^{-1}$) |
|---------|----------------------------------------|---------|----------------------------------------|
| −8 | 77.00 | 25 | 71.97 |
| −5 | 76.40 | 30 | 71.18 |
| 0 | 75.60 | 40 | 69.56 |
| 5 | 74.90 | 50 | 67.91 |
| 10 | 74.22 | 60 | 66.18 |
| 15 | 73.49 | 70 | 64.40 |
| 18 | 73.05 | 80 | 62.60 |
| 20 | 72.75 | 100 | 58.90 |

**Table C.5**  Vapour pressure of water

| $T(°C)$ | $p_{vap}$ (Pa) | $T(°C)$ | $p_{vap}$ (Pa) |
|---|---|---|---|
| −15 | 191.45 | 55 | 15 737.32 |
| −10 | 286.50 | 60 | 19 915.64 |
| −5 | 421.69 | 65 | 25 003.20 |
| 0 | 610.48 | 70 | 31 157.35 |
| 5 | 872.32 | 75 | 38 543.39 |
| 10 | 1227.76 | 80 | 47 342.64 |
| 15 | 1704.92 | 85 | 57 808.41 |
| 20 | 2337.80 | 90 | 70 095.37 |
| 25 | 3167.19 | 95 | 84 512.81 |
| 30 | 4242.83 | 100 | 101 324.72 |
| 35 | 5622.85 | 105 | 120 799.06 |
| 40 | 7375.90 | 110 | 143 262.48 |
| 45 | 9583.18 | 115 | 169 049.62 |
| 50 | 12 333.61 | 120 | 198 535.12 |

**Fig. C.2**  Dynamic viscosity of water as a function of temperature

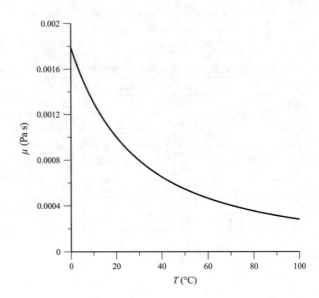

**Table C.6** Dynamic viscosity of water

| $T\,(°C)$ | $\mu\,(\times 10^{-3}\ \text{Pa s})$ | $T\,(°C)$ | $\mu\,(\times 10^{-3}\ \text{Pa s})$ |
|---|---|---|---|
| 0 | 1.7870 | 55 | 0.5040 |
| 5 | 1.5190 | 60 | 0.4665 |
| 10 | 1.3070 | 65 | 0.4335 |
| 15 | 1.1390 | 70 | 0.4042 |
| 20 | 1.0020 | 75 | 0.3781 |
| 25 | 0.8904 | 80 | 0.3547 |
| 30 | 0.7975 | 85 | 0.3337 |
| 35 | 0.7194 | 90 | 0.3147 |
| 40 | 0.6529 | 95 | 0.2975 |
| 45 | 0.5960 | 100 | 0.2818 |
| 50 | 0.5468 | | |

Interpolating functions:

$$\log_{10} \mu = \frac{1301}{998.333 + 8.1855\,(T - 20) + 0.005\,85\,(T - 20)^2} - 4.302\,33,$$

with $T$ in °C and $\mu$ in Pa s, for $0\,°C < T < 20\,°C$

$$\log_{10} \frac{\mu}{\mu_{20}} = \frac{1.3272\,(20 - T) - 0.001\,053\,(20 - T)^2}{T + 105},$$

for $20\,°C < T < 100\,°C$.

**Table C.7** Dynamic viscosity of air

| $T\,(°C)$ | $\mu\,(\times 10^{-5}\ \text{Pa s})$ | $T\,(°C)$ | $\mu\,(\times 10^{-5}\ \text{Pa s})$ |
|---|---|---|---|
| $-100$ | 1.16 | 20 | 1.81 |
| $-50$ | 1.45 | 30 | 1.86 |
| $-20$ | 1.61 | 40 | 1.90 |
| $-10$ | 1.66 | 50 | 1.95 |
| 0 | 1.71 | 100 | 2.18 |
| 10 | 1.76 | | |

**Table C.8** Isentropic bulk modulus of water

| $T$ (°C) | $\varepsilon$ (GPa) | $T$ (°C) | $\varepsilon$ (GPa) |
|------|-------|------|-------|
| −10  | 1.801 | 25   | 2.21  |
| −5   | 1.885 | 30   | 2.235 |
| 0    | 1.962 | 35   | 2.251 |
| 5    | 2.029 | 40   | 2.261 |
| 10   | 2.088 | 45   | 2.265 |
| 15   | 2.138 | 50   | 2.264 |
| 20   | 2.178 |      |       |

**Table C.9** Isentropic bulk modulus of some liquids at 20 °C

| Liquid   | $\varepsilon$ (GPa) | Liquid          | $\varepsilon$ (GPa) |
|----------|-------|-----------------|-------|
| Benzene  | 1.48  | Lubricating oil | 1.44  |
| Glycerin | 4.59  | Mercury         | 28.5  |
| Kerosene | 1.43  | Sea water       | 2.42  |

**Table C.10** Relative density of some manometric liquids (referred to water density at 4 °C, $\rho = 1000$ kg m$^{-3}$)

| Liquid         | $s$    |
|----------------|--------|
| Benzene        | 0.879  |
| Mercury        | 13.55  |
| Meriam red oil | 0.827  |
| Meriam blue    | 1.75   |

**Table C.11** Relative density of some materials (referred to water density at 4 °C, $\rho = 1000$ kg m$^{-3}$)

| Material        | $s$   |
|-----------------|-------|
| Steel           | 7.83  |
| Aluminium       | 2.64  |
| Concrete        | 2.4   |
| Liquid concrete | 2.5   |
| Cast iron       | 7.08  |
| Light pine wood | 0.43  |
| Oak wood        | 0.77  |
| Lead            | 11.4  |
| Copper          | 8.91  |

# Appendix D
# Losses in Pipes and Channels

© Springer Nature Switzerland AG 2021
S. Longo et al., *Problems in Hydraulics and Fluid Mechanics*, Springer Tracts
in Civil Engineering, https://doi.org/10.1007/978-3-030-51387-0

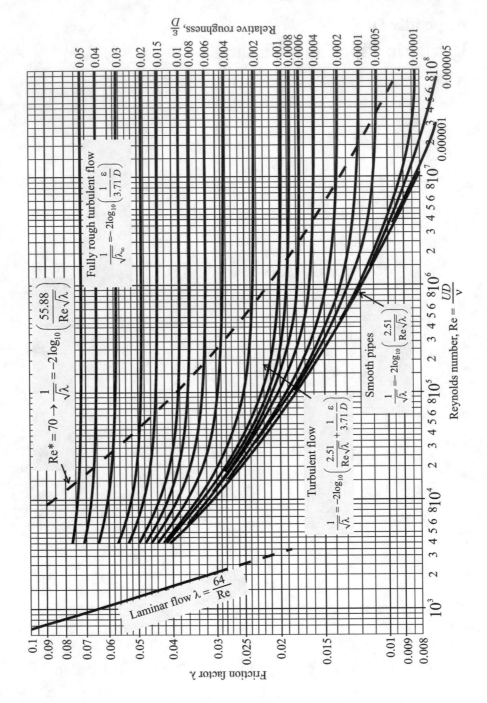

## Modern Formulas for Circular Cross-Section Pipes

Colebrook–White formula:

$$J = \lambda \frac{U^2}{2g} \frac{1}{D}, \quad \frac{1}{\sqrt{\lambda}} = -2\log_{10}\left(\frac{2.51}{Re\sqrt{\lambda}} + \frac{1}{3.71}\frac{\varepsilon}{D}\right).$$

Smooth pipes:

$$\frac{1}{\sqrt{\lambda}} = -2\log_{10}\left(\frac{2.51}{Re\sqrt{\lambda}}\right) \qquad \text{(Prandtl)},$$

$$\lambda = 0.316\,Re^{-0.25} \qquad Re \leq 10^5 \quad \text{(Blasius)},$$

$$\lambda = 0.0031 + 0.221\,Re^{-0.237} \;\; Re > 10^5 \;\; \text{(Nikuradse)}.$$

Fully rough turbulent flow:

$$\frac{1}{\sqrt{\lambda_\infty}} = -2\log_{10}\left(\frac{1}{3.71}\frac{\varepsilon}{D}\right) \;\; \text{(Prandtl–Nikuradse)}.$$

## Classical Formulas for Circular Cross-Section Pipes

$$J = \frac{V^2}{\chi^2 R} \quad \text{with} \quad R = \frac{D}{4}.$$

where

$$\chi = \frac{87}{1 + \dfrac{\gamma}{\sqrt{D/4}}} \quad \text{(Bazin)},$$

or

$$\chi = \frac{100}{1 + \dfrac{m}{\sqrt{D/4}}} \quad \text{(Kutter)},$$

or

$$\chi = k\left(\frac{D}{4}\right)^{1/6} \quad \text{(Gauckler–Strickler)}.$$

In all these formulas $D$ is in metres and $\chi$ is in $m^{1/2}\,s^{-1}$.

**Table D.1** Coefficients of roughness for cylindrical circular pipes

| Type of pipeline | Equivalent roughness $\varepsilon$ (mm) | Bazin $\gamma$ (m$^{1/2}$) | Kutter $m$ (m$^{1/2}$) | Gauckler–Strickler $k$ (m$^{1/3}$ s$^{-1}$) |
|---|---|---|---|---|
| Smooth pipes (glass, brass, extruded copper, resin) | 0–0.02 | | | |
| Steel pipes with coatings degradable over time | 0.05–3.0 | 0.06 | 0.12 | 70–120 |
| Steel pipes with coatings that do not degrade over time | 0.05–0.15 | 0.10–0.23 | 0.15–0.35 | 120 |
| Welded sheet metal pipes in good condition | 0.2–0.3 | 0.06 | 0.12 | 90 |
| Welded sheet metal pipes in current service, with incrustations | 0.4–1.0 | 0.16 | 0.20–0.25 | 75–87 |
| New bolted sheet metal pipes | 0.3–3.0 | 0.10–0.30 | 0.18–0.35 | 70–90 |
| Pipes in bolted sheet metal with incrustations | up to 5.0 | 0.36 | 0.45 | 65 |
| New cast iron pipes | 0.1–0.4 | 0.06–0.10 | 0.12–0.15 | 90–100 |
| Encrusted cast iron pipes | 0.4–5.0 | 0.16–0.36 | 0.20–0.45 | 65–85 |
| New concrete pipes | 0.1–0.15 | 0.06 | 0.12 | 100–105 |
| Cement pipes in current service | 2.0–5.0 | 0.23–0.36 | 0.30–0.45 | 65–70 |

**Fig. D.1** Convergent pipe, energy loss

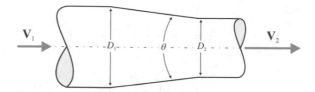

## Energy Loss for a Gradual Contraction

Energy loss expressed as (Fig. D.1):

$$H_1 - H_2 = \xi \frac{V_1^2}{2g},$$

with

$$\xi = \frac{0.8 \sin(\theta/2)\left[1 - (D_2/D_1)^2\right]}{(D_2/D_1)^4} \quad \text{if} \quad \theta \leq 45°,$$

$$\xi = \frac{0.5\sqrt{\sin(\theta/2)}\left[1 - (D_2/D_1)^2\right]}{(D_2/D_1)^4} \quad \text{if} \quad 45° < \theta \leq 180°.$$

**Fig. D.2** Divergent pipe, energy loss

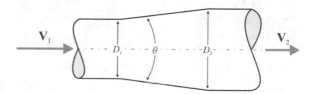

## Energy Loss for a Gradual Expansion

Energy loss expressed as (Fig. D.2):

$$H_1 - H_2 = \xi \frac{V_2^2}{2g},$$

**Table D.2** Pressure drop coefficients for inlet and outlet of circular ducts

| Inlet | $\xi$ | Outlet | $\xi$ |
|---|---|---|---|
| Sharp-edged | 0.5 | Sharp-edged | 1 |
| Re-entering conduct | 1.16 | Conical outlet | 0.20–0.50 |
| With invitation mouthpiece | 0.06 | | |
| Rounded inlet | 0.06–0.10 | | |
| Conical inlet | 0.10–0.25 | | |

with

$$\xi = \frac{2.6\sin{(\theta/2)}\left[1 - (D_1/D_2)^2\right]^2}{(D_1/D_2)^4} \quad \text{if} \ \ \theta \le 45°,$$

$$\xi = \frac{\left[1 - (D_1/D_2)^2\right]^2}{(D_1/D_2)^4} \quad \text{if} \ \ 45° < \theta \le 180°.$$

### Modern Formulas for Channels

$$J = \lambda \frac{U^2}{2g}\frac{1}{4R}, \quad \frac{1}{\sqrt{\lambda}} = -2\log_{10}\left(\frac{2.83}{f\mathrm{Re}\sqrt{\lambda}} + \frac{1}{13.3}\frac{\varepsilon}{fR}\right)$$

where $R$ = hydraulic radius, $f$ = shape factor (Table D.3).

**Table D.3**  Shape factor

| Cross-section shape | $f$ |
| --- | --- |
| Equilateral triangular | 1.30–1.25 |
| Right triangular | 1.20–1.15 |
| Semicircular | 0.90 |
| Semi-hexagonal | 1.00–0.90 |
| Very wide trapeze | 0.80 |
| Rectangular ($b = 2y$) | 0.95 |
| Very wide rectangular | 0.80 |

Smooth walls:

$$\frac{1}{\sqrt{\lambda}} = -2\log_{10}\left(\frac{2.83}{f\mathrm{Re}\sqrt{\lambda}}\right),$$

Fully rough turbulent flow:

$$\frac{1}{\sqrt{\lambda_\infty}} = -2\log_{10}\left(\frac{1}{13.3}\frac{\varepsilon}{fR}\right).$$

### Classical Formulas for Channels

$$J = \frac{Q^2}{\chi^2 \Omega^2 R},$$

with

$$\chi = \frac{87}{1 + \dfrac{\gamma}{\sqrt{R}}} \quad \text{(Bazin)},$$

or

$$\chi = \frac{100}{1 + \dfrac{m}{\sqrt{R}}} \quad \text{(Kutter)},$$

or

$$\chi = kR^{1/6} \quad \text{(Gauckler–Strickler)},$$

where $\Omega$ = cross-section area of the current. In all these formulas $D$ is in metres and $\chi$ is in $\mathrm{m}^{1/2}\,\mathrm{s}^{-1}$.

**Table D.4** Coefficients of roughness for channels

| Type of channel | Equivalent roughness $\varepsilon$ (mm) | Bazin $\gamma$ (m$^{1/2}$) | Kutter $m$ (m$^{1/2}$) | Gauckler–Strickler $k$ (m$^{1/3}$ s$^{-1}$) |
|---|---|---|---|---|
| Partially coated concrete walls, irregular masonry, regular soil without vegetation | 8 | 0.46 | 0.55–0.75 | 60 |
| Fairly regular earth, old masonry | 15–30 | 0.60–0.85 | 0.75–1.25 | 50 |
| Earth with grass, regular watercourses | 70 | 1.3 | 1.5 | 40 |
| Earth in poor condition, natural watercourses with pebbles | 120–200 | 1.75 | 2.0 | 35 |
| Abandoned canals with vegetation, watercourses with gravel beds or dug into rock | 300–400 | 2.0–2.3 | 3.0 | 30 |

Printed in the United States
by Baker & Taylor Publisher Services